AF355957

Advances in Breast Cancer Research and Treatment

Advances in Breast Cancer Research and Treatment

Guest Editors

Taobo Hu
Mengping Long
Lei Wang
Riccardo Autelli

Basel • Beijing • Wuhan • Barcelona • Belgrade • Novi Sad • Cluj • Manchester

Guest Editors

Taobo Hu	Mengping Long	Lei Wang
Breast Surgery	Pathology	Oncology
Peking University	Peking University	Shenzhen University
People's Hospital	Cancer Hospital	Shenzhen
Beijing	Beijing	China
China	China	

Riccardo Autelli
Dipartimento di Scienze
Cliniche e Biologiche
Università degli Studi di Torino
Turin
Italy

Editorial Office
MDPI AG
Grosspeteranlage 5
4052 Basel, Switzerland

This is a reprint of the Special Issue, published open access by the journal *Life* (ISSN 2075-1729), freely accessible at: www.mdpi.com/journal/life/special_issues/I1TMBYR98I.

For citation purposes, cite each article independently as indicated on the article page online and using the guide below:

Lastname, A.A.; Lastname, B.B. Article Title. *Journal Name* **Year**, *Volume Number*, Page Range.

ISBN 978-3-7258-3078-7 (Hbk)
ISBN 978-3-7258-3077-0 (PDF)
https://doi.org/10.3390/books978-3-7258-3077-0

Contents

Editorial

Bridging Discoveries and Treatments: The New Landscape of Breast Cancer Research

Taobo Hu [1],* , **Lei Wang [2],***, **Riccardo Autelli [3],*** and **Mengping Long [4],***

[1] Department of Breast Surgery, Peking University People's Hospital, Beijing 100033, China
[2] International Cancer Center, Shenzhen University, Shenzhen 518060, China
[3] Department of Clinical and Biological Sciences, University of Turin, 10125 Turin, Italy
[4] Department of Pathology, Peking University Cancer Hospital, Beijing 100191, China
* Correspondence: thuac@bjmu.edu.cn (T.H.); lei.wang@szu.edu.cn (L.W.); riccardo.autelli@unito.it (R.A.);
mplong@bjmu.edu.cn (M.L.)

Citation: Hu, T.; Wang, L.; Autelli, R.; Long, M. Bridging Discoveries and Treatments: The New Landscape of Breast Cancer Research. *Life* **2024**, *14*, 747. https://doi.org/10.3390/life14060747

Received: 28 May 2024
Accepted: 3 June 2024
Published: 12 June 2024

Welcome to our Special Issue, "Advances in Breast Cancer Research and Treatment" of *Life*, where we have embarked on a comprehensive exploration of groundbreaking studies that advance our understanding and management of breast cancer. Each paper contributes uniquely to the evolving landscape of breast cancer research, ranging from the clinicopathological characteristics of breast carcinoma with neuroendocrine features [1], the identification of hub genes in the context of non-alcoholic fatty liver disease and triple-negative breast cancer [2], to advancements in diagnostic techniques [3], and the exploration of immunotherapy biomarkers [4]. Some of the topics covered include innovative methods for breast cancer classification combining transfer learning and attention mechanisms [5] and the role of diffusion-weighted imaging in breast cancer diagnosis among young patients [6]. The collective insights presented here not only underscore the complexity of this disease but also highlight the promising pathways toward more effective treatments and improved patient outcomes.

1. Highlights from This Special Issue

1.1. Application of Deep Learning in Breast Cancer Pathology Image Classification

The article "Improved Breast Cancer Classification through Combining Transfer Learning and Attention Mechanism" introduces a novel approach that enhances the accuracy and interpretability of breast cancer histopathological image classification [5]. This method utilizes modified pre-trained Convolutional Neural Network (CNN) models and attention mechanisms to emphasize localized features and enable accurate discrimination in complex cases.

1.2. The Use of Diffusion-Weighted Imaging (DWI) in Young Breast Cancer Patients

"The Role of Diffusion-Weighted Imaging Based on Maximum-Intensity Projection in Young Patients with Marked Background Parenchymal Enhancement on Contrast-Enhanced Breast MRI" explores the application of DWI, particularly in young patients with significant background parenchymal enhancement (BPE) on contrast-enhanced MRI (CE-MRI) [6]. This study found that DWI outperforms CE-MRI in terms of lesion detection.

1.3. Association between Non-Alcoholic Fatty Liver Disease (NAFLD) and Triple-Negative Breast Cancer (TNBC)

The article titled "Identification of Hub Genes and Biological Mechanisms Associated with Non-Alcoholic Fatty Liver Disease and Triple-Negative Breast Cancer" identified hub genes associated with NAFLD and TNBC by analyzing publicly available transcriptomic data [2]. This study also explored the potential co-pathogenesis and prognostic linkage between these two diseases.

                    1

1.4. Breast Cancer Exposomics

The review titled "Breast Cancer Exposomics" discusses the impact of environmental exposures on the development of breast cancer, including the roles of environmental toxins, dietary components, psychosocial stressors, and their associated biological processes and molecular pathways [7]. This review emphasized the role of food and nutrition, as well as endocrine-disrupting chemicals (EDCs), in breast cancer development.

1.5. The Foundational Role of Breast Cancer Cell Lines in Cancer Research

"Molecular, Cellular, and Technical Aspects of Breast Cancer Cell Lines as a Foundational Tool in Cancer Research" reviews the history and origins of breast cancer cell lines and analyzes the molecular pathways that pharmaceutical drugs apply to these cell lines in vitro and in vivo [8]. This review also discussed controversies regarding the use of patient-derived xenografts (PDXs) versus cell-derived xenograft (CDXs) and 2D versus 3D cell culturing techniques.

1.6. Progress and Challenges of Immunotherapy Predictive Biomarkers for Triple-Negative Breast Cancer

"Progress and Challenges of Immunotherapy Predictive Biomarkers for Triple Negative Breast Cancer in the Era of Single-Cell Multi-Omics" discusses the advancements in single-cell sequencing techniques that have allowed for a deeper exploration of the complex and heterogeneous TNBC tumor microenvironment [4]. This review highlighted the potential of single-cell multi-omics analysis for identifying more effective biomarkers and personalized treatment strategies for TNBC patients.

1.7. Adverse Events of PD-1 or PD-L1 Inhibitors in Triple-Negative Breast Cancer

"Adverse Events of PD-1 or PD-L1 Inhibitors in Triple-Negative Breast Cancer: A Systematic Review and Meta-Analysis" provides a comprehensive understanding of treatment-related adverse events when using PD-1 or PD-L1 inhibitors in TNBC [9]. This study included an analysis of the incidence of serious immune-related adverse events and suggested considerations for their management.

1.8. In Silico Analysis of Triple-Negative Breast Cancer-Specific Biomarkers

"In Silico Analysis of Publicly Available Transcriptomic Data for the Identification of Triple-Negative Breast Cancer-Specific Biomarkers" employed in silico analyses to identify biomarkers for triple-negative breast cancer (TNBC), a subtype with limited treatment options [10]. Using publicly available transcriptomic data, the researchers of this study identified 34 differentially expressed genes (DEGs) associated with TNBC. These findings could help in developing targeted therapies and improving diagnostic accuracy.

1.9. Neuroendocrine Breast Carcinoma: Characteristics and Prognosis

"Clinicopathological Characteristics and Prognostic Profiles of Breast Carcinoma with Neuroendocrine Features" examined the clinicopathological characteristics and prognostic outcomes of breast carcinoma with neuroendocrine features [1]. This study found that these tumors are generally hormone receptor-positive and have a higher prevalence among postmenopausal women. Factors such as diabetes and advanced disease stage were associated with poorer progression-free survival.

1.10. Advancements in Post-Mastectomy Breast Reconstruction

"Breast Reconstruction following Mastectomy for Breast Cancer or Prophylactic Mastectomy: Therapeutic Options and Results" discusses various reconstructive options following mastectomy for breast cancer or as a preventive measure [11]. It highlights the evolution of techniques and materials that offer women more choices for breast restoration, aiming to improve psychological outcomes and quality of life after surgery.

1.11. Artificial Intelligence in Breast Cancer Diagnosis: Patient Perspectives

"Patients' Perceptions and Attitudes to the Use of Artificial Intelligence in Breast Cancer Diagnosis: A Narrative Review" synthesizes patient perspectives on the use of artificial intelligence (AI) in breast cancer diagnostics [3]. It reveals that while there is interest in AI's potential to improve diagnostic accuracy, there is also significant concern regarding trust and the desire for human oversight in the diagnostic process.

1.12. Investigating the Role of Eosinophils in Reactive Breast Stroma

"Eosinophilic Dermatoses: Cause of Non-Infectious Erythema after Volume Replacement with Diced Acellular Dermal Matrix in Breast Cancer?" explores the role of eosinophils in reactive breast stroma, particularly in the context of inflammation and tumor microenvironment interactions [12]. The findings of this study suggested that eosinophils may play a part in the breast's response to tumor presence, although their exact role remains to be fully understood.

2. Advancing Frontlines: New Perspectives in Breast Cancer Research

Currently, several critical areas in breast cancer research are drawing considerable attention. Among them, significant advancements in immunotherapy, particularly for TNBC, are at the forefront [13]. TNBC is known for its aggressive nature and lack of targeted therapies, which makes the development of effective immunotherapy treatments especially crucial [14]. These treatments aim to harness the body's immune system to better recognize and combat cancer cells, offering new hope for improving survival rates in a subgroup of breast cancer that has traditionally been challenging to treat [15]. Recent studies have highlighted the effectiveness of treatments like pembrolizumab, which, when combined with chemotherapy, has shown to improve survival rates in patients with high-risk early-stage TNBC [16,17].

Another major area of focus is the management of HER2-positive breast cancer. This subtype, characterized by the overexpression of the HER2 protein, has seen transformative treatments in recent decades, such as targeted therapies that significantly improve patient outcomes [18]. Research is ongoing to enhance these therapies' efficacy and reduce side effects, ensuring more patients can benefit from these advanced treatments [19,20].

Additionally, the role and optimization of radiotherapy in breast cancer treatment protocols remain critical [21]. Radiotherapy is a cornerstone of breast cancer management, used both in the early and more advanced stages of this disease [22]. Innovations in radiotherapy techniques aim to increase the precision and effectiveness of radiation delivery, minimize damage to surrounding healthy tissues, and enhance its cancer-killing capabilities [23,24].

These research topics reflect a concerted and multidisciplinary effort to improve patient survival rates, manage risk factors more effectively, and refine surgical and chemotherapy strategies to offer tailored and less invasive treatment options. By pushing the boundaries in these key areas, researchers hope to not only extend the lives of those diagnosed with breast cancer but also improve their quality of life during and after treatment [25].

3. Final Reflections

This Special Issue embodies our collective quest to understand the complexities of breast cancer through cutting-edge research and to translate these discoveries into actionable treatments that improve patient outcomes. Through a multidisciplinary lens, we explore innovative diagnostic tools [5,10], breakthrough therapies [4,9], and pioneering surgical techniques [11] that are reshaping the way we approach this disease. Our contributors, leading experts in their fields, offer insights into the evolving paradigms of breast cancer management, from molecular genetics to personalized medicine. Their work not only reflects the current state of knowledge but also charts a course for future research directions. We invite you to delve into these pages, where the synergy of scientific discovery and clinical excellence illuminates the path toward a world with more effective breast cancer treatments.

Conflicts of Interest: The authors declare no conflict of interest.

References

1. Qiu, Y.; Dai, Y.; Zhu, L.; Hao, X.; Zhang, L.; Bao, B.; Chen, Y.; Wang, J. Clinicopathological Characteristics and Prognostic Profiles of Breast Carcinoma with Neuroendocrine Features. *Life* **2023**, *13*, 532. [CrossRef]
2. Zhu, J.; Min, N.; Gong, W.; Chen, Y.; Li, X. Identification of Hub Genes and Biological Mechanisms Associated with Non-Alcoholic Fatty Liver Disease and Triple-Negative Breast Cancer. *Life* **2023**, *13*, 998. [CrossRef] [PubMed]
3. Pesapane, F.; Giambersio, E.; Capetti, B.; Monzani, D.; Grasso, R.; Nicosia, L.; Rotili, A.; Sorce, A.; Meneghetti, L.; Carriero, S.; et al. Patients' Perceptions and Attitudes to the Use of Artificial Intelligence in Breast Cancer Diagnosis: A Narrative Review. *Life* **2024**, *14*, 454. [CrossRef] [PubMed]
4. Yu, J.; Guo, Z.; Wang, L. Progress and Challenges of Immunotherapy Predictive Biomarkers for Triple Negative Breast Cancer in the Era of Single-Cell Multi-Omics. *Life* **2023**, *13*, 1189. [CrossRef] [PubMed]
5. Ashurov, A.; Chelloug, S.A.; Tselykh, A.; Muthanna, M.S.A.; Muthanna, A.; Al-Gaashani, M. Improved Breast Cancer Classification through Combining Transfer Learning and Attention Mechanism. *Life* **2023**, *13*, 1945. [CrossRef] [PubMed]
6. Park, G.E.; Kang, B.J.; Kim, S.H.; Jung, N.Y. The Role of Diffusion-Weighted Imaging Based on Maximum-Intensity Projection in Young Patients with Marked Background Parenchymal Enhancement on Contrast-Enhanced Breast MRI. *Life* **2023**, *13*, 1744. [CrossRef] [PubMed]
7. Neagu, A.N.; Jayaweera, T.; Corrice, L.; Johnson, K.; Darie, C.C. Breast Cancer Exposomics. *Life* **2024**, *14*, 402. [CrossRef] [PubMed]
8. Witt, B.L.; Tollefsbol, T.O. Molecular, Cellular, and Technical Aspects of Breast Cancer Cell Lines as a Foundational Tool in Cancer Research. *Life* **2023**, *13*, 2311. [CrossRef] [PubMed]
9. Zhang, Y.; Wang, J.; Hu, T.; Wang, H.; Long, M.; Liang, B. Adverse Events of PD-1 or PD-L1 Inhibitors in Triple-Negative Breast Cancer: A Systematic Review and Meta-Analysis. *Life* **2022**, *12*, 1990. [CrossRef]
10. Kaddoura, R.; Alqutami, F.; Asbaita, M.; Hachim, M. In Silico Analysis of Publicly Available Transcriptomic Data for the Identification of Triple-Negative Breast Cancer-Specific Biomarkers. *Life* **2023**, *13*, 422. [CrossRef]
11. Simion, L.; Petrescu, I.; Chitoran, E.; Rotaru, V.; Cirimbei, C.; Ionescu, S.O.; Stefan, D.C.; Luca, D.; Stanculeanu, D.L.; Gheorghe, A.S.; et al. Breast Reconstruction following Mastectomy for Breast Cancer or Prophylactic Mastectomy: Therapeutic Options and Results. *Life* **2024**, *14*, 138. [CrossRef] [PubMed]
12. Schneider, J.; Lim, S.T.; Yi An, Y.; Suh, Y.J. Eosinophilic Dermatoses: Cause of Non-Infectious Erythema after Volume Replacement with Diced Acellular Dermal Matrix in Breast Cancer? *Life* **2024**, *14*, 608. [CrossRef] [PubMed]
13. Onkar, S.S.; Carleton, N.M.; Lucas, P.C.; Bruno, T.C.; Lee, A.V.; Vignali, D.A.A.; Oesterreich, S. The Great Immune Escape: Understanding the Divergent Immune Response in Breast Cancer Subtypes. *Cancer Discov.* **2023**, *13*, 23–40. [CrossRef] [PubMed]
14. Liu, Y.; Hu, Y.; Xue, J.; Li, J.; Yi, J.; Bu, J.; Zhang, Z.; Qiu, P.; Gu, X. Advances in immunotherapy for triple-negative breast cancer. *Mol. Cancer* **2023**, *22*, 145. [CrossRef] [PubMed]
15. Leon-Ferre, R.A.; Goetz, M.P. Advances in systemic therapies for triple negative breast cancer. *BMJ* **2023**, *381*, e071674. [CrossRef] [PubMed]
16. Tarantino, P.; Corti, C.; Schmid, P.; Cortes, J.; Mittendorf, E.A.; Rugo, H.; Tolaney, S.M.; Bianchini, G.; Andre, F.; Curigliano, G. Immunotherapy for early triple negative breast cancer: Research agenda for the next decade. *NPJ Breast Cancer* **2022**, *8*, 23. [CrossRef] [PubMed]
17. Schmid, P.; Cortes, J.; Dent, R.; Pusztai, L.; McArthur, H.; Kummel, S.; Bergh, J.; Denkert, C.; Park, Y.H.; Hui, R.; et al. Event-free Survival with Pembrolizumab in Early Triple-Negative Breast Cancer. *N. Engl. J. Med.* **2022**, *386*, 556–567. [CrossRef] [PubMed]
18. Marra, A.; Chandarlapaty, S.; Modi, S. Management of patients with advanced-stage HER2-positive breast cancer: Current evidence and future perspectives. *Nat. Rev. Clin. Oncol.* **2024**, *21*, 185–202. [CrossRef]
19. Hurvitz, S.A.; Hegg, R.; Chung, W.P.; Im, S.A.; Jacot, W.; Ganju, V.; Chiu, J.W.Y.; Xu, B.; Hamilton, E.; Madhusudan, S.; et al. Trastuzumab deruxtecan versus trastuzumab emtansine in patients with HER2-positive metastatic breast cancer: Updated results from DESTINY-Breast03, a randomised, open-label, phase 3 trial. *Lancet* **2023**, *401*, 105–117. [CrossRef]
20. Squifflet, P.; Saad, E.D.; Loibl, S.; van Mackelenbergh, M.T.; Untch, M.; Rastogi, P.; Gianni, L.; Schneeweiss, A.; Conte, P.; Piccart, M.; et al. Re-Evaluation of Pathologic Complete Response as a Surrogate for Event-Free and Overall Survival in Human Epidermal Growth Factor Receptor 2-Positive, Early Breast Cancer Treated With Neoadjuvant Therapy Including Anti-Human Epidermal Growth Factor Receptor 2 Therapy. *J. Clin. Oncol.* **2023**, *41*, 2988–2997. [CrossRef]
21. Coles, C.E.; Haviland, J.S.; Kirby, A.M.; Griffin, C.L.; Sydenham, M.A.; Titley, J.C.; Bhattacharya, I.; Brunt, A.M.; Chan, H.Y.C.; Donovan, E.M.; et al. Dose-escalated simultaneous integrated boost radiotherapy in early breast cancer (IMPORT HIGH): A multicentre, phase 3, non-inferiority, open-label, randomised controlled trial. *Lancet* **2023**, *401*, 2124–2137. [CrossRef] [PubMed]
22. Meattini, I.; Becherini, C.; Caini, S.; Coles, C.E.; Cortes, J.; Curigliano, G.; de Azambuja, E.; Isacke, C.M.; Harbeck, N.; Kaidar-Person, O.; et al. International multidisciplinary consensus on the integration of radiotherapy with new systemic treatments for breast cancer: European Society for Radiotherapy and Oncology (ESTRO)-endorsed recommendations. *Lancet Oncol.* **2024**, *25*, e73–e83. [CrossRef] [PubMed]
23. Demircan, N.V.; Bese, N. New Approaches in Breast Cancer Radiotherapy. *Eur. J. Breast Health* **2024**, *20*, 1–7. [CrossRef] [PubMed]

24. Sigurdson, S.; Thibodeau, S.; Korzeniowski, M.; Moraes, F.Y. A Precise Approach for Radiotherapy of Breast Cancer. *Cancer Treat. Res.* **2023**, *188*, 175–198. [CrossRef]
25. Coles, C.E.; Earl, H.; Anderson, B.O.; Barrios, C.H.; Bienz, M.; Bliss, J.M.; Cameron, D.A.; Cardoso, F.; Cui, W.; Francis, P.A.; et al. The Lancet Breast Cancer Commission. *Lancet* **2024**, *403*, 1895–1950. [CrossRef]

Review

Molecular, Cellular, and Technical Aspects of Breast Cancer Cell Lines as a Foundational Tool in Cancer Research

Brittany L. Witt [1] and Trygve O. Tollefsbol [1,2,3,4,5,6,*]

[1] Department of Biology, University of Alabama at Birmingham, 902 14th Street, Birmingham, AL 35228, USA; bwitt95@uab.edu

[2] Integrative Center for Aging Research, University of Alabama at Birmingham, 1530 3rd Avenue South, Birmingham, AL 35294, USA

[3] O'Neal Comprehensive Cancer Center, University of Alabama at Birmingham, 1802 6th Avenue South, Birmingham, AL 35294, USA

[4] Nutrition Obesity Research Center, University of Alabama at Birmingham, 1675 University Boulevard, Birmingham, AL 35294, USA

[5] Comprehensive Diabetes Center, University of Alabama at Birmingham, 1825 University Boulevard, Birmingham, AL 35294, USA

[6] University Wide Microbiome Center, University of Alabama at Birmingham, 845 19th Street South, Birmingham, AL 35294, USA

[*] Correspondence: trygve@uab.edu; Tel.: +1-205-934-4573; Fax: +1-205-975-6097

Abstract: Breast cancer comprises about 30% of all new female cancers each year and is the most common malignant cancer in women in the United States. Breast cancer cell lines have been harnessed for many years as a foundation for in vitro analytic studies to understand the use of cancer prevention and therapy. There has yet to be a compilation of works to analyze the pitfalls, novel discoveries, and essential techniques for breast cancer cell line studies in a scientific context. In this article, we review the history of breast cancer cell lines and their origins, as well as analyze the molecular pathways that pharmaceutical drugs apply to breast cancer cell lines in vitro and in vivo. Controversies regarding the origins of certain breast cancer cell lines, the benefits of utilizing Patient-Derived Xenograft (PDX) versus Cell-Derived Xenograft (CDX), and 2D versus 3D cell culturing techniques will be analyzed. Novel outcomes from epigenetic discovery with dietary compound usage are also discussed. This review is intended to create a foundational tool that will aid investigators when choosing a breast cancer cell line to use in multiple expanding areas such as epigenetic discovery, xenograft experimentation, and cancer prevention, among other areas.

Keywords: breast cancer; cell lines; molecular pathways; epigenetics; xenograft; cancer prevention; cancer therapy

Citation: Witt, B.L.; Tollefsbol, T.O. Molecular, Cellular, and Technical Aspects of Breast Cancer Cell Lines as a Foundational Tool in Cancer Research. *Life* **2023**, *13*, 2311. https://doi.org/10.3390/life13122311

Academic Editors: Riccardo Autelli, Taobo Hu, Mengping Long and Lei Wang

Received: 19 September 2023
Revised: 30 November 2023
Accepted: 2 December 2023
Published: 8 December 2023

1. Introduction

Breast cancer cells have been utilized for over 50 years to establish various techniques to give prognosis and treatment for breast cancer. Many well-known mechanisms have been utilized and analyzed with the use of breast cancer cell line research. For example, Trastuzumab monotherapy (brand name Herceptin) is used to treat breast cancer patients with amplified or over-expressed human epidermal growth factor 2 (HER2+) in the body [1]. Researchers used Herceptin, which can also be combined with chemotherapy, to analyze the pathophysiologies of over 51 breast cancer cell lines with abnormalities in the genome which mirror over 145 primary breast cancer tumor types [2]. The process of cultivating human cells for research received impetus in 1951 when George Gey established the HeLa cell line named after Henrietta Lacks who had cervical carcinoma [3]. About eight years later, HeLa cells were established [4]. The first breast cancer cell line, referred to as BT-20, was established in the 1950's and, since then, many other breast cancer cell lines have been established, such as MCF7 and MDA-MB-231 cells [5]. By analyzing these breast cancer cell lines, researchers can distinguish between different molecular aberrations to facilitate the identification of mechanisms that can help prevent or decrease the occurrence of breast cancer through therapeutical practices. Even though various breast cancer cell lines have been vital, challenges can occur, such as contamination due to poor technique or genetic drift due to over-passaging of breast cancer cells. Researchers have documented this instance of contamination from HeLa cells to other cultures that caused a large amount of 'false cell lines' to be produced [6].

Since then, many advancements have been made to categorize, illustrate, and utilize various breast cancer cell types for use by investigators. Notably, the MD Anderson series of various breast cancer cell lines that are known worldwide was established in 1973 at the Michigan Cancer Foundation [7]. T47D, MCF7, MDA-MB-231, and MDA-MB-157 are just a few breast cancer cell lines that have been used in vitro over the years to translate to in vivo studies. They have even found utility in studies of treatment for human breast cancer, e.g., Herceptin was found to predict the therapeutic response in many HER2+ cell lines such as SK-BR-3 [2]. Many have found that breast cancer cell lines are also modified by changes in gene expression that can be heritable from cell to cell, but do not modify DNA sequences [8]. This aspect of epigenetics has been shown through many different mechanisms such as chromatin remodeling, DNA methylation, effects from non-coding RNA, and histone modifications [9]. Some enzymes perform essential tasks within the epigenome to silence (e.g., DNA methyltransferase and histone deacetylase) or activate (e.g., histone acetyltransferase) genes in cells during development. Other methyltransferases can impact epigenetics by either altering the concentration of the methyl donors or creating a metabolic methylation sink. An example of this is the enzyme Nicotinamide N-methyltransferase (NNMT) which promotes epigenetic remodeling in breast cancer and affects all other NAD+ dependent enzymes such as poly-ADP ribose polymerases (PARPs) and sirtuins [10,11]. In addition, research has been performed using combinations of various dietary phytochemicals that induce inhibition of these silencing enzymes to analyze a decrease in the growth of cancerous cell lines [12].

There are remaining issues regarding cross-contamination when handling cell lines that could cause a crisis during experimentation [13]. This can occur, although proper sterilization and techniques are always helpful to combat inconsistencies. Cell line research can also translate to in vivo experiments when new connections are being constructed. For example, the gut microbiome has been analyzed to investigate any contribution to breast cancer prevention and treatment. Short-chain fatty acids such as sodium butyrate can be generated in the gut microbiome by digestion of food, and these compounds have enzyme-silencing properties that may contribute to a reduction in breast cancer development [14]. There has been a considerable amount of work performed with breast cancer cell lines, so much so that it can be difficult to find the origin of certain cell line types and the reasoning behind using a certain subtype in an experiment. Researchers need a foundational guide to organize the many categories of breast cancer cell lines, techniques to sustain them, and information on how to prevent contamination of these breast cancer cell lines. There have been publications that list the classifications of breast cancer cell lines and how they reflect heterogeneity from breast cancer [15]. However, there is a major gap in previous research to show a comprehensive overview of modern cell culturing techniques and example breast cancer cell lines that may apply when creating an experimental idea.

The purpose of this review is to analyze and organize previous research performed on breast cancer cell lines to create a foundational starting point for investigators who are involved with cell-culturing experiments. If a certain cell line type is needed, researchers will be able to use this review for an overview of the origin of the cell line. Many investigators have made a large impact in this scientific field by classifying multiple types of breast cancer cell lines and describing the morphologies of each [16]. This review will extend such research by adding the techniques used to culture these categorized breast cancer cell types and how to maintain the lines without contamination. Having a foundational piece of work to guide researchers through the process of finding an ideal breast cancer cell line to work with during experimentation and key proteins or genes that they may be able to study is a valuable aspect that this review will bring to this research field. Even examples of techniques able to provide an ideal environment analogous to breast cancer tissue in vivo, cell-derived xenograft versus patient-derived xenograft, and how breast cancer cellular lines participate in creating pharmaceutical drugs that can execute functions in molecular pathways will be discussed. Also, understanding key components that are essential and may be detrimental to the growth of breast cancer cells will be explored. Through this review, many aspects of cell line research will be considered in detail. The goal is to categorize the various cell lines in such a way that investigators will be able to use this resource as a tool to help focus experimentation and further study.

2. Methods

Multiple empirical and review articles were accessed to investigate previous research on the history of breast cancer cell lines. PubMed and Google Scholar databases were searched using the keywords breast cancer, cell lines, molecular pathways, epigenetics, xenograft, cancer prevention, and cancer therapy. There was no meta-analysis, quantitative, or qualitative synthesis performed. Some references were published more than five years ago for the basis of grasping a history of cell line usage.

3. Results

3.1. Epigenetics and Breast Cancer

Analyses have been performed to discover the components of dietary phytochemicals such as epigallocatechin and sulforaphane that have been shown to decrease breast cancer risk [17]. These discoveries were originally performed with the use of in vitro cell lines and have significantly advanced the field. Analyses such as cell viability assay [(MTT(4,5-Dimethylthiazol-2-yl))] can reveal the number of cells that can survive from 3–7 days of treatment with these dietary phytochemicals that have been studied to show DNA methyltransferase (DNMT) and histone deacetylase (HDAC) inhibition. Gene-silencing enzymes such as DNMTs can attach a methyl group to a portion of DNA called a CpG (Cytosine–Phosphate–Guanine) island and prevent the transcription of that portion of DNA from being expressed. The other silencing enzyme HDAC has the role of removing an acetyl group from a histone structure, thus allowing the protein to be tightly coiled and unable to proceed with transcription [9]. If the dietary phytochemicals mentioned can inhibit the function of these silencing enzymes, more genes will undergo transcription and expression in the cells. Thus, knowledge of DNMT and HDAC inhibition properties of certain dietary phytochemicals is important to breast cancer cell line treatment and prevention in epigenetics.

Other experiments, such as western blot, PCR (Polymerase Chain Reaction), and enzymatic activity assays when a proper concentration of treatment (e.g., dietary phyto-chemicals) has been established, can be used to identify the optimal dosage to treat cancer cells and decrease their growth without toxicity to the control cell line of non-transformed cells. Studies have shown that some environmental factors can cause certain genes to be either silenced or expressed in the human body. As evidence of this, twins can have the same DNA but expressed differently based on the internal and external environment [18]. A common issue that often arises is the uncertainty of the exact number of polyphenols in the foods that humans consume to associate with the impacts they may have on various cancer types, specifically breast cancer [19]. This is why the use of breast cancer cell lines is so essential for this purpose; an exact number of dietary phytochemicals can be tested on cancerous cell lines to determine if there is a decrease in growth. Although research has accumulated over the years, there is still a limited number of studies on breast cancer cell lines [15]. In Table 1, various cell lines are shown that have been discovered to have either estrogen receptor (ER), progesterone receptor (PR), luminal–HER2, HER2, or triple-negative breast cancer basal-like/normal-like (TNBCA/B) properties along with the most beneficial medium in which they can be grown. The type of cancer cell line is also listed along with the morphology of the cancerous cell.

Table 1. Breast cancer cell line classification.

Cell Line	[1] ER	PR	HER2	Type	Protein Status	TP53 Status	Additional Mutated Genes	Medium	Morphology	References
ZR-75-1	+	+/−	−	Luminal A	Low Ki-67	Wild type	PTEN	RPMI	Grape-like	[16,20–25]
T-47D	+	+	−	Luminal A	Low Ki-67	L194F	PIK3CA, SPEN	RPMI, DMEM	Mass	[16,20,22,24,25]
MCF7	+	+	−	Luminal A	Low Ki-67	Wild type	PIK3CA	RPMI, DMEM	Mass	[16,20–25]
MDA-MB-415	+	+/−	−	Luminal A	Low Ki-67	Y236C	MAP2K4, PTEN	DMEM	Round	[7,16,20,22,24,25]
MDA-MB-330	+/−	−	+	Luminal B	High Ki-67	Y220C	CTNNA1	RPMI	Grape-like	[7,16,22,25–27]
ZR-75-30	+	−	+	Luminal B	High Ki-67	Wild type	BRCA2, AKT1	RPMI	Grape-like	[16,25,28,29]
BT-474	+/−	+	+	Luminal HER2	High Ki-67	E285K	PIK3CA, BRCA2	RPMI	Mass	[16,20–22,24,25]
MDA-MB-361	+	+/−	+	Luminal HER2	High Ki-67	E56 *	PIK3CA, SPEN	RPMI, DMEM	Grape-like	[7,16,22,24,25,30]
UACC-812	+/−	+/−	+	Luminal HER2	High Ki-67	Wild type	BRCA1	L-15, RPMI, DMEM	Grape-like	[16,24,25,30–33]
AU565	−	−	+	HER2+	High Ki-67	R175H	CDH1	RPMI	Grape-like	[16,21,24,25,30]
MDA-MB-453	−	−	+	HER2+	High Ki-67	T382S	BRCA2, PTEN, PIK3CA	RPMI, DMEM	Grape-like	[7,16,20,21,24,25]
HCC1569	−	−	+	HER2+	High Ki-67	E294 *	PTEN, PIK3CA	RPMI	Mass	[20,24,25,34,35]
SUM190PT	−	−	+	HER2+	High Ki-67	Q317 *	PIK3CA	Ham's F12	Mass	[2,12,16,25,36]
SK-BR-3	−	−	+	HER2+	High Ki-67	R175H	CDK2NA	DMEM, McCoys	Grape-like	[16,20,21,24,25]
HCC70	−	−	−	TNBCA	High Ki-67	R248Q	FANCA, PIK3CA	RPMI	Mass	[16,20,24,25]
BT-20	−	−	−	TNBCA	High Ki-67	K132Q	RB1, PIK3CA	EMEM, RPMI, DMEM	Stellate	[16,20,21,25,37,38]
CAL148	−	−	−	TNBCA	High Ki-67	E224K	RB1, PTEN, PIK3CA	DMEM	Round	[16,25,30,38]
SUM229PE	−	−	−	TNBCA	High Ki-67	R273C	CDK2NA, PIK3CA	RPMI, Ham's F12	Spindle	[2,16,25,30,37,39]
BT-549	−	−	−	TNBCB	High Ki-67	R249S	RB1, PTEN	RPMI	Stellate	[16,20,24,25]
MDA-MB-157	−	−	−	TNBCB	High Ki-67	A88fs	MED12, SPEN	RPMI, DMEM	Stellate	[7,16,20,25,40]

Table 1. *Cont.*

Cell Line	[1] ER	PR	HER2	Type	Protein Status	TP53 Status	Additional Mutated Genes	Medium	Morphology	References
MDA-MB-231	–	–	–	TNBCB	High Ki-67	R280K	BRAF, TERT, KRAS	RPMI, DMEM	Stellate	[7,16,20,21,24,25]
Hs 578T	–	–	–	TNBCB	High Ki-67	V157F	PIK3CA, MED12, CDKN2A	RPMI, DMEM	Stellate	[16,24,25,30]
MDA-MB-435	–	–	–	MEL	High Ki-67	G266E	BRAF, CDKN2A	L15, RPMI, DMEM	Spindle	[16,20,25]

[1] ER (estrogen receptor), PR (progesterone receptor), HER2 (human epidermal growth factor receptor 2), and MEL (Melanoma). Medium is the culture media that the breast cancer cells can be grown in, but this is not limited to what is shown. Options for media include EMEM (Eagle's Minimum Essential Medium), L-15 (Leibovitz's L-15 Medium), RPMI (Roswell Park Memorial Institute-1640 Medium), DMEM (Dulbecco's Modified Eagle Medium), McCoys (McCoys 5a Medium) and Ham's F12 (Ham's F-12 (Kaighn's) Medium). All media have nutrients added to ensure proper cell viability (see Section 3.3.1 for more information). +/– represents the positive or negative receptor that the breast cancer cell line is classified to have The type consists of the receptors that are present or not present within the breast cancer cell line such as Luminal A, Luminal B, Luminal HER2, HER2+, and Triple-Negative Breast Cancer (TNBC) that can be further divided into Basal-like (TNBCA) and Normal-like (TNBCB). Protein status consists of the level of Ki-67 proliferation index of the breast cancer cell line type. TP53 status indicates the protein sequence mutation for each breast cancer cell line listed. Some of the amino acid alterations have an asterisk symbol "*" that indicates a premature stop codon that is produced from the *TP53* mutation. Wild type indicates that the breast cancer cell line does not have a mutation within the *TP53* gene. Morphology consists of stellate, mass, grape-like, spindle, and round.

These breast cancer cell lines were chosen based on being commonly used and well-known with respect to various breast cancer cell line types. The *TP53* and protein status are also listed for the breast cancer cell lines. Multiple articles discuss *p53* mutations and how this gene is often called "the guardian of the genome", as it can activate genes that play a role in cell cycle arrest, apoptosis, and DNA repair. Mutations within this gene may cause several aberrations to be multiplied and eventually produce proteins that are not supposed to be expressed in the cells of the body [41]. The *TP53* status of many breast cancer cell lines can be identified to highlight the actual mutation within the DNA binding domain. There has been much controversy over the years about the *TP53* status and how it can cause incorrect experimental interpretations. There is a collection of widely used cell lines from the NCI-60 panel that have the *TP53* mutant status collected for the ICGC/TCGA Pan-cancer Analysis of Whole Genomes databases [20]. This Pan-Cancer Project was established in 2006 and is an international collaboration for cancer genomics and the molecular characterization of over 20,000 primary cancers to match over 50 cancer types. By utilizing these mutation databases, analysis of *TP53* gene variants can be executed along with other important mutated genes in breast cancer research that are listed in Table 1.

3.2. Breast Cancer Cell Line Classification

There are many ways in which to classify breast cancer cell lines, although the most common strategy that has been used over the past few decades entails immunohistochemistry (IHC) methods to recognize different hormone receptors such as ER, PR, and androgen (AR) [21]. IHC has been utilized to distinguish the expression through the phenotype of breast cancer cell lines with tyrosine kinase receptors such as HER2 and the Epidermal Growth Factor receptor (EGFR) [21]. Figure 1 shows a simplified schematic of breast cancer cell line types and their common characteristics according to the category. The first published work using IHC techniques dates to 1941, while the discovery of these influential hormone receptors to classify breast cancer cells dates back to 1940 (AR) and 1977 (ER and PR) [42,43]. Some investigators also include a normal-like subtype to classify about 2–8% of breast cancer that is also associated with Luminal A for being targeted with tamoxifen. Additionally, there is an intermediate subtype, Luminal HER2+, that is closely related to the HER2+ subtype owing to an overexpression of HER2 which has been linked to ER downregulation [16]. Immunohistochemistry is a technique that utilizes various types of antibodies to detect specific antigens in the tissue. The hormones estrogen, progesterone, and androgen have been recognized to be more prevalent in breast cancer tissue, and efforts have been made to target such a tissue through therapy or surgical approaches. By deriving the tissue from patients, breast cancer cell lines have been used to perform further tests that have become useful to treat different breast cancer types and reduce the risk of breast cancer in general [44–46]. These hormonal receptors can be found embedded in the phospholipid bilayer or free floating in the cytoplasm of many cells and function by allowing estrogen, progesterone, or androgen to bind. This then causes translocation to the nucleus or a cascade of enzymes to react and eventually silence or activate important genes that regulate transcription and produce genes that cause breast cancer to progress.

The origin of the name for most of the breast cancer cell lines discussed below is based on the scientists who derived the cell line, such as Zeida Rae for ZR-75-1 and Iafa Keydar who established T-47D at Tel Aviv University in the 47th petri dish [47]. Most of the other breast cancer cell line names originated from organizations such as the 'M.D. Anderson Hospital and Tumor Institute' for the MD Anderson-Metastatic Breast series, 'Streamlined University of Michigan' for the SUM series, 'Hamon Cancer Center' for HCC1569 and HCC70. Certain breast cancer cell lines were named after the number of cells that grew from a particular dish, such as the 'BT' series (BT-20, BT-549, BT-474), which includes the first breast cancer cell line derived and which was named by the researcher that isolated them, Etienne Y. Lasfargues, and the UACC-812 named by researchers from the 'University of Arizona Cancer Center' to create the University of Arizona Cell Culture series [47]. The MCF7 breast cancer cell line was named after the organization from which it was derived, the 'Michigan Cancer Foundation'. The name AU565 originates from Adenocarcinoma Unknown 565 [47], while SK-BR-3 is named after the Memorial-Sloan Kettering Cancer Center, more specifically after the Sloan-Kettering Breast Cancer Cell Line 3 [47]. CAL148 breast cancer cell line owes its name to the Centre Antoine Lacassagne [47]. Hs 578T breast cancer cell line has a name that originates from *Homo sapiens* or Human sarcoma based on multiple sources [47].

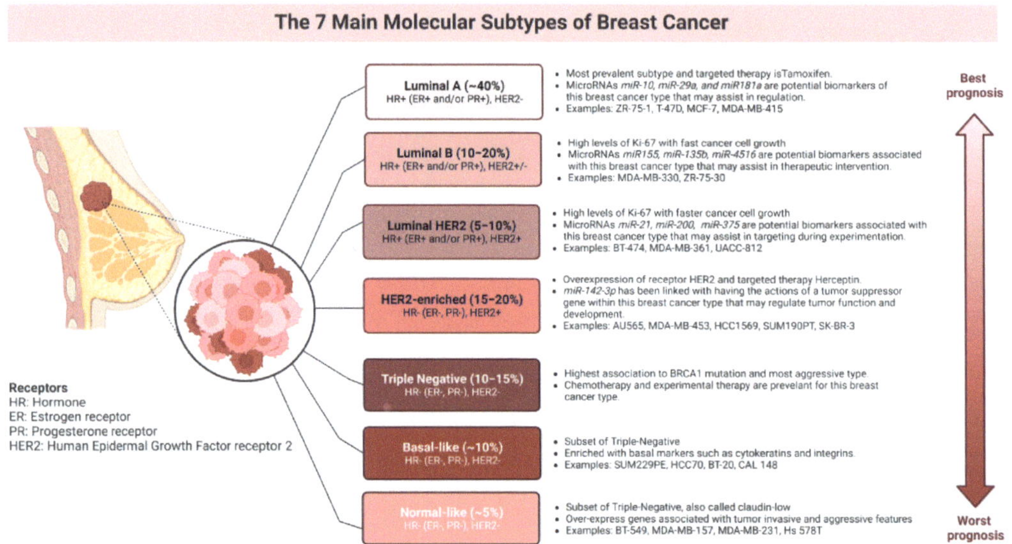

Figure 1. Types of breast cancer are grouped by Luminal A, Luminal B, HER2, and Triple—Negative. BC cell line examples; additional characteristics are also mentioned.

3.2.1. Hormonal Receptors

Breast cancer may be caused by the mutation of multiple genes that are usually silenced or activated for expression. Many of these breast cancer cell lines are categorized by the hormonal receptors ER, PR, and AR that are derived from the breast cancer patient of origin. The absence or presence of the estrogen, progesterone, and androgen hormonal receptors can determine what kind of breast cancer cell line investigators would obtain for breast cancer experimentation. MDA-MB-453 breast cancer cell line, in particular, has a high level of AR, which has been shown to play a vital role in breast cancer pathogenesis [48]. AR expression is prevalent in breast cancer subtypes and has been found to be about 50% expressed in ER breast cancer types. The ER− or PR− status in breast cancer cell lines can determine if AR-targeted hormone therapy would be beneficial [49]. There has been treatment for ER through aromatase inhibition; however, this therapy technique alone may increase androgens, so usage of the hormonal therapy tamoxifen is mostly recommended to treat ER+ breast cancer in women and men [49]. As seen in Figure 2, the presence of ER+ in the patient's breast cancer tissue can be treated with tamoxifen therapy, considering that ER+ may place breast cancer cell lines under the type Luminal A, the most prevalent type that also has low levels of Ki-67. One of the many important protein biomarkers for breast cancer is Ki-67 (derived from Kiel city and the number of the original clone 67). This biomarker has been studied to show an association with tumor cell proliferation and growth. When different types of tissue are tested through IHC monoclonal antibodies and tested for the expression of Ki-67, a prognosis may be made about the threshold and treatment available for patients [50]. Ki-67 was first studied by Scholzen to show that this protein is vital for the progression of the cell cycle, since it is present in each cell phase (G1, S, and G2, but not G0) [31,51,52].

PR+ breast cancer cell lines can be associated with the Luminal A type along with the Luminal B type that may be treated with tamoxifen hormonal therapy. The Luminal B subtype has a higher level of Ki-67 expression with faster cancer cell growth. In Figure 2, the HER2+ receptors are membrane-bound, showing that the targeted therapy for HER2+, Herceptin, will need the ability to enter the cell to cause homo- or heterodimerization through the cell signaling pathway [36]. HER2+ breast cancer cell type is also associated with receptor tyrosine kinases (RTKs), which play a role in cellular functions such as cell growth and survival. Other proteins that are RTKs are within the Erythoblastic oncogene B (ErbB) family, such as ErbB-1 (also referred to as epidermal growth factor receptor EGFR), ErbB-2 (also referred to as HER2), ErbB-3 (also referred to as HER3), and ErbB-4 (also referred to as HER4) [53]. These ErbB receptors are essential for the normal functioning of the body, but when present in excessive quantities cause dysregulation and can potentially lead to breast cancer [54]. Examples such as the ones reported above illustrate how vital these hormones and membrane receptors are for the overall advancement of breast cancer cell line research.

Drug Regulation Mechanism Comparison in Breast Cancer

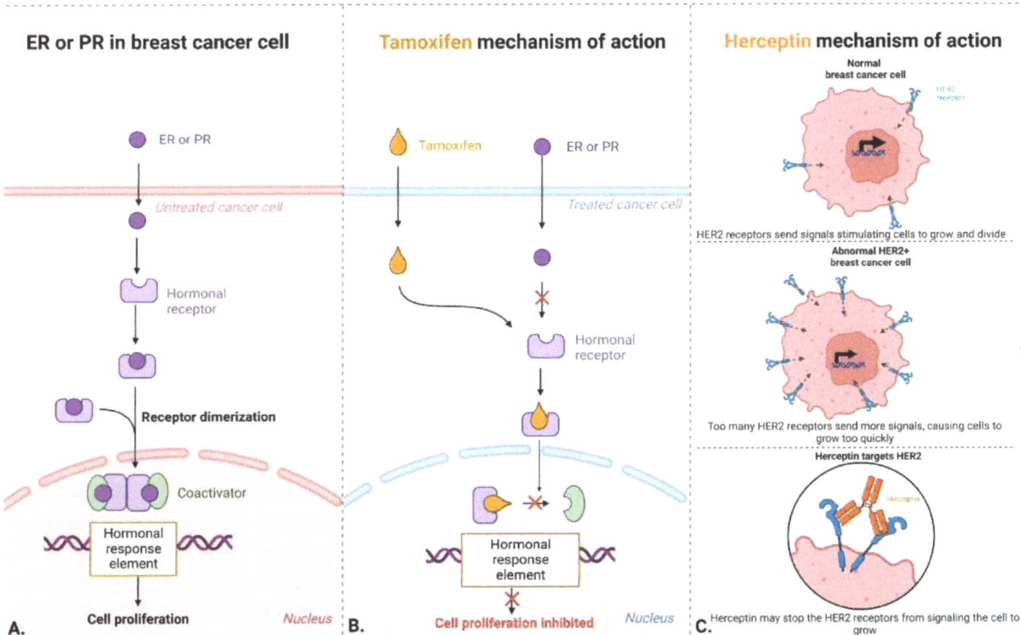

Figure 2. This schematic exemplifies the drug regulation mechanisms that occur in breast cancer when pharmaceutical drugs are administered. Part (**A**) shows an untreated cancer cell that receives either ER or PR ligands to bind to an intracellular hormonal receptor that will cause dimerization and increase breast cancer cell proliferation. Part (**B**) gives an example of tamoxifen affecting the molecular pathway by preventing the ER or PR ligand from binding to the hormonal receptor which inhibits breast cancer cell proliferation. The arrow with an red X represents the process not taking place due to binding of Tamoxifen during the pathway. Part (**C**) gives a schematic of Herceptin preventing HER2+ receptors from being able to express more signals in breast cancer cell.

3.2.2. Luminal Breast Cancer Cell Lines

ZR-75-1, T-47D, MCF7, and MDA-MB-415 all represent the Luminal A subtype of breast cancer, and all were established in the 1970s. The nomenclature for each of these breast cancer cell lines is unique and most have a mass structure according to Table 1.

The ZR-75-1 breast cancer cell line was first derived from the isolated metastatic ascites of a 63-year-old white female patient and has been used to study different radioactive diagnostic agents that are used with PET (positron emission tomography) imaging, such as fluoroestradiol [22]. Being able to visualize ER+ breast cancer gene expression in ZR-75-1 cell lines was vital to further clinical applications. The T-47D breast cancer cell line was originally isolated from the metastatic pleural effusion from a 54-year-old female patient [55]. Studies have shown that T-47D is more susceptible to progesterone compared to the breast cancer cell line MCF7, progesterone being a hormone that is known to be prevalent in Luminal A breast cancer. The MCF7 breast cancer cell line was derived in the 1970s from a 69-year-old white female patient with metastatic pleural effusion and has since been one of the most well-known breast cancer cell lines [56]. Sources show that MCF7 has been searched the most (up to about 22,000 times depending on whether the hyphen is included) [57]. This ER+ breast cancer cell line has been used for 50 years and has many sub-clones, such as LM-MCF7, that have demonstrated high metastasis potential, downregulation of p27 (cyclin-dependent kinase inhibitor, also called KIP1, which is a tumor suppressor) expression, and upregulation of bcl-2 (b-cell lymphoma-2 is a regulator protein of cell death) protein expression when injected into severe combined immunodeficiency mice (SCID) [58]. Finally, Luminal A breast cancer cell line MDA-MB-415 was derived from a 38-year-old white female patient with pleural effusion at the metastatic site in 1975 [7]. MDA-MB-415 was used in a study to show that overexpression of the tumor suppressor p53 regulates apoptosis-inducing protein 1 (TP53AIP1) and can decrease cell viability in the breast cancer cell line [59].

All these Luminal A breast cancer cell line examples have in common a low percentage of Ki-67, meaning the recovery from breast cancer in the clinical setting is manageable. Some studies show the opposite of not having enough Ki-67 expression for Luminal A breast cancer, rendering the prognosis difficult without having that marker present [21]. Ki-67 was first discovered as an antigen in 1983 and has since been used as a proliferation marker for tumor growth in breast cancer cells [60]. Being able to analyze the expression of Ki-67 through all phases of the cell cycle, with the exception of G0, makes it an ideal tool to investigate the regulation of breast cancer growth and apoptosis [23]. There has been some controversy around the usefulness of the Ki-67 marker for Luminal A breast cancer prognosis based on the inconsistent cut-off range of 10–20% within the specimen for different research findings [61]. Immunohistochemistry staining of breast cancer tissue seems to be the primary tool for the identification of Ki-67 expression in breast cancer patients and is continuously being used for the prognosis of breast cancer progression status. Luminal A breast cancer subtypes have the advantage of being characterized by the presence of the ER+ receptor which can be manipulated in experiments involving the antiestrogen tamoxifen, the latter believed to be able to treat breast cancer long-term [62]. However, there are also disadvantages associated with the usage of Luminal A type breast cancer cell lines, such as the high levels of differentiation that can render the interpretation of the morphology difficult for various breast cancer cell line types within the category [16].

MDA-MB-330 and ZR-75-30 are the breast cancer cell lines of choice to represent the Luminal B subtype of breast cancer for this review. BT-474, MDA-MB-361, and UACC-812 are the breast cancer cell lines of choice that represent the Luminal-HER2 subtype of breast cancer cell lines for this review. Many of these Luminal B cell line types have a grape-like structure, as shown in Table 1. MDA-MB-330 was derived from a 43-year-old white female patient with metastatic pleural effusion in 1974 [7]. This breast cancer cell line is associated with invasive lobular carcinoma (ILC), rendering the transition from in vitro work to clinical therapy practices more specific and beneficial for patients with respect to treatment [26]. There has also been controversy with the ER+ status for this breast cancer cell line, hence why Table 1 shows +/− for ER in MDA-MB-330 cells [27]. ZR-75-30 is a breast cancer cell line derived from a 47-year-old black female patient in 1999 via metastatic ascites [28]. This breast cancer cell line is often associated with metastasis-associated proteins (MTA) MTA1 and MTA2 that depict overexpression or reduced expression to contribute to the metastasis of the ZR-75-30 breast cancer cell line [63]. Additionally, dietary compounds with epigenetic mechanisms such as berberine within *Coptis chinensis* have been shown to inhibit growth within the ZR-75-30 breast cancer cell line [64]. BT-474 was derived from a 60-year-old white female patient in 1978 who had invasive ductal breast carcinoma [65]. This breast cancer cell line has been used to improve the therapeutic efficiency against breast cancer with the use of the chemotherapy drug Hydroxyurea (HU) [66]. MDA-MB-361 was one of the nineteen cell lines derived at the M.D. Anderson Hospital and Tumor Institute in Houston, Texas. This breast cancer cell line was derived from metastases in the brain of a 40-year-old white female patient in 1975 [7]. The M.D. Anderson's series has been influential in cancer discovery and prevention. For example, by knocking out the *MALAT1* (metastasis associated lung adenocarcinoma transcript 1) gene with the use of clustered regularly interspaced short palindromic repeats (CRISPR) editing in the MDA-MB-361 breast cancer cell line, decreased proliferation and increased apoptosis has been observed, a phenomenon which can be further studied in vivo in a clinical setting for breast cancer patients [67]. In 1986, UACC-812 was derived from a 42-year-old white female patient who had ductal carcinoma [32]. Some researchers have shown this breast cancer cell line as a HER2 subtype because only ErbB-2 is expressed in experiments completed via PCR, culture assays, or other methods [68].

The Ki-67 protein biomarker for proliferation in cancerous cells is high in Luminal B type breast cancers [51]. This means that the cell lines have a larger number of proliferating cells that cause the cancerous cells to divide more quickly [69]. This is not the only gene that can be used to characterize Luminal B type breast cancer. *BRCA1* was first discovered by Dr. Mary-Claire King in 1994 and the gene is associated with hereditary breast cancer [70]. *BRCA1*, along with the *BRCA2* gene which is more prevalent in male breast cancer, is a tumor suppressor that helps repair DNA double-strand breaks and decrease the amount of rapid cancerous growth in cells. Studies have shown that, when comparing Luminal type breast cancer to TNBC, *BRCA1* is equally downregulated in mRNA expression for Luminal type and TNBC [71]. This does not explain why luminal breast cancers have a better prognosis and are an easy target for certain hormonal therapies like tamoxifen. Another gene that is highly studied within different types of breast cancer is *TP53*, which has responsibilities in the endocrine response and resistance pathway [72]. Mutations within the *TP53* gene have shown that this tumor suppressor can be a great biological marker for therapeutic strategies in breast cancer treatment. The prevalence of certain genes in these breast cancer cells can be utilized to create different medications in clinical settings. As seen in Figure 2, there is a specific pathway that these pharmaceutical drugs take to be able to intercept the molecular mechanisms of breast cancer with hormonal receptors (ER and PR) or membrane receptors (HER2). However, much work remains to be done to maintain these breast cancer cell lines and ensure no genotypic variants affect the results being produced. This may be seen as a disadvantage associated with the utilization of Luminal B breast cancer cell line types, as the more invasive characteristics weaken the luminal phenotype that is associated with high expression of HER2+ expression which, in turn, downregulates the expression of ER+ [16].

3.2.3. HER2+ Breast Cancer Cell Lines

MDA-MB-453, HCC1569, SUM190PT, AU565, and SK-BR-3 are all HER2+ breast cancer cell lines that do not have hormonal receptors for ER or PR. Grape-like structures and mass morphology are common within this type of breast cancer. MDA-MB-453 was first derived in 1976 from a 48-year-old white female patient with metastatic pericardial effusion [7]. This first example of HER2+ breast cancer cell lines has been a vital tool for the identification of molecular mechanisms with certain compounds used to treat cancer, such as hirsuteine [73]. Hirsuteine is an alkaloid that can be extracted from *Uncaria rhynchophylla* to treat different diseases, but the mechanism of downregulating BCL-2 and promoting apoptosis was not discovered until the use of the MDA-MB-453 cell line. MDA-MB-453 cells are an androgen-responsive breast carcinoma cell line with high-level AR expression. The HCC1569 breast cancer cell line was derived from a 70-year-old black female patient by isolating the mammary gland with metaplastic carcinoma in 1995 [74]. Since then, this breast cancer cell line has been used in the analysis of synergistic effects within the combination of PI3K (plasma membrane-associated lipid kinases) inhibitors and Herceptin to develop a less toxic implementation protocol for cancer patient treatment [75]. Unfortunately, researchers have yet to find any significant alteration within PTEN (phosphatase and tendon homolog that is a tumor suppressor) through the combination of treatment from PI3K and Herceptin applied to the HCC1569 breast cancer cell line [75], so more studies are required to find better concentrations or methodologies.

SUM190PT was first derived from a 40-year-old woman's primary tumor in 1996 and has been shown to lack ER and PR [76]. The SUM breast cancer cell line was established in the 1990s and has such a following that the SUM breast cancer knowledge base (SLKBase) was created for researchers to comment and post about different experiences, research attributes, and findings concerning this cell line and others [76]. SUM190PT has been used to observe the effectiveness of histone deacetylase (HDAC) inhibitors in creating new therapy treatments for cancer patients. Findings have shown that the HDAC inhibitor CG-1521 does induce apoptosis, although it targeted tubulin (non-histone protein used for spindle assembly in the cell). Therefore, more work needs to be performed to ensure the safety of CG-1521 for treatment purposes [77]. AU565 was first derived in 1970 from a 43-year-old white female patient with breast adenocarcinoma from the metastatic pleural effusion [78]. This breast cancer cell line is known to overexpress the ErbB2 protein [79]. The SK-BR-3 breast cancer cell line was derived in the same year as AU565, from the same 43-year-old white female through the pleural effusion metastasis that has similar characteristics of overexpression of the ErbB-2 gene product. Investigators discovered that SK-BR-3 was stimulated by the activation of the ninth EGF-like domain that contained GST (glutathione *S*-transferase) fusion proteins to indicate that no other ErbB2 ligand was derived from that specific EGF-like domain. There have also been studies conducted using flaxseed lignans in combination with pharmaceutical drugs that caused a decrease in cell viability in SK-BR-3, suggesting a potential improvement in chemotherapeutic applications in the clinical setting [80].

The ErbB receptor family is important for this type of breast cancer cell line, since HER2 is included as a receptor. A common characteristic of HER2+ breast cancer is that the prognosis is poor and the ability to treat this type of breast cancer is limited. Studies have been conducted with genes such as *EGF* and *hMena* (cytoskeleton regulatory protein) to analyze the phosphorylation and cell proliferation of HER2+ breast cancer cells [81]. This study did show that the downstream results of the knockout of *hMena* affecting activity of EGF in the cell may be a great attribute to identify new methods for the prognosis in breast cancer patients. An advantage of working with HER2+ breast cancer cell lines is that the HER2+ receptor allows certain drug targets to attach when conducting experimentation, e.g., Herceptin. The translation from in vitro to in vivo formed poorly differentiated tumors in immunocompromised mice with HER2+ SK-BR-3 cell line, so this may be a disadvantage [16].

3.2.4. Triple-Negative Breast Cancer Cell Lines

SUM229PE, BT-549, HCC70, BT-20, CAL148, MDA-MB-157, MDA-MB-231, and Hs 578T are all breast cancer cell lines that have been classified as triple-negative breast cancer (TNBC) [16,24,40]. Figure 3 illustrates a timeline of the cell line derivations along with key elements that have taken place in breast cancer cell line research. TNBC cell lines lack ER, PR, and HER2 and have been known to be the most aggressive/worst prognosis breast cancer type [16]. This breast cancer category is also known to be most common in black women who have a familial history of cancer [82]. Stellate is the most common structure among TNBC cells, along with a few that show round, mass, and spindle structures, as shown in Table 1. SUM229PE is a breast cancer cell line that was retrieved in 1996 from the pleural effusion fluid of a female patient [76]. This breast cancer cell line has also been used to study resistance to inhibitors such as MEK1/2 (mitogen-activated protein kinase 1/2) and as a control to test the vulnerabilities in drugs that may affect heterogeneity amongst the different cancerous cell types [39]. In 1978, BT-549 was derived from a 72-year-old white female patient by isolating the papillary of metastasized portions of the lymph node [83]. This breast cancer cell line showed downregulation of HDAC7 and established the possibility of this gene being a tumor suppressor in TNBC [84]. HCC70 was first derived from a 49-year-old black female patient in 1992 by isolating a primary ductal carcinoma [75]. This TNBC cell line has been used to study the effectiveness of cancer treatment drug conjugates on breast cancer cells, such as the peptide TH1902 for docetaxel [85]. BT-20 was the first breast cancer cell line to be derived in 1958 from a 74-year-old white female patient by isolating the tumor within her breast [5]. This hallmark breast cancer cell line has been used in studies to show alterations in mitochondria due to the function of key elements of the mTOR (mammalian target of rapalycin) pathway [86]. The CAL148 breast cancer cell line was derived from a 58-year-old French female patient with pleural effusion in 1994 [87]. In 2019, CAL148 was used to discover if two drugs, palbociclib and MLN0128, could work synergistically to inhibit cell proliferation, with results revealing that further investigation in the clinical setting would be beneficial [88]. MDA-MB-157 was first derived in 1972 from a 44-year-old black female patient with metastatic breast cancer and pleural effusion [7]. Both MDA-MB-157 and MDA-MB-231 have been used to study the effects of the HDAC inhibitor Panobinostat. Research has shown that this HDAC inhibitor is toxic to TNBC and could be a potential tool for treatment [89]. In 1973, MDA-MB-231 was derived from a 51-year-old white female patient by isolating the mammary gland of an adenocarcinoma via metastatic pleural effusion [7]. This TNBC cell line is known to be very aggressive and is associated with a poor diagnosis. The Hs 578T cell was first derived in 1977 from the breast of a 74-year-old white female patient [90]. This cell line has since been used as a tool to evaluate HMGA1 (high mobility group A) protein expression, as it relates to mitochondrial mutation in cancer cell research [91].

There has been controversy over the years concerning the MDA-MB-435 cell line and its derived origin. Investigators have shown that the MDA-MB-435 cell line was derived from a 31-year-old white female patient in 1976 with metastatic, ductal adenocarcinoma of the breast [92]. However, studies through gene analysis have depicted the clustering of this cell line with melanoma-origin cell lines. This discovery was made in 2000 by DNA microarray analysis and, following debate in 2007, the MDA-MB-435 cell line was determined to have originated from melanoma [93,94]. Even though this issue has been settled, there have still been scientific articles published that categorize the cell line as originating from breast cancer.

TNBC constitutes 10% to 15% of all breast cancers, and an important component to distinguish the presence of TNBC is retinoblastoma (RB1) status. RB1 is a tumor suppressor that is not found in TNBC and is currently being studied to understand its effects within specific therapies [95]. Research has shown that the presence of RB1 in TNBC lines is more sensitive to gamma radiation and that few RB1 are present in TNBC. All TNBC cell lines discussed above are examples from many studies aiming to reach the goal of cancer cure and prevention. More in vivo work is necessary as well as in the clinical setting to ensure proper verification of methodologies and techniques and prevent and eventually cure cancer. In vitro studies have been essential to understand breast cancer and will continue to inform and stimulate future research. TNBC cells also have subsections that are basal-like for enrichment with basal markers and claudin-low, which is associated with genes that are tumor-invasive and aggressive [16]. An advantage of TNBC cell line use is the characterization with *BRCA1* gene mutation for the basal-like subsection that can be analyzed with cell lines for the translation to clinical settings. A disadvantage of utilizing TNBC cell lines is the lack of receptors that may show relatability to immunotherapy that is done in the clinical setting [96]. However, dietary compounds have been shown to restore ER+ gene expression in TBNC through epigenetic mechanisms that may be applicable to chemotherapeutic approaches with additional findings [97].

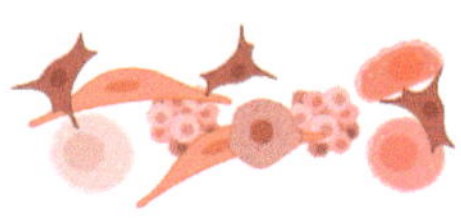

Breast Cancer Cell Line Derivation Timeline

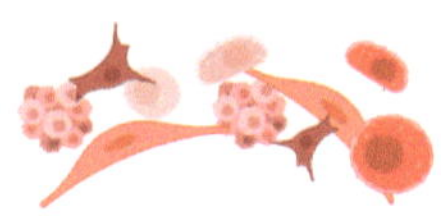

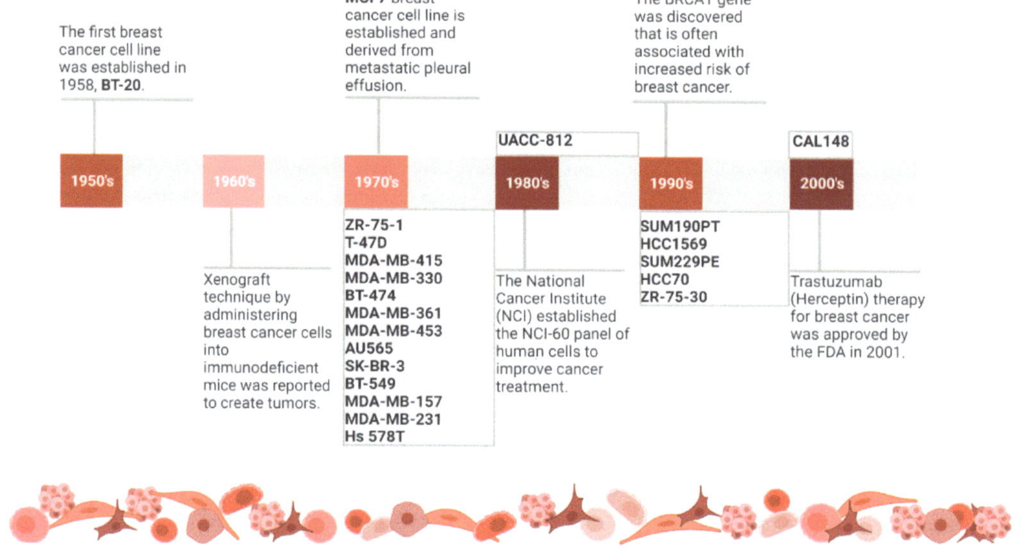

Figure 3. Time of breast cancer cell line derivation along with additional novel findings from breast cancer cell line research. Abbreviations for each breast cancer cell line origin can be found in Section 3.2.

3.3. Common Requirements, Techniques, and Approaches to Cell Culture Maintenace

3.3.1. Medium Choice and Control Cell Lines

These breast cancer cell lines must have a mixture of components for each medium type to ensure the proper nutrients are available to support cell growth and maintain cells viable for experimental studies. In Table 2, the components for each of the breast cancer cell lines from Table 1 are outlined. These are examples for each breast cancer cell line and the addition of antibiotics is based on laboratory preference. Some investigators also recommend the addition of a DMEM high-glucose medium instead of a DMEM normal-glucose medium for cell culture components to ensure higher maintenance efficiency in breast cancer cell lines [98]. Investigators can choose various medium types and add the necessary nutrients for the breast cancer cell lines being utilized. Common media of choice are Human Plasma-Like Medium (HPLM), Minimal Essential Medium (MEM), Iscove's Modified Dulbecco's Medium (IMDM), Dulbecco's modified Eagles Medium (DMEM, high glucose), DMEM/F-12 nutrient media, Ham's F-12 Nutrient Medium, and Roswell Park Memorial Institute 1640 Medium (RPMI 1640, low glucose) [99]. Often, Fetal Bovine Serum (FBS), a growth supplement, is used to promote growth through proteins and growth factors in a cell culture environment. Antibiotics such as streptomycin, amphotericin B, and penicillin are used to prevent cell wall synthesis and interfere with cell permeability and bacterial development in cell cultures. On the other hand, antibiotics such as penicillin and streptomycin have been shown to also alter gene expression, cell regulation, and drug response. This is an aspect that scientists must consider when designing an experiment. In addition to the medium of choice, certain additives can be placed within the medium to ensure proper breast cancer cell line maintenance, such as pyruvate that stabilizes the hypoxia-inducible factors in TNBC cells [100]. The environment within which all these breast cancer cell lines must remain to grow is a laboratory-validated incubator at 37 °C. While a 5–10% CO_2 and air mixture is used in association with most available culture media, Leibovitz's L-15 medium (used for the UACC-812 cell line) is detrimental to cell cultivation in a CO_2 and air mixture environment [32].

Table 2. Optional medium choice for breast cancer cell line maintenance.

Breast Cancer Cell Line	Foundational Media	Supplement Additives	References
ZR-75-1	RPMI	10% FBS, 10 mL penicillin and streptomycin	[56]
T-47D	RPMI/DMEM	100% FBS, 100 IU/mL penicillin, 100 μg/mL streptomycin	[55,101]
MCF7	RPMI/DMEM	10% FBS, 100 IU/mL penicillin, 100 μg/mL streptomycin	[55,101]
MDA-MB-415	DMEM	10% FBS, 100 U/mL streptomycin and penicillin	[60]
MDA-MB-330	RPMI	10% FBS, non-essential amino acids and insulin	[26,27]
ZR-75-30	RPMI	10% FBS [1], 10 μg/mL insulin	[28]
BT-474	RPMI	10% FBS, Hybri-Care Medium, 1 L cell-culture0grade-water, 1.5 g/L sodium bicarbonate	[102]
MDA-MB-361	RPMI/DMEM	8–10% FBS, 100 U/mL penicillin and 100 μg/mL streptomycin	[101]
UACC-812	L-15/RPMI/DMEM	10–20% FBS, 2 mmol/L glutamine, 1% PSF	[103,104]
MDA-MB-453	RPMI/DMEM	10% FBS, penicillin (100 U/mL), streptomycin (100 μg/mL), 200 mM L-glutamine	[73,105]
HCC1569	RPMI	10% FBS, 2 mM L-glutamine, 100 U/mL penicillin, and 100 μg/mL streptomycin	[75,106]
SUM190PT	Ham's F12	2% FBS, 1 g/L BSA, 5 mM ethanolamine, 10 mM HEPES, 0.1% hydrocortisone, 5 μg/mL insulin, 50 nM sodium selenite, 5 μg/mL transferrin, 10 nM T3	[107]
AU565	RPMI	10% FBS, 10 mM HEPES, 1 mM sodium pyruvate, 1% penicillin/streptomycin, 2.5 g/L glucose	[107]
SK-BR-3	DMEM/MCCOYS	10% heat-inactivated FBS, 100 μg/mL penicillin G, and 80 μg/mL streptomycin	[108]
SUM229E	RPMI/Ham's F12	5% FBS, 10 μg/mL, penicillin-streptomycin, 0.5 μg/mL hydrocortisone	[55]
BT-549	RPMI	10% FBS, 100 μg/mL streptomycin, 100 U/mL penicillin, 10 μg/mL insulin	[109]
HCC70	RPMI	10% FBS	[110]
BT-20	EMEM/RPMI/DMEM	10% FBS, penicillin, and streptomycin	[38]
CAL148	DMEM	10% FBS, 1% penicillin-streptomycin, 1% sodium pyruvate	[88]
MDA-MB-157	RPMI/DMEM	10% FBS, 1% 100× penicillin-streptomycin-amphotericin B, 1% 100× nonessential amino acid solution	[7]
MDA-MB-231	RPMI/DMEM	10% FBS, 100 IU/mL penicillin, 100 μg/mL streptomycin	[101]
MDA-MB-435	L-15/RPMI/DMEM	10% heat-inactivated FBS, 100 μg/mL penicillin G, and 80 μg/mL streptomycin	[108]
Hs 578T	RPMI/DMEM	10% FBS, 0.01 mg/mL human insulin	[7,111]

[1] Foundational media must be used for each breast cancer cell line: EMEM (Eagle's Minimum Essential Medium), L-15 (Leibovitz's L-15 Medium), RPMI (Roswell Park Memorial Institute-1640 medium), DMEM (Dulbecco's Modified Eagle Medium), and Ham's F12 (Ham's F-12 (Kaighn's) Medium). CO2 and air mixture are detrimental to UACC-812 cells when using an L-15 medium for cultivation. Additives are used to ensure the breast cancer cell lines have nutrients to support growth in a controlled environment. Antibiotics are optional for laboratories that would like to prevent bacterial growth. The source for the HCC70 cell line did not add anything to the FBS. FBS: Fetal Bovine Serum, PSF: Penicillin G-streptomycin–fungizone solution, BSA: Bovine Serum Albumin, HEPES (4-(2-hydroxyethyl)-1-piperazineethanesulfonic acid) is a zwitterionic sulfonic acid buffering agent, T3 (Triiodo Thyronine) is a hormone produced by the thyroid gland.

The MCF10A cell line is non-cancerous and has different strains that have been used by many investigators. Breast cancer cell line maintenance is vital for the success of experimental trials, leading, hopefully, to clinical trials. The presence of control cell lines is a requirement to ensure the data acquired are reliable and accurate. At least two, preferably three, cell lines and one control cell line is what most consider to be the optimal number of cell lines to obtain reliable data. For example, an investigator used [ER− PR−] MDA-MB-231 and ER+ MCF7 breast cancer cell lines along with MCF10A control cells in their study for combinatorial epigenetic mechanisms of sulforaphane, genistein, and sodium butyrate in breast cancer inhibition [12]. The MCF10 cell line was derived in 1990 by Soule and colleagues as the first non-transformed, human mammary epithelial cell line derived from normal breast tissue [112]. Multiple sublines including MCF10AneoT, MCF10AT, MCF10DCIS, and MCF10CA1 have been used as excellent models to help analyze and classify many breast cancer cell line types. Although MCF10 cell lines are widely known and the most commonly used benign proliferative breast cancer tissue-derived cell line, there have been studies that show MCF10A cells may not represent luminal, basal, and normal cells, phenotypically, when placed in different culture systems (2D versus 3D) [113].

The use of human mammary epithelial cells (hMECs) has also played a vital role in breast cancer cell line research, representing another control cell line for the comparison with breast cancer cell line progression and development [114]. hMECs are normal epithelial cells that have been utilized to monitor, in vitro, the early stages of tumorigenesis along with the ability to reprogram to a previous state (neoplastic) [115]. There have been issues with the short, allotted time for passages during the cell culture process. hMECs are already destroyed embryos from stillbirths that are difficult to grow in a lab setting. Based on research conducted previously, investigators have found that the passage time is limited to 5–8 times before the cell line is unable to be chosen as a control cell line for breast cancer research. On the other hand, MCF10A cell lines have been utilized within so many sublines and, even though the growth phase is slower than most breast cancer cells, the cell line is an optimal choice for breast cancer cell line work.

3.3.2. Cellular Techniques and Morphology

The discovery of histological differences in origin can be stated to have started in 1906 when Histology was recognized as a biomedical discipline [98]. Advances in microscope technology have enabled investigators to observe inside organs, tissues, and even cells. The ability to distinguish between different cell types was not possible until the invention of the microscope in the 1600s and. Since then, the evolution of histology has hastened. The classification of different shapes for breast cancer types in Table 1 was established using the first 3D models of cell cultures in microenvironments in 2007 [77]. The grape-like morphology of breast cancer cell types is mostly associated with Luminal B breast cancer, but AU565, which is an example of triple-negative breast cancer, ZR-75-1, associated with Luminal A, and SK-BR-3, which is an example of HER2+, have also been classified as exhibiting a grape-like morphology.

The round morphology is associated with MDA-MB-415 and CAL148 breast cancer cell lines. These breast cancer cell types are from two different categories (Luminal A and TNBC, respectively), but there is a common protein that is associated with round morphology breast cancer cell types. Moderate levels of ErbB-2 are required for the formation of round morphology cell lines, while higher expression levels are required for HER2+ cell lines [66]. Mass morphology is very diverse and incorporates breast cancer cell types from all four categories. Mass morphology breast cancer cell types include T-47D, MCF7, BT-474, HCC1569, and HCC70, all of which have the highest level of proton ErbB-2 expression from western blot analysis [66]. Stellate morphology is characteristic of all TNBC cell lines, including BT-549, BT-20, MDA-MB-157, MDA-MB-231, and Hs 578T. MDA-MB-435 is also classified as having stellate morphology despite being a melanoma-derived cell line [88]. The morphology of each breast cancer cell line has been described, including mass, grape-like, stellate, round, and spindle which are depicted in Figure 4.

Breast Cancer Cell Culture Classification by Morphological Structure

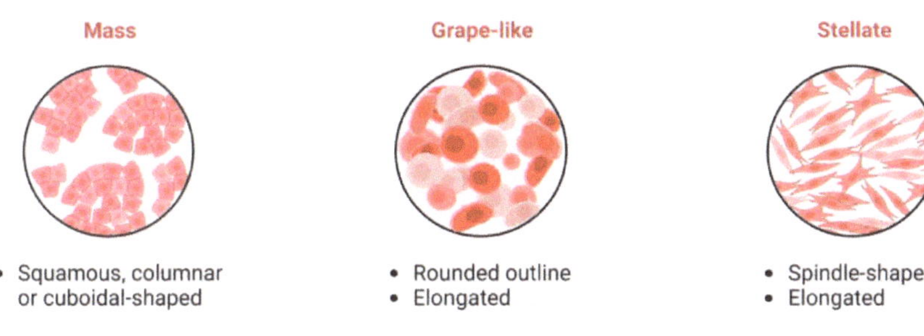

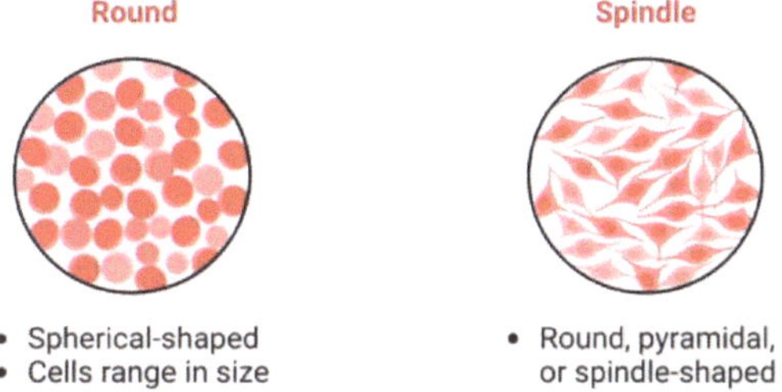

Figure 4. Types of morphology to group the Luminal A, Luminal B, Luminal-HER2, HER2, TNBC, and other cancerous cell lines. Each breast cancer cell line morphology is explained under Sections 3.2.2–3.2.4 and 3.3.2.

The process of creating 2D versus 3D cell culture models is summarized in Figure 5. Maintaining 2D cell cultures has the advantages of being cost-effective and convenient while allowing diffusion of soluble factors into the media. Disadvantages of 2D cell culturing include reduced cell-to-cell interactions and the less translatable models [116]. Three-dimensional culturing has been expanded with an additional technique of co-culturing between the MCF7 breast cancer cell line and MRC-5 fibroblast spheroids to study the many mechanisms that contribute to a 3D environment [117]. The utilization of 3D cell culture techniques allows a more accurate representation for translation to in vivo studies, increases cell-to-cell ECM interactions, and the matrix fibrils can restrict cell spreading [118]. The disadvantages of using 3D culture include the handling of a more complex culture system, as seen in Figure 5, which includes ECM layers to incorporate an aspect of the basement membrane (matrigel, collagen, polydimethylsiloxane, and laminin) [119]. Breast cancer cell lines that have been utilized as spheroid models cultured in a 3D system include, but are not limited to, ZR-75-1, T-47D, MCF7, MDA-MB-415, BT-474, MDA-MB-361, UACC-812, MDA-MB-453, HCC1569, AU565, BT-540, HCC70, BT-20, MDA-MB-231, and Hs578T cells [24]. Additionally, cellular techniques can be utilized to maintain various breast cancer cell line types.

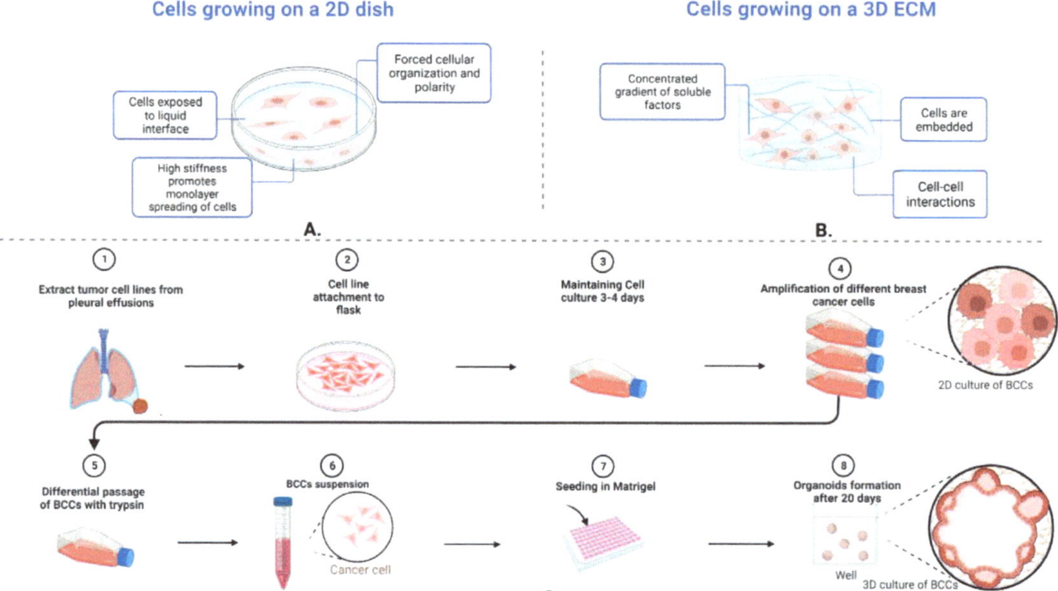

Figure 5. There are key differences between the utilization of 2D cell cultures and 3D cell cultures. Part (**A**) shows that breast cancer cells grown on a 2D dish have a forced cellular organization and polarity, high stiffness to promote monolayer of spreading cells, and no cell–cell interaction. Part (**B**) illustrates 3D cell culturing techniques to embed breast cancer cells in ECM, incorporate cell–cell interactions, and form a more complex culturing system. Part (**C**) illustrates most breast cancer cell line derivation from pleural effusions to plating in a 2D culture flask for amplification of the differentiated cells. While organoid formation can then be formed after proper breast cancer proliferation and embedding in the proper ECM material, see Section 3.3.2 for more information.

4. Additional Concepts Involving Breast Cancer Cell Line Research

4.1. Cross-Contamination

Cross-contamination of breast cancer cells can occur through multiple means such as culturing techniques and mislabeling of containers [120]. HeLa cells were the first to be developed in 1952 from glandular cancer of the cervix [6]. The HeLa cell confusion caused a catastrophe in many laboratories over 60 years ago, but there are still breast cancer cell lines in use that derive from that very discovery [3,121]. Sterilization of the area within the Biosafety Cabinet (BSC) and gloves when handling breast cancer cells with 70% ethanol is essential to keep contamination at a minimum. Studies have shown that benzalkonium chloride with corrosive inhibition and distilled water in wet conditions would be the optimal combination for the sterilization of the BSC area with respect to a specific bacterium [122]. Considering that benzalkonium chloride with corrosive inhibition may not be the optimal option for most laboratories, 70% ethanol is a common choice for use when sterilizing surfaces. Freezing cells for future use can ensure the longevity of a breast cancer cell line, but care must be taken when labeling and opening the cryovial for the re-suspension of cell lines. If a 37 °C water bath is not properly maintained to thaw frozen cell lines for <1 min, this can cause contamination. Removing cell lines from liquid nitrogen storage and placing them directly in the water bath is essential to maintain cell viability. If placed on ice temporarily, breast cancer cells can thaw and die before being plated. UV radiation is a great method before and after cell culturing to maintain a sterile environment and not cause any cross-contamination. Investigators have discovered that UV radiation can cause DNA methylation alteration in cells; therefore, proper protocols should be in place in a laboratory environment to prevent this from influencing cellular morphology [123].

4.2. Genetic Mutations

Studies have shown that breast cancer cell lines may have more mutations than the tumor from which they are derived, and this raises concerns for researchers who are interested in translating conclusions found from breast cancer cell line studies [16]. History has shown that an increase in mutated BRCA1 or BRCA2 can put patients at higher risk of developing cancer, but the origin of many of these breast cancer cell lines studied may have mutated. An example of a breast cancer cell line that is widespread and has been shown to have diverse genetic uncertainties through multiple analyses is MCF7. This breast cancer cell line has been shown to react differently to drugs used over a length of time from the various subculturing processes to the freezing and thawing of the cells [124]. Prevention of genetic mutations of breast cancer cell lines is possible, with helpful techniques including maintenance of a sterile environment when subculturing, minimization of the freeze–thaw process, and documentation of the cell line passage number to prevent overpassaging [100]. Additional genetic mutations that have been utilized and used as tools to target breast cancer are cyclin-dependent kinase inhibitor 2A (*CDKN2A*), phosphatidylinositol 3-kinase Catalytic Subunit Alpha (*PIK3CA*), phosphate and tensin homolog (*PTEN*), and *TP53*, which is a tumor suppressor protein p53 that was previously mentioned [125–128]. Another novel molecular biomarker that has been influential in oncological decision-making is programmed death ligand 1 (*PD-L1*), which has been shown to have intrinsic capabilities concerning triple-negative breast cancer cells. The knockdown of this immunosuppressive protein has been shown to decrease cell proliferation and tumor growth in the model organism chosen for the experiment [129]. Examples such as these illustrate the essential role breast cancer cell lines play in the development of therapeutic techniques for breast cancer prevention and therapy. There are numerous genes that can be either overexpressed or underexpressed in breast cancer cell lines, and a few of those that have been studied are mentioned in Table 1, e.g., cadherin 1 (*CDH1*) and FA complementation group A (*FANCA*) that has been shown to be associated with lobular breast cancer [130–132]. Spen family transcriptional repressor (*SPEN*) has been associated with poor prognosis of breast cancer and is involved with chromatin remodeling [130,133]. Catenin alpha 1 (*CTNNA1*) and mediator complex subunit 12 (*MED12*) have been shown to associate with breast cancer [130,131,134].

4.3. Cell-Derived Xenograft and Gut Microbiota in Breast Cancer

Utilization of patients with breast cancer cells has been employed with respect to many techniques to diagnose and investigate various methods for better treatment of the disease. The first xenograft technique was implemented in 1962 from human breast cancer to an immunodeficient mouse [135]. The technique of patient-derived xenografts (PDXs) has been conducted by numerous institutions for therapeutic and clinical trials [136]. This process involves the removal of cells from a patient with a known status of breast cancer and the injection of such cells into an immunodeficient mouse in a specific area that will show the growth of a tumor for analysis, as seen in Figure 6. The exact location of injection could be, for example, intraperitoneally or orthotopically, although it depends on the the type of cancer. Injection of breast cancer cells into the mammary fat pad of immunocompromised mice is the most common example of PDXs. Many strains of immunodeficient mice have been utilized since the discovery of the 'nude' mice model in 1962 by Grist [137]. An interesting concept that has been discovered is that the difference between PDXs and cell-derived xenografts (CDXs) may not be as obvious according to pathologists viewing histology slides [138]. Viewing immunohistological quality control (QC) slides is a daily routine in pathology; therefore, utilization of CDXs instead of PDXs may be beneficial if there is no clear difference.

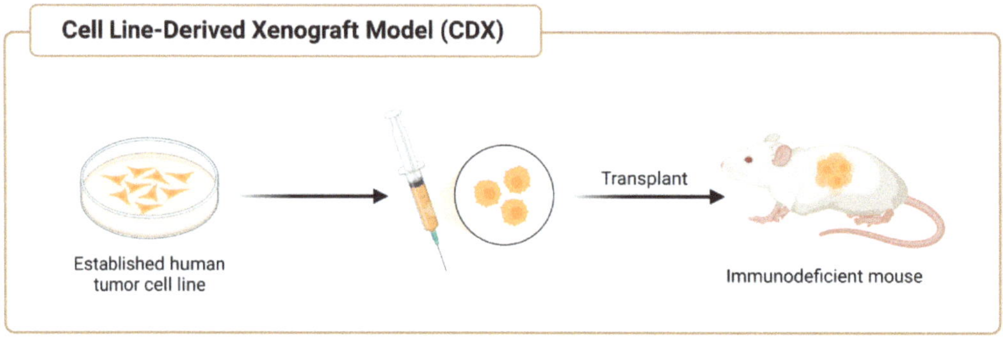

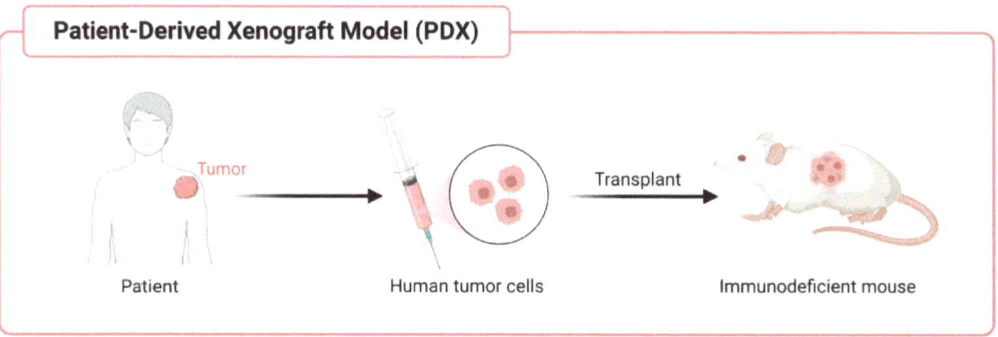

Figure 6. Comparison of the Cell Line-Derived Xenograft (CDX) model process versus Patient-Dervied Xenograft (PDX) model process.

One connection that has been investigated but needs further investigation is the relationship between breast cancer and the gut microbiota environment. Studies have established that as a result of the introduction of (GE) genistein into mice diet, there were microbial alterations in members of the family Lachnospiraceae and Ruminococcaceae [139]. No significant difference in microbiota composition was found between pre-chemotherapy and post-chemotherapy fecal samples of breast cancer patients, although introduction of GE into their diet did induce epigenetic changes resulting in reduced tumor size and increased tumor latency. The use of known epigenetic factors from dietary compounds that can be included into the diet of immunocompromised mice may become vital for breast cancer research. The fact that the mice used in the pilot study were humanized mice and not injected directly with breast cancer cells shows how the heterogeneity of breast cancer cells translated appropriately. Utilization of CDXs over PDXs could uncover novel findings in breast cancer research and prevention, but the data to analyze translatable evidence that CDXs can be applied to human clinical trials are still being gathered and investigated. The Luminal A breast cancer cell lines MCF7 and T-47D have been frequently used to inoculate mice into xenograft models for further examination because of the presence of estrogen [55]. However, Luminal A or Luminal B breast cancer cells engrafted from PDXs are difficult to grow and maintain due to ER+ tumors [140]. Investigators have used HER2+ breast cancer cell lines such as MDA-MB-453 to discover metastasis ability through intravenous injection (IVI) for CDX models [141]. The TNBC cell line MDA-MB-231 has been utilized with CDXs to investigate a decrease in tumor growth and induce G1 cell cycle arrest when using targeting agents [142].

Researchers have analyzed the immunohistological stains for β-catenin, Ki-67, and E-cadherin in human cancer types versus cell-derived xenografts. The expression of β-catenin in tissue can result from aberrations in the Wingless-related integration site (Wnt) signaling pathway. Levels of Ki-67 in breast cancer tissue have been used as a biomarker protein for cell proliferation in breast cancer. E-cadherin expression in breast cancer cells can distinguish invasive ductal or lobular cancer. Being able to investigate these protein markers as they relate to the gut microbiome can show how estrogen in breast cancer reacts in the body. Utilization of CDXs may lower costs for experimentation and develop connections that are lacking in breast cancer research. Studies have uncovered phenotypical similarities that suggest that CDXs can play a role in further investigation of tumor budding in colorectal cancer [143]. Realizing how essential CDXs have been used in many cancer types to facilitate deeper understanding is important. On the other hand, studies have shown that six breast cancer cell lines (ER+:[UCD4, UCD12, and UCD65] and ER−:[UCD46, UCD115, and UCD178]) have been created from PDXs to increase the overall ER+ number of cell lines within the archive to manipulate for breast cancer research [144]. These findings may bridge the gap in recognizing the connection between breast cancer and the gut microbiota environment.

5. Conclusions

Breast cancer cell lines have been utilized for over 50 years to establish prognosis, protein biomarkers, morphological differences, and genetic mutations. There is much to be discovered still, but, through this review, a researcher can take advantage of the established knowledge that has been produced. By categorizing breast cancer cell lines into Luminal A, Luminal B, Luminal-HER2, HER2+, and TNBC, the investigator can be very precise when analyzing data and creating experiments to make novel advancements in science. By utilizing the unique techniques described, such as 2D versus 3D subculturing, xenograft experimentation, and, possibly, the application of various dietary compound concentrations to analyze epigenetic regulation within breast cancer cell line growth inhibition, investigators may discover creative applications that can be practiced through analyzing the breast cancer cell lines discussed in this review. Being able to understand and appreciate the components of breast cancer cell line types that include, but are not limited to, Ki-67 expression level and various genetic mutation statuses, the investigator may carefully analyze and choose the correct cell line that can coincide with their field of expertise. This review is intended to assist the researcher when creating new ideas, and can be used as a guidance resource when additional background information can be useful for the origin of various breast cancer cell types.

Author Contributions: Conceptualization, B.L.W. and T.O.T.; writing—original draft preparation, B.L.W.; writing—review and editing, B.L.W. and T.O.T.; supervision, T.O.T.; funding acquisition, T.O.T. All authors have read and agreed to the published version of the manuscript.

Funding: This research was funded by grants from the National Cancer Institute (R01 CA178441 and R01 CA204346).

Acknowledgments: All figures were created with BioRender.com (accessed on 17 September–5 December 2023). The authors acknowledge support from the Tollefsbol lab members. Also, thanks are extended to Huixin Wu and Sebanti Ganguly for their helpful comments.

Conflicts of Interest: The authors declare no conflict of interest.

References

1. Boekhout, A.H.; Beijnen, J.H.; Schellens, J.H. Trastuzumab. *Oncologist* **2011**, *16*, 800–810. [CrossRef] [PubMed]
2. Neve, R.M.; Chin, K.; Fridlyand, J.; Yeh, J.; Baehner, F.L.; Fevr, T.; Clark, L.; Bayani, N.; Coppe, J.P.; Tong, F.; et al. A collection of breast cancer cell lines for the study of functionally distinct cancer subtypes. *Cancer Cell* **2006**, *10*, 515–527. [CrossRef]
3. Masters, J.R. HeLa cells 50 years on: The good, the bad and the ugly. *Nat. Rev. Cancer* **2002**, *2*, 315–319. [CrossRef]
4. Holliday, D.L.; Speirs, V. Choosing the right cell line for breast cancer research. *Breast Cancer Res. BCR* **2011**, *13*, 215. [CrossRef]
5. Lasfargues, E.Y.; Ozzello, L. Cultivation of human breast carcinomas. *J. Natl. Cancer Inst.* **1958**, *21*, 1131–1147. [PubMed]

6. Burdall, S.E.; Hanby, A.M.; Lansdown, M.R.; Speirs, V. Breast cancer cell lines: Friend or foe? *Breast Cancer Res. BCR* **2003**, *5*, 89–95. [CrossRef]
7. Cailleau, R.; Olivé, M.; Cruciger, Q.V. Long-term human breast carcinoma cell lines of metastatic origin: Preliminary characterization. *In Vitro* **1978**, *14*, 911–915. [CrossRef]
8. Jovanovic, J.; Rønneberg, J.A.; Tost, J.; Kristensen, V. The epigenetics of breast cancer. *Mol. Oncol.* **2010**, *4*, 242–254. [CrossRef]
9. Mahgoub, M.; Monteggia, L.M. Epigenetics and psychiatry. *Neurother. J. Am. Soc. Exp. NeuroTher.* **2013**, *10*, 734–741. [CrossRef]
10. Couto, J.P.; Vulin, M.; Jehanno, C.; Coissieux, M.M.; Hamelin, B.; Schmidt, A.; Ivanek, R.; Sethi, A.; Bräutigam, K.; Frei, A.L.; et al. Nicotinamide N-methyltransferase sustains a core epigenetic program that promotes metastatic colonization in breast cancer. *EMBO J.* **2023**, *42*, e112559. [CrossRef]
11. Campagna, R.; Vignini, A. NAD+ Homeostasis and NAD+-Consuming Enzymes: Implications for Vascular Health. *Antioxidants* **2023**, *12*, 376. [CrossRef] [PubMed]
12. Sharma, M.; Tollefsbol, T.O. Combinatorial epigenetic mechanisms of sulforaphane, genistein and sodium butyrate in breast cancer inhibition. *Exp. Cell Res.* **2022**, *416*, 113160. [CrossRef]
13. Mirabelli, P.; Coppola, L.; Salvatore, M. Cancer Cell Lines Are Useful Model Systems for Medical Research. *Cancers* **2019**, *11*, 1098. [CrossRef]
14. Wu, H.; Ganguly, S.; Tollefsbol, T.O. Modulating Microbiota as a New Strategy for Breast Cancer Prevention and Treatment. *Microorganisms* **2022**, *10*, 1727. [CrossRef] [PubMed]
15. Lacroix, M.; Leclercq, G. Relevance of breast cancer cell lines as models for breast tumours: An update. *Breast Cancer Res. Treat.* **2004**, *83*, 249–289. [CrossRef]
16. Dai, X.; Cheng, H.; Bai, Z.; Li, J. Breast Cancer Cell Line Classification and Its Relevance with Breast Tumor Subtyping. *J. Cancer* **2017**, *8*, 3131–3141. [CrossRef] [PubMed]
17. Hui, C.; Qi, X.; Qianyong, Z.; Xiaoli, P.; Jundong, Z.; Mantian, M. Flavonoids, flavonoid subclasses and breast cancer risk: A meta-analysis of epidemiologic studies. *PLoS ONE* **2013**, *8*, e54318. [CrossRef]
18. Fraga, M.F.; Ballestar, E.; Paz, M.F.; Ropero, S.; Setien, F.; Ballestar, M.L.; Heine-Suñer, D.; Cigudosa, J.C.; Urioste, M.; Benitez, J.; et al. Epigenetic differences arise during the lifetime of monozygotic twins. *Proc. Natl. Acad. Sci. USA* **2005**, *102*, 10604–10609. [CrossRef]
19. Zhou, Y.; Zheng, J.; Li, Y.; Xu, D.P.; Li, S.; Chen, Y.M.; Li, H.B. Natural Polyphenols for Prevention and Treatment of Cancer. *Nutrients* **2016**, *8*, 515. [CrossRef]
20. Leroy, B.; Girard, L.; Hollestelle, A.; Minna, J.D.; Gazdar, A.F.; Soussi, T. Analysis of TP53 mutation status in human cancer cell lines: A reassessment. *Hum. Mutat.* **2014**, *35*, 756–765. [CrossRef]
21. Subik, K.; Lee, J.F.; Baxter, L.; Strzepek, T.; Costello, D.; Crowley, P.; Xing, L.; Hung, M.C.; Bonfiglio, T.; Hicks, D.G.; et al. The Expression Patterns of ER, PR, HER2, CK5/6, EGFR, Ki-67 and AR by Immunohistochemical Analysis in Breast Cancer Cell Lines. *Breast Cancer Basic Clin. Res.* **2010**, *4*, 35–41. [CrossRef]
22. Ding, Z.; Xu, X.; Li, T.; Wang, J.; Sun, J.; Tang, L. ZR-75-1 breast cancer models to study the utility of 18F-FES by PET imaging. *Transl. Cancer Res.* **2021**, *10*, 1430–1438. [CrossRef] [PubMed]
23. Li, L.T.; Jiang, G.; Chen, Q.; Zheng, J.N. Ki67 is a promising molecular target in the diagnosis of cancer (review). *Mol. Med. Rep.* **2015**, *11*, 1566–1572. [CrossRef] [PubMed]
24. Kenny, P.A.; Lee, G.Y.; Myers, C.A.; Neve, R.M.; Semeiks, J.R.; Spellman, P.T.; Lorenz, K.; Lee, E.H.; Barcellos-Hoff, M.H.; Petersen, O.W.; et al. The morphologies of breast cancer cell lines in three-dimensional assays correlate with their profiles of gene expression. *Mol. Oncol.* **2007**, *1*, 84–96. [CrossRef] [PubMed]
25. Blenkiron, C.; Goldstein, L.D.; Thorne, N.P.; Spiteri, I.; Chin, S.F.; Dunning, M.J.; Barbosa-Morais, N.L.; Teschendorff, A.E.; Green, A.R.; Ellis, I.O.; et al. MicroRNA expression profiling of human breast cancer identifies new markers of tumor subtype. *Genome Biol.* **2007**, *8*, R214. [CrossRef] [PubMed]
26. Tasdemir, N.; Bossart, E.A.; Li, Z.; Zhu, L.; Sikora, M.J.; Levine, K.M.; Jacobsen, B.M.; Tseng, G.C.; Davidson, N.E.; Oesterreich, S. Comprehensive Phenotypic Characterization of Human Invasive Lobular Carcinoma Cell Lines in 2D and 3D Cultures. *Cancer Res.* **2018**, *78*, 6209–6222. [CrossRef]
27. Sflomos, G.; Schipper, K.; Koorman, T.; Fitzpatrick, A.; Oesterreich, S.; Lee, A.V.; Jonkers, J.; Brunton, V.G.; Christgen, M.; Isacke, C.; et al. Atlas of Lobular Breast Cancer Models: Challenges and Strategic Directions. *Cancers* **2021**, *13*, 5396. [CrossRef]
28. Engel, L.W.; Young, N.A.; Tralka, T.S.; Lippman, M.E.; O'Brien, S.J.; Joyce, M.J. Establishment and characterization of three new continuous cell lines derived from human breast carcinomas. *Cancer Res.* **1978**, *38*, 3352–3364.
29. Koseoglu, S.; Lu, Z.; Kumar, C.; Kirschmeier, P.; Zou, J. AKT1, AKT2 and AKT3-dependent cell survival is cell line-specific and knockdown of all three isoforms selectively induces apoptosis in 20 human tumor cell lines. *Cancer Biol. Ther.* **2007**, *6*, 755–762. [CrossRef]
30. Depmap. Available online: https://depmap.org/portal/home/#/ (accessed on 8 September 2023).
31. Davey, M.G.; Hynes, S.O.; Kerin, M.J.; Miller, N.; Lowery, A.J. Ki-67 as a Prognostic Biomarker in Invasive Breast Cancer. *Cancers* **2021**, *13*, 4455. [CrossRef]
32. Meltzer, P.; Leibovitz, A.; Dalton, W.; Villar, H.; Kute, T.; Davis, J.; Nagle, R.; Trent, J. Establishment of two new cell lines derived from human breast carcinomas with HER-2/neu amplification. *Br. J. Cancer* **1991**, *63*, 727–735. [CrossRef] [PubMed]

33. Liu, X.; Ye, Y.; Zhu, L.; Xiao, X.; Zhou, B.; Gu, Y.; Si, H.; Liang, H.; Liu, M.; Li, J.; et al. Niche stiffness sustains cancer stemness via TAZ and NANOG phase separation. *Nat. Commun.* **2023**, *14*, 238. [CrossRef]

34. Ruiz-Saenz, A.; Dreyer, C.; Campbell, M.R.; Steri, V.; Gulizia, N.; Moasser, M.M. HER2 Amplification in Tumors Activates PI3K/Akt Signaling Independent of HER3. *Cancer Res.* **2018**, *78*, 3645–3658. [CrossRef] [PubMed]

35. Barnabas, N.; Cohen, D. Phenotypic and Molecular Characterization of MCF10DCIS and SUM Breast Cancer Cell Lines. *Int. J. Breast Cancer* **2013**, *2013*, 872743. [CrossRef] [PubMed]

36. Hollestelle, A.; Elstrodt, F.; Timmermans, M.; Sieuwerts, A.M.; Klijn, J.G.; Foekens, J.A.; den Bakker, M.A.; Schutte, M. Four human breast cancer cell lines with biallelic inactivating alpha-catenin gene mutations. *Breast Cancer Res. Treat.* **2010**, *122*, 125–133. [CrossRef] [PubMed]

37. Castles, C.G.; Fuqua, S.A.; Klotz, D.M.; Hill, S.M. Expression of a constitutively active estrogen receptor variant in the estrogen receptor-negative BT-20 human breast cancer cell line. *Cancer Res.* **1993**, *53*, 5934–5939. [PubMed]

38. Papadakos, K.S.; Ekström, A.; Slipek, P.; Skourti, E.; Reid, S.; Pietras, K.; Blom, A.M. Sushi domain-containing protein 4 binds to epithelial growth factor receptor and initiates autophagy in an EGFR phosphorylation independent manner. *J. Exp. Clin. Cancer Res. CR* **2022**, *41*, 363. [CrossRef]

39. Gu, Y.; Helenius, M.; Väänänen, K.; Bulanova, D.; Saarela, J.; Sokolenko, A.; Martens, J.; Imyanitov, E.; Kuznetsov, S. BRCA1-deficient breast cancer cell lines are resistant to MEK inhibitors and show distinct sensitivities to 6-thioguanine. *Sci. Rep.* **2016**, *6*, 28217. [CrossRef]

40. Ziperstein, M.J.; Guzman, A.; Kaufman, L.J. Breast Cancer Cell Line Aggregate Morphology Does Not Predict Invasive Capacity. *PLoS ONE* **2015**, *10*, e0139523. [CrossRef]

41. Chiang, Y.T.; Chien, Y.C.; Lin, Y.H.; Wu, H.H.; Lee, D.F.; Yu, Y.L. The Function of the Mutant p53-R175H in Cancer. *Cancers* **2021**, *13*, 4088. [CrossRef]

42. Duraiyan, J.; Govindarajan, R.; Kaliyappan, K.; Palanisamy, M. Applications of immunohistochemistry. *J. Pharm. Bioallied Sci.* **2012**, *4* (Suppl. 2), S307–S309. [CrossRef]

43. Riva, C.; Dainese, E.; Caprara, G.; Rocca, P.C.; Massarelli, G.; Tot, T.; Capella, C.; Eusebi, V. Immunohistochemical study of androgen receptors in breast carcinoma. Evidence of their frequent expression in lobular carcinoma. *Virchows Arch. Int. J. Pathol.* **2005**, *447*, 695–700. [CrossRef]

44. Lumachi, F.; Santeufemia, D.A.; Basso, S.M. Current medical treatment of estrogen receptor-positive breast cancer. *World J. Biol. Chem.* **2015**, *6*, 231–239. [CrossRef] [PubMed]

45. Gajria, D.; Chandarlapaty, S. HER2-amplified breast cancer: Mechanisms of trastuzumab resistance and novel targeted therapies. *Expert Rev. Anticancer Ther.* **2011**, *11*, 263–275. [CrossRef] [PubMed]

46. Li, Z.; Wei, H.; Li, S.; Wu, P.; Mao, X. The Role of Progesterone Receptors in Breast Cancer. *Drug Des. Dev. Ther.* **2022**, *16*, 305–314. [CrossRef]

47. Cellusaurus. Available online: https://www.cellosaurus.org/ (accessed on 25 November 2023).

48. Nahleh, Z. Androgen receptor as a target for the treatment of hormone receptor-negative breast cancer: An unchartered territory. *Future Oncol.* **2008**, *4*, 15–21. [CrossRef]

49. Niță, I.; Nițipir, C.; Toma, Ș.A.; Limbău, A.M.; Pîrvu, E.; Bădărău, I.A.; Suciu, I.; Suciu, G.; Manolescu, L.S.C. Correlation between Androgen Receptor Expression and Immunohistochemistry Type as Prognostic Factors in a Cohort of Breast Cancer Patients: Result from a Single-Center, Cross Sectional Study. *Healthcare* **2021**, *9*, 277. [CrossRef] [PubMed]

50. Gucalp, A.; Traina, T.A.; Eisner, J.R.; Parker, J.S.; Selitsky, S.R.; Park, B.H.; Elias, A.D.; Baskin-Bey, E.S.; Cardoso, F. Male breast cancer: A disease distinct from female breast cancer. *Breast Cancer Res. Treat.* **2019**, *173*, 37–48. [CrossRef]

51. Healey, M.A.; Hirko, K.A.; Beck, A.H.; Collins, L.C.; Schnitt, S.J.; Eliassen, A.H.; Holmes, M.D.; Tamimi, R.M.; Hazra, A. Assessment of Ki67 expression for breast cancer subtype classification and prognosis in the Nurses' Health Study. *Breast Cancer Res. Treat.* **2017**, *166*, 613–622. [CrossRef]

52. Scholzen, T.; Gerdes, J. The Ki-67 protein: From the known and the unknown. *J. Cell. Physiol.* **2000**, *182*, 311–322. [CrossRef]

53. Mueller, C.; Haymond, A.; Davis, J.B.; Williams, A.; Espina, V. Protein biomarkers for subtyping breast cancer and implications for future research. *Expert Rev. Proteom.* **2018**, *15*, 131–152. [CrossRef] [PubMed]

54. Hsu, J.L.; Hung, M.C. The role of HER2, EGFR, and other receptor tyrosine kinases in breast cancer. *Cancer Metastasis Rev.* **2016**, *35*, 575–588. [CrossRef] [PubMed]

55. Yu, S.; Kim, T.; Yoo, K.H.; Kang, K. The T47D cell line is an ideal experimental model to elucidate the progesterone-specific effects of a luminal A subtype of breast cancer. *Biochem. Biophys. Res. Commun.* **2017**, *486*, 752–758. [CrossRef] [PubMed]

56. Soule, H.D.; Vazquez, J.; Long, A.; Albert, S.; Brennan, M. A human cell line from a pleural effusion derived from a breast carcinoma. *J. Natl. Cancer Inst.* **1973**, *51*, 1409–1416. [CrossRef]

57. Yu, M.; Selvaraj, S.K.; Liang-Chu, M.M.; Aghajani, S.; Busse, M.; Yuan, J.; Lee, G.; Peale, F.; Klijn, C.; Bourgon, R.; et al. A resource for cell line authentication, annotation and quality control. *Nature* **2015**, *520*, 307–311. [CrossRef]

58. Ye, L.H.; Wu, L.Y.; Guo, W.; Ma, H.T.; Zhang, X.D. Screening of a sub-clone of human breast cancer cells with high metastasis potential. *Zhonghua Yi Xue Za Zhi* **2006**, *86*, 61–65.

59. Liang, Y.; Wang, S.; Liu, J. Overexpression of Tumor Protein p53-regulated Apoptosis-inducing Protein 1 Regulates Proliferation and Apoptosis of Breast Cancer Cells through the PI3K/Akt Pathway. *J. Breast Cancer* **2019**, *22*, 172–184. [CrossRef]

60. Sun, X.; Kaufman, P.D. Ki-67: More than a proliferation marker. *Chromosoma* **2018**, *127*, 175–186. [CrossRef]

61. Gallardo, A.; Garcia-Valdecasas, B.; Murata, P.; Teran, R.; Lopez, L.; Barnadas, A.; Lerma, E. Inverse relationship between Ki67 and survival in early luminal breast cancer: Confirmation in a multivariate analysis. *Breast Cancer Res. Treat.* **2018**, *167*, 31–37. [CrossRef]

62. Badia, E.; Oliva, J.; Balaguer, P.; Cavaillès, V. Tamoxifen resistance and epigenetic modifications in breast cancer cell lines. *Curr. Med. Chem.* **2007**, *14*, 3035–3045. [CrossRef]

63. Zhang, L.; Wang, Q.; Zhou, Y.; Ouyang, Q.; Dai, W.; Chen, J.; Ding, P.; Li, L.; Zhang, X.; Zhang, W.; et al. Overexpression of MTA1 inhibits the metastatic ability of ZR-75-30 cells in vitro by promoting MTA2 degradation. *Cell Commun. Signal. CCS* **2019**, *17*, 4. [CrossRef] [PubMed]

64. Ma, W.; Zhu, M.; Zhang, D.; Yang, L.; Yang, T.; Li, X.; Zhang, Y. Berberine inhibits the proliferation and migration of breast cancer ZR-75-30 cells by targeting Ephrin-B2. *Phytomed. Int. J. Phytother. Phytopharm.* **2017**, *25*, 45–51. [CrossRef] [PubMed]

65. Lasfargues, E.Y.; Coutinho, W.G.; Redfield, E.S. Isolation of two human tumor epithelial cell lines from solid breast carcinomas. *J. Natl. Cancer Inst.* **1978**, *61*, 967–978. [PubMed]

66. Akbari, A.; Akbarzadeh, A.; Rafiee Tehrani, M.; Ahangari Cohan, R.; Chiani, M.; Mehrabi, M.R. Development and Characterization of Nanoliposomal Hydroxyurea Against BT-474 Breast Cancer Cells. *Adv. Pharm. Bull.* **2020**, *10*, 39–45. [CrossRef] [PubMed]

67. Ahmadi-Baloutaki, S.; Doosti, A.; Jaafarinia, M.; Goudarzi, H. Editing of the MALAT1 Gene in MDA-MB-361 Breast Cancer Cell Line using the Novel CRISPR Method. *J. Ilam Univ. Med. Sci.* **2022**, *30*, 18–31. [CrossRef]

68. Watrowski, R.; Castillo-Tong, D.C.; Obermayr, E.; Zeillinger, R. Gene Expression of Kallikreins in Breast Cancer Cell Lines. *Anticancer Res.* **2020**, *40*, 2487–2495. [CrossRef] [PubMed]

69. Kilickap, S.; Kaya, Y.; Yucel, B.; Tuncer, E.; Babacan, N.A.; Elagoz, S. Higher Ki67 expression is associates with unfavorable prognostic factors and shorter survival in breast cancer. *Asian Pac. J. Cancer Prev. APJCP* **2014**, *15*, 1381–1385. [CrossRef]

70. Hurst, J.H. Pioneering geneticist Mary-Claire King receives the 2014 Lasker~Koshland Special Achievement Award in Medical Science. *J. Clin. Investig.* **2014**, *124*, 4148–4151. [CrossRef]

71. Darbeheshti, F.; Izadi, P.; Emami Razavi, A.N.; Yekaninejad, M.S.; Tavakkoly Bazzaz, J. Comparison of BRCA1 Expression between Triple-Negative and Luminal Breast Tumors. *Iran. Biomed. J.* **2018**, *22*, 210–214.

72. Ades, F.; Zardavas, D.; Bozovic-Spasojevic, I.; Pugliano, L.; Fumagalli, D.; de Azambuja, E.; Viale, G.; Sotiriou, C.; Piccart, M. Luminal B breast cancer: Molecular characterization, clinical management, and future perspectives. *J. Clin. Oncol. Off. J. Am. Soc. Clin. Oncol.* **2014**, *32*, 2794–2803. [CrossRef]

73. Meng, J.; Yuan, Y.; Li, Y.; Yuan, B. Effects of hirsuteine on MDA-MB-453 breast cancer cell proliferation. *Oncol. Lett.* **2022**, *25*, 4. [CrossRef]

74. Gazdar, A.F.; Kurvari, V.; Virmani, A.; Gollahon, L.; Sakaguchi, M.; Westerfield, M.; Kodagoda, D.; Stasny, V.; Cunningham, H.T.; Wistuba, I.I.; et al. Characterization of paired tumor and non-tumor cell lines established from patients with breast cancer. *Int. J. Cancer* **1998**, *78*, 766–774. [CrossRef]

75. Chung, W.P.; Huang, W.L.; Lee, C.H.; Hsu, H.P.; Huang, W.L.; Liu, Y.Y.; Su, W.C. PI3K inhibitors in trastuzumab-resistant HER2-positive breast cancer cells with PI3K pathway alterations. *Am. J. Cancer Res.* **2022**, *12*, 3067–3082. [PubMed]

76. Ethier, S.P.; Duchinsky, K.; Couch, D. Abstract P4-05-13: The SUM breast cancer cell line knowledge base (SLKBase): A knowledge base and functional genomics platform for breast cancer cell lines. *Cancer Res.* **2020**, *80* (Suppl. 4), P4–05–13. [CrossRef]

77. Chatterjee, N.; Wang, W.L.; Conklin, T.; Chittur, S.; Tenniswood, M. Histone deacetylase inhibitors modulate miRNA and mRNA expression, block metaphase, and induce apoptosis in inflammatory breast cancer cells. *Cancer Biol. Ther.* **2013**, *14*, 658–671. [CrossRef] [PubMed]

78. Bacus, S.S.; Kiguchi, K.; Chin, D.; King, C.R.; Huberman, E. Differentiation of cultured human breast cancer cells (AU-565 and MCF-7) associated with loss of cell surface HER-2/neu antigen. *Mol. Carcinog.* **1990**, *3*, 350–362. [CrossRef]

79. Bacus, S.S.; Huberman, E.; Chin, D.; Kiguchi, K.; Simpson, S.; Lippman, M.; Lupu, R. A ligand for the erbB-2 oncogene product (gp30) induces differentiation of human breast cancer cells. *Cell Growth Differ. Mol. Biol. J. Am. Assoc. Cancer Res.* **1992**, *3*, 401–411.

80. Di, Y.; De Silva, F.; Krol, E.S.; Alcorn, J. Flaxseed Lignans Enhance the Cytotoxicity of Chemotherapeutic Agents against Breast Cancer Cell Lines MDA-MB-231 and SKBR3. *Nutr. Cancer* **2018**, *70*, 306–315. [CrossRef]

81. Di Modugno, F.; Mottolese, M.; DeMonte, L.; Trono, P.; Balsamo, M.; Conidi, A.; Melucci, E.; Terrenato, I.; Belleudi, F.; Torrisi, M.R.; et al. The cooperation between hMena overexpression and HER2 signalling in breast cancer. *PLoS ONE* **2010**, *5*, e15852. [CrossRef]

82. Jogalekar, M.P.; Serrano, E.E. Morphometric analysis of a triple negative breast cancer cell line in hydrogel and monolayer culture environments. *PeerJ* **2018**, *6*, e4340. [CrossRef]

83. Katayose, Y.; Kim, M.; Rakkar, A.N.; Li, Z.; Cowan, K.H.; Seth, P. Promoting apoptosis: A novel activity associated with the cyclin-dependent kinase inhibitor p27. *Cancer Res.* **1997**, *57*, 5441–5445.

84. Zhu, M.; Liu, N.; Lin, J.; Wang, J.; Lai, H.; Liu, Y. HDAC7 inhibits cell proliferation via NudCD1/GGH axis in triple-negative breast cancer. *Oncol. Lett.* **2022**, *25*, 33. [CrossRef]

85. Demeule, M.; Charfi, C.; Currie, J.C.; Larocque, A.; Zgheib, A.; Kozelko, S.; Béliveau, R.; Marsolais, C.; Annabi, B. TH1902, a new docetaxel-peptide conjugate for the treatment of sortilin-positive triple-negative breast cancer. *Cancer Sci.* **2021**, *112*, 4317–4334. [CrossRef]

86. Pelicano, H.; Zhang, W.; Liu, J.; Hammoudi, N.; Dai, J.; Xu, R.H.; Pusztai, L.; Huang, P. Mitochondrial dysfunction in some triple-negative breast cancer cell lines: Role of mTOR pathway and therapeutic potential. *Breast Cancer Res. BCR* **2014**, *16*, 434. [CrossRef] [PubMed]

87. Lehmann, B.D.; Bauer, J.A.; Chen, X.; Sanders, M.E.; Chakravarthy, A.B.; Shyr, Y.; Pietenpol, J.A. Identification of human triple-negative breast cancer subtypes and preclinical models for selection of targeted therapies. *J. Clin. Investig.* **2011**, *121*, 2750–2767. [CrossRef] [PubMed]

88. Yamamoto, T.; Kanaya, N.; Somlo, G.; Chen, S. Synergistic anti-cancer activity of CDK4/6 inhibitor palbociclib and dual mTOR kinase inhibitor MLN0128 in pRb-expressing ER-negative breast cancer. *Breast Cancer Res. Treat.* **2019**, *174*, 615–625. [CrossRef] [PubMed]

89. Tate, C.R.; Rhodes, L.V.; Segar, H.C.; Driver, J.L.; Pounder, F.N.; Burow, M.E.; Collins-Burow, B.M. Targeting triple-negative breast cancer cells with the histone deacetylase inhibitor panobinostat. *Breast Cancer Res. BCR* **2012**, *14*, R79. [CrossRef] [PubMed]

90. Hay, R.; Park, J.; Gadzar, A. *Atlas of Human Tumor Cell Lines*; Academic Press: Cambridge, UK, 1994.

91. Mao, L.; Wertzler, K.J.; Maloney, S.C.; Wang, Z.; Magnuson, N.S.; Reeves, R. HMGA1 levels influence mitochondrial function and mitochondrial DNA repair efficiency. *Mol. Cell. Biol.* **2009**, *29*, 5426–5440. [CrossRef] [PubMed]

92. Sheng, S.; Carey, J.; Seftor, E.A.; Dias, L.; Hendrix, M.J.; Sager, R. Maspin acts at the cell membrane to inhibit invasion and motility of mammary and prostatic cancer cells. *Proc. Natl. Acad. Sci. USA* **1996**, *93*, 11669–11674. [CrossRef] [PubMed]

93. Rae, J.M.; Creighton, C.J.; Meck, J.M.; Haddad, B.R.; Johnson, M.D. MDA-MB-435 cells are derived from M14 melanoma cells--a loss for breast cancer, but a boon for melanoma research. *Breast Cancer Res. Treat.* **2007**, *104*, 13–19. [CrossRef]

94. Scherf, U.; Ross, D.T.; Waltham, M.; Smith, L.H.; Lee, J.K.; Tanabe, L.; Kohn, K.W.; Reinhold, W.C.; Myers, T.G.; Andrews, D.T.; et al. A gene expression database for the molecular pharmacology of cancer. *Nat. Genet.* **2000**, *24*, 236–244. [CrossRef]

95. Robinson, T.J.; Liu, J.C.; Vizeacoumar, F.; Sun, T.; Maclean, N.; Egan, S.E.; Schimmer, A.D.; Datti, A.; Zacksenhaus, E. RB1 status in triple negative breast cancer cells dictates response to radiation treatment and selective therapeutic drugs. *PLoS ONE* **2013**, *8*, e78641. [CrossRef]

96. Li, L.; Zhang, F.; Liu, Z.; Fan, Z. Immunotherapy for Triple-Negative Breast Cancer: Combination Strategies to Improve Outcome. *Cancers* **2023**, *15*, 321. [CrossRef]

97. Gianfredi, V.; Vannini, S.; Moretti, M.; Villarini, M.; Bragazzi, N.L.; Izzotti, A.; Nucci, D. Sulforaphane and Epigallocatechin Gallate Restore Estrogen Receptor Expression by Modulating Epigenetic Events in the Breast Cancer Cell Line MDA-MB-231: A Systematic Review and Meta-Analysis. *J. Nutrigenet. Nutrigenom.* **2017**, *10*, 126–135. [CrossRef] [PubMed]

98. Ackermann, T.; Tardito, S. Cell Culture Medium Formulation and Its Implications in Cancer Metabolism. *Trends Cancer* **2019**, *5*, 329–332. [CrossRef] [PubMed]

99. Weiskirchen, S.; Schröder, S.K.; Buhl, E.M.; Weiskirchen, R. A Beginner's Guide to Cell Culture: Practical Advice for Preventing Needless Problems. *Cells* **2023**, *12*, 682. [CrossRef] [PubMed]

100. Soule, H.D.; Maloney, T.M.; Wolman, S.R.; Peterson, W.D.; Brenz, R., Jr.; McGrath, C.M.; Russo, J.; Pauley, R.J.; Jones, R.F.; Brooks, S.C. Isolation and characterization of a spontaneously immortalized human breast epithelial cell line, MCF-10. *Cancer Res.* **1990**, *50*, 6075–6086. [PubMed]

101. Wawruszak, A.; Luszczki, J.J.; Grabarska, A.; Gumbarewicz, E.; Dmoszynska-Graniczka, M.; Polberg, K.; Stepulak, A. Assessment of Interactions between Cisplatin and Two Histone Deacetylase Inhibitors in MCF7, T47D and MDA-MB-231 Human Breast Cancer Cell Lines-An Isobolographic Analysis. *PLoS ONE* **2015**, *10*, e0143013. [CrossRef] [PubMed]

102. Grigoriadis, A.; Mackay, A.; Noel, E.; Wu, P.J.; Natrajan, R.; Frankum, J.; Reis-Filho, J.S.; Tutt, A. Molecular characterisation of cell line models for triple-negative breast cancers. *BMC Genom.* **2012**, *13*, 619. [CrossRef] [PubMed]

103. Kao, J.; Salari, K.; Bocanegra, M.; Choi, Y.L.; Girard, L.; Gandhi, J.; Kwei, K.A.; Hernandez-Boussard, T.; Wang, P.; Gazdar, A.F.; et al. Molecular profiling of breast cancer cell lines defines relevant tumor models and provides a resource for cancer gene discovery. *PLoS ONE* **2009**, *4*, e6146. [CrossRef]

104. Finn, R.S.; Dering, J.; Conklin, D.; Kalous, O.; Cohen, D.J.; Desai, A.J.; Ginther, C.; Atefi, M.; Chen, I.; Fowst, C.; et al. PD 0332991, a selective cyclin D kinase 4/6 inhibitor, preferentially inhibits proliferation of luminal estrogen receptor-positive human breast cancer cell lines in vitro. *Breast Cancer Res. BCR* **2009**, *11*, R77. [CrossRef] [PubMed]

105. Albi, E.; Mandarano, M.; Cataldi, S.; Ceccarini, M.R.; Fiorani, F.; Beccari, T.; Sidoni, A.; Codini, M. The Effect of Cholesterol in MCF7 Human Breast Cancer Cells. *Int. J. Mol. Sci.* **2023**, *24*, 5935. [CrossRef] [PubMed]

106. Morales Torres, C.; Wu, M.Y.; Hobor, S.; Wainwright, E.N.; Martin, M.J.; Patel, H.; Grey, W.; Grönroos, E.; Howell, S.; Carvalho, J.; et al. Selective inhibition of cancer cell self-renewal through a Quisinostat-histone H1.0 axis. *Nat. Commun.* **2020**, *11*, 1792. [CrossRef] [PubMed]

107. Jernström, S.; Hongisto, V.; Leivonen, S.K.; Due, E.U.; Tadele, D.S.; Edgren, H.; Kallioniemi, O.; Perälä, M.; Mælandsmo, G.M.; Sahlberg, K.K. Drug-screening and genomic analyses of HER2-positive breast cancer cell lines reveal predictors for treatment response. *Breast Cancer (Dove Med. Press)* **2017**, *9*, 185–198. [CrossRef] [PubMed]

108. Riaz, M.; van Jaarsveld, M.T.; Hollestelle, A.; Prager-van der Smissen, W.J.; Heine, A.A.; Boersma, A.W.; Liu, J.; Helmijr, J.; Ozturk, B.; Smid, M.; et al. miRNA expression profiling of 51 human breast cancer cell lines reveals subtype and driver mutation-specific miRNAs. *Breast Cancer Res. BCR* **2013**, *15*, R33. [CrossRef] [PubMed]

109. Yamashita, N.; Morimoto, Y.; Fushimi, A.; Ahmad, R.; Bhattacharya, A.; Daimon, T.; Haratake, N.; Inoue, Y.; Ishikawa, S.; Yamamoto, M.; et al. MUC1-C Dictates PBRM1-Mediated Chronic Induction of Interferon Signaling, DNA Damage Resistance, and Immunosuppression in Triple-Negative Breast Cancer. *Mol. Cancer Res. MCR* **2023**, *21*, 274–289. [CrossRef]
110. Singha, M.; Pu, L.; Stanfield, B.A.; Uche, I.K.; Rider, P.J.F.; Kousoulas, K.G.; Ramanujam, J.; Brylinski, M. Artificial intelligence to guide precision anticancer therapy with multitargeted kinase inhibitors. *BMC Cancer* **2022**, *22*, 1211. [CrossRef]
111. Twomey, J.D.; Zhang, B. Exploring the Role of Hypoxia-Inducible Carbonic Anhydrase IX (CAIX) in Circulating Tumor Cells (CTCs) of Breast Cancer. *Biomedicines* **2023**, *11*, 934. [CrossRef]
112. Puleo, J.; Polyak, K. The MCF10 Model of Breast Tumor Progression. *Cancer Res.* **2021**, *81*, 4183–4185. [CrossRef]
113. Sabarinathan, D.; Chandrika, S.P.; Venkatraman, P.; Easwaran, M.; Sureka, C.S.; Preethi, K. Production of polyhydroxybutyrate (PHB) from Pseudomonas plecoglossicida and its application towards cancer detection. *Inf. Med. Unlocked* **2018**, *11*, 61–67. [CrossRef]
114. DeAngelis, J.T.; Li, Y.; Mitchell, N.; Wilson, L.; Kim, H.; Tollefsbol, T.O. 2D difference gel electrophoresis analysis of different time points during the course of neoplastic transformation of human mammary epithelial cells. *J. Proteome Res.* **2011**, *10*, 447–458. [CrossRef]
115. Mazzarini, M.; Falchi, M.; Bani, D.; Migliaccio, A.R. Evolution and new frontiers of histology in bio-medical research. *Microsc. Res. Tech.* **2021**, *84*, 217–237. [CrossRef]
116. Cailleau, R.; Young, R.; Olivé, M.; Reeves, W.J., Jr. Breast tumor cell lines from pleural effusions. *J. Natl. Cancer Inst.* **1974**, *53*, 661–674. [CrossRef] [PubMed]
117. Yakavets, I.; Francois, A.; Benoit, A.; Merlin, J.L.; Bezdetnaya, L.; Vogin, G. Advanced co-culture 3D breast cancer model for investigation of fibrosis induced by external stimuli: Optimization study. *Sci. Rep.* **2020**, *10*, 21273. [CrossRef] [PubMed]
118. Ryan, S.L.; Baird, A.M.; Vaz, G.; Urquhart, A.J.; Senge, M.; Richard, D.J.; O'Byrne, K.J.; Davies, A.M. Drug Discovery Approaches Utilizing Three-Dimensional Cell Culture. *Assay Drug Dev. Technol.* **2016**, *14*, 19–28. [CrossRef] [PubMed]
119. Katt, M.E.; Placone, A.L.; Wong, A.D.; Xu, Z.S.; Searson, P.C. In Vitro Tumor Models: Advantages, Disadvantages, Variables, and Selecting the Right Platform. *Front. Bioeng. Biotechnol.* **2016**, *4*, 12. [CrossRef]
120. Masters, J.R. Human cancer cell lines: Fact and fantasy. Nature reviews. *Mol. Cell Biol.* **2000**, *1*, 233–236. [CrossRef]
121. Engel, L.W.; Young, N.A. Human breast carcinoma cells in continuous culture: A review. *Cancer Res.* **1978**, *38 Pt 2*, 4327–4339.
122. Mizuno, M.; Matsuda, J.; Watanabe, K.; Shimizu, N.; Sekiya, I. Effect of disinfectants and manual wiping for processing the cell product changeover in a biosafety cabinet. *Regen. Ther.* **2023**, *22*, 169–175. [CrossRef]
123. de Oliveira, N.F.P.; de Souza, B.F.; de Castro Coêlho, M. UV Radiation and Its Relation to DNA Methylation in Epidermal Cells: A Review. *Epigenomes* **2020**, *4*, 23. [CrossRef]
124. Ben-David, U.; Siranosian, B.; Ha, G.; Tang, H.; Oren, Y.; Hinohara, K.; Strathdee, C.A.; Dempster, J.; Lyons, N.J.; Burns, R.; et al. Genetic and transcriptional evolution alters cancer cell line drug response. *Nature* **2018**, *560*, 325–330. [CrossRef] [PubMed]
125. Borg, A.; Sandberg, T.; Nilsson, K.; Johannsson, O.; Klinker, M.; Måsbäck, A.; Westerdahl, J.; Olsson, H.; Ingvar, C. High frequency of multiple melanomas and breast and pancreas carcinomas in CDKN2A mutation-positive melanoma families. *J. Natl. Cancer Inst.* **2000**, *92*, 1260–1266. [CrossRef]
126. Janku, F.; Wheler, J.J.; Westin, S.N.; Moulder, S.L.; Naing, A.; Tsimberidou, A.M.; Fu, S.; Falchook, G.S.; Hong, D.S.; Garrido-Laguna, I.; et al. PI3K/AKT/mTOR inhibitors in patients with breast and gynecologic malignancies harboring PIK3CA mutations. *J. Clin. Oncol.* **2012**, *30*, 777–782. [CrossRef] [PubMed]
127. Cizkova, M.; Susini, A.; Vacher, S.; Cizeron-Clairac, G.; Andrieu, C.; Driouch, K.; Fourme, E.; Lidereau, R.; Bièche, I. PIK3CA mutation impact on survival in breast cancer patients and in ERalpha, PR and ERBB2-based subgroups. *Breast Cancer Res* **2012**, *14*, R28. [CrossRef] [PubMed]
128. May, P.; May, E. Twenty years of p53 research: Structural and functional aspects of the p53 protein. *Oncogene* **1999**, *18*, 7621–7636. [CrossRef] [PubMed]
129. Alkaabi, D.; Arafat, K.; Sulaiman, S.; Al-Azawi, A.M.; Attoub, S. PD-1 Independent Role of PD-L1 in Triple-Negative Breast Cancer Progression. *Int. J. Mol. Sci.* **2023**, *24*, 6420. [CrossRef] [PubMed]
130. Jiang, G.; Zhang, S.; Yazdanparast, A.; Li, M.; Pawar, A.V.; Liu, Y.; Inavolu, S.M.; Cheng, L. Comprehensive comparison of molecular portraits between cell lines and tumors in breast cancer. *BMC Genom.* **2016**, *17* (Suppl. 7), 525. [CrossRef]
131. Wijshake, T.; Zou, Z.; Chen, B.; Zhong, L.; Xiao, G.; Xie, Y.; Doench, J.G.; Bennett, L.; Levine, B. Tumor-suppressor function of Beclin 1 in breast cancer cells requires E-cadherin. *Proc. Natl. Acad. Sci. USA* **2021**, *118*, e2020478118. [CrossRef]
132. Haitjema, A.; Mol, B.M.; Kooi, I.E.; Massink, M.P.; Jørgensen, J.A.; Rockx, D.A.; Rooimans, M.A.; de Winter, J.P.; Meijers-Heijboer, H.; Joenje, H.; et al. Coregulation of FANCA and BRCA1 in human cells. *SpringerPlus* **2014**, *3*, 381. [CrossRef]
133. Légaré, S.; Chabot, C.; Basik, M. SPEN, a new player in primary cilia formation and cell migration in breast cancer. *Breast Cancer Res. BCR* **2017**, *19*, 104. [CrossRef]
134. Lin, J.; Ye, S.; Ke, H.; Lin, L.; Wu, X.; Guo, M.; Jiao, B.; Chen, C.; Zhao, L. Changes in the mammary gland during aging and its links with breast diseases. *Acta Biochim. Biophys. Sin.* **2023**, *55*, 1001–1019. [CrossRef] [PubMed]
135. Kim, J.B.; O'Hare, M.J.; Stein, R. Models of breast cancer: Is merging human and animal models the future? *Breast Cancer Res. BCR* **2004**, *6*, 22–30. [CrossRef] [PubMed]
136. Souto, E.P.; Dobrolecki, L.E.; Villanueva, H.; Sikora, A.G.; Lewis, M.T. In Vivo Modeling of Human Breast Cancer Using Cell Line and Patient-Derived Xenografts. *J. Mammary Gland Biol. Neoplasia* **2022**, *27*, 211–230. [CrossRef] [PubMed]

137. Okada, S.; Vaeteewoottacharn, K.; Kariya, R. Application of Highly Immunocompromised Mice for the Establishment of Patient-Derived Xenograft (PDX) Models. *Cells* **2019**, *8*, 889. [CrossRef] [PubMed]
138. Hasan, T.; Carter, B.; Denic, N.; Gai, L.; Power, J.; Voisey, K.; Kao, K.R. Evaluation of cell-line-derived xenograft tumours as controls for immunohistochemical testing for ER and PR. *J. Clin. Pathol.* **2015**, *68*, 746–751. [CrossRef]
139. Paul, B.; Royston, K.J.; Li, Y.; Stoll, M.L.; Skibola, C.F.; Wilson, L.S.; Barnes, S.; Morrow, C.D.; Tollefsbol, T.O. Impact of genistein on the gut microbiome of humanized mice and its role in breast tumor inhibition. *PLoS ONE* **2017**, *12*, e0189756. [CrossRef]
140. Whittle, J.R.; Lewis, M.T.; Lindeman, G.J.; Visvader, J.E. Patient-derived xenograft models of breast cancer and their predictive power. *Breast Cancer Res. BCR* **2015**, *17*, 17. [CrossRef]
141. Han, Y.; Azuma, K.; Watanabe, S.; Semba, K.; Nakayama, J. Metastatic profiling of HER2-positive breast cancer cell lines in xenograft models. *Clin. Exp. Metastasis* **2022**, *39*, 467–477. [CrossRef]
142. Liu, B.; Fan, Z.; Edgerton, S.M.; Deng, X.S.; Alimova, I.N.; Lind, S.E.; Thor, A.D. Metformin induces unique biological and molecular responses in triple negative breast cancer cells. *Cell Cycle* **2009**, *8*, 2031–2040. [CrossRef]
143. Georges, L.M.C.; De Wever, O.; Galván, J.A.; Dawson, H.; Lugli, A.; Demetter, P.; Zlobec, I. Cell Line Derived Xenograft Mouse Models Are a Suitable in vivo Model for Studying Tumor Budding in Colorectal Cancer. *Front. Med.* **2019**, *6*, 139. [CrossRef]
144. Finlay-Schultz, J.; Jacobsen, B.M.; Riley, D.; Paul, K.V.; Turner, S.; Ferreira-Gonzalez, A.; Harrell, J.C.; Kabos, P.; Sartorius, C.A. New generation breast cancer cell lines developed from patient-derived xenografts. *Breast Cancer Res. BCR* **2020**, *22*, 68. [CrossRef] [PubMed]

Article

The Role of Diffusion-Weighted Imaging Based on Maximum-Intensity Projection in Young Patients with Marked Background Parenchymal Enhancement on Contrast-Enhanced Breast MRI

Ga-Eun Park [1], Bong-Joo Kang [1], Sung-hun Kim [1] and Na-Young Jung [2,*]

[1] Department of Radiology, Seoul Saint Mary's Hospital, College of Medicine, The Catholic University of Korea, Seoul 06591, Republic of Korea; hoonhoony@naver.com (G.-E.P.); lionmain@catholic.ac.kr (B.-J.K.); rad-ksh@catholic.ac.kr (S.-h.K.)

[2] Department of Radiology, Uijeongbu Eulji Medical Center, College of Medicine, Eulji University, Uijeongbu 11759, Republic of Korea

* Correspondence: nyjung1230@gmail.com; Tel.: +82-31-951-1739

Abstract: Diffusion-weighted imaging (DWI) utilizing maximum-intensity projection (MIP) was suggested as a cost-effective alternative tool without the risk of gadolinium-based contrast agents. The purpose of this study was to investigate whether DWI MIPs played a supportive role in young ($\leq$60) patients with marked background parenchymal enhancement (BPE) on contrast-enhanced MRI (CE-MRI). The research included 1303 patients with varying degrees of BPE, and correlations between BPE on CE-MRI, the background diffusion signal (BDS) on DWI, and clinical parameters were analyzed. Lesion detection scores were compared between CE-MRI and DWI, with DWI showing higher scores. Among the 186 lesions in 181 patients with marked BPE on CE-MRI, the main lesion on MIPs of CE-MRI was partially or completely seen in 88.7% of cases, while it was not seen in 11.3% of cases. On the other hand, the main lesion on MIPs of DWI was seen in 91.4% of cases, with only 8.6% of cases showing no visibility. DWI achieved higher scores for lesion detection compared to CE-MRI. The presence of a marked BDS was significantly associated with a lower likelihood of a higher DWI score ($p < 0.001$), and non-mass lesions were associated with a decreased likelihood of a higher DWI score compared with mass lesions ($p = 0.196$). In conclusion, the inclusion of MIPs of DWI in the preoperative evaluation of breast cancer patients, particularly young women with marked BPE, proved highly beneficial in improving the overall diagnostic process.

Keywords: breast cancer; MRI; diffusion-weighted

Citation: Park, G.-E.; Kang, B.-J.; Kim, S.-h.; Jung, N.-Y. The Role of Diffusion-Weighted Imaging Based on Maximum-Intensity Projection in Young Patients with Marked Background Parenchymal Enhancement on Contrast-Enhanced Breast MRI. *Life* **2023**, *13*, 1744. https://doi.org/10.3390/life13081744

Academic Editors: Taobo Hu, Mengping Long, Lei Wang and Riccardo Autelli

Received: 19 July 2023
Revised: 3 August 2023
Accepted: 11 August 2023
Published: 14 August 2023

1. Introduction

Contrast-enhanced magnetic resonance imaging (CE-MRI) of the breast has the highest sensitivity for breast cancer detection among several imaging modalities [1,2]. Despite its highest degree of sensitivity, breast MRI does not result in false-negative cases in comparison to other imaging modalities. False-negative cases may be attributed to perceptive errors in the absence of radiological detection at the time of screening; interpretation errors, where the cases are recognized but mistaken for benign lesions; and various technical errors [3]. A recent study reported three main causes of undetected breast malignancy in CE-MRI: (1) non-enhancing histologic features; (2) location; and (3) significant background parenchymal enhancement (BPE) [4]. BPE significantly affects breast MRI interpretation and is a valuable imaging marker for assessing breast cancer risk [5]. BPE is widely recognized for increasing the recall, false-positive, and false-negative rates in breast MR readings. In particular, substantial BPE may prevent the clear demarcation of lesions from the breast parenchyma [6].

Several earlier studies have demonstrated that the combination of a diffusion-weighted imaging (DWI) protocol with CE-MRI leads to higher specificity compared to CE-MRI alone [7–10]. Furthermore, recent research has indicated its potential for breast cancer detection and the characterization of breast lesions [11,12].

The abbreviated protocol (AP) for breast MRI, utilizing a single pre-contrast and a single post-contrast acquisition along with maximum-intensity projection (MIP) images, has become increasingly popular. The advantages include a shorter acquisition and reading time, lower cost, and diagnostic accuracy comparable to the full protocol [13,14]. So, DWI MIPs could be proposed as a cost-effective alternative tool, eliminating the risk associated with gadolinium-based contrast agents in this study.

Therefore, the objective of this study was to assess the effectiveness of DWI utilizing MIPs in distinguishing lesions compared with the conventional protocol of using MIPs of CE-MRI for patients with preoperative breast cancer, with a specific focus on young women presenting marked BPE.

2. Materials and Methods

2.1. Study Population

This study was approved by our Institutional Review Board, and informed consent was waived due to its retrospective nature. The study period spanned from 1 July 2020 to 30 September 2022, and a total of 4199 MRI scans were included. Among these, 1712 scans were performed preoperatively. A total of 409 cases were excluded such as those associated with neoadjuvant chemotherapy (NAC) (309 cases), post-excision scans (33 cases), and specific criteria such as old age (>60 years) (51 cases), inflammatory cancers (9 cases), implants (5 cases), and absence of surgical confirmation (2 cases).

2.2. MRI Protocol

Breast MRI was performed using 3T MR machines (Verio and Vida, Siemens Healthcare, Erlangen, Germany). Breast MRI scans were conducted in the prone position using a specialized breast surface coil. The enrolled patients underwent the following MRI sequences for the Verio system: (1) Axial T2-weighted imaging with a turbo spin-echo technique, using a TR/TE of 4530/93, a flip angle of 80, 34 slices, a 320 mm field of view, a matrix size of 576×403, 1 excitation, a 4 mm slice thickness, and an acquisition time of 2 min 28 s. (2) Axial DWI with a readout-segmented echoplanar image, employing b values of 0 and 1000 s/mm^2, a TR/TE of 5200/53 ms, a field of view of 340×205 mm^2, a matrix size of 192×116, a 4 mm slice thickness, and an acquisition time of 2 min 31 s with 5 readout segments. The apparent diffusion coefficient (ADC) maps were automatically calculated using software. (3) Pre- and post-contrast axial T1-weighted 3D volumetric interpolated brain examination (VIBE) sequences with a TR/TE of 2.7/0.8, a flip angle of 10, and a 1.2 mm slice thickness. The images were acquired before and at 10, 70, 130, 190, 250, and 310 s after the injection of gadolinium DTPA (0.1 mmol/kg of Gadovist; Bayer Schering Pharma, Berlin, Germany). For the Vida system, the MRI sequences were as follows: (1) Axial T2-weighted imaging with a turbo spin-echo DIXON sequence, using a TR/TE of 5000/96 ms, a flip angle of 120, 50 slices, a 320 mm field of view, a matrix size of 448×314, a 3 mm slice thickness, and an acquisition time of 3 min 23 s. (2) Axial DWI with readout-segmented long variable echo trains, employing b values of 0 and 1000 s/mm^2, a TR/TE of 4720/60 ms, a field of view of 350×210 mm^2, a matrix size of 256×154, a 3 mm slice thickness, and an acquisition time of 3 min 29 s with 9 readout segments. The ADC maps were automatically calculated using software. (3) Pre- and post-contrast axial T1-weighted 3D fast low-angle-shot (FLASH) sequences with a TR/TE of 4.7/2.27 ms, a flip angle of 10, and a 1 mm slice thickness. The images were acquired before and at 10, 93, 176, 259, 342, and 425 s after the injection of gadolinium DTPA.

Digital Imaging and Communications in Medicine (DICOM) files from DCE-MRI and DWI were transferred to a computer software program (Syngovia; Simens healthcare,

Erlangen, Germany) in order to generate MIP images using high b-value DWI and first postcontrast subtracted images.

2.3. Image Analysis

Among the 1303 enrolled breast MRI results, breast parenchymal enhancement (BPE) and background diffusion signal (BDS) were assessed by one of three radiologists with 5–20 years of experience in breast imaging. The degrees of BPE and BDS were categorized as minimal, mild, moderate, or marked according to the American College of Radiology Breast Imaging and Reporting Data System (BI-RADS) [4]. In cases where BPE or BDS exhibited asymmetry between bilateral scans, the breast with the highest BPE or BDS was utilized for categorization purposes.

Two breast radiologists with experiences of 18 years and 20 years reviewed the two sets of images and arrived at a consensus based on MIPs of DWI and CE-MRI. The readers were blinded to the women's clinical histories and other imaging sets. The images were transferred to numbered folders containing anonymized image data on a Picture Archiving and Communication System (PACS) and read according to the following standardized protocol.

First, the readers reviewed the MIP of DWI to identify significant lesions: (1) definitively seen group, based on the consensus of two readers' findings of suspicious lesions; partial or retrospectively seen group, based on findings of suspicious lesion by one out of two readers; (2) unseen or undetected group, if neither reader found a true lesion. The readers characterized (mass/non-mass) and scored (1 to 10) the detected lesions on DWI MIPs.

Second, the readers reviewed the MIPs of CE-MRI obtained under early enhancement similar to DWI MIPs. The MIPs did not allow a full assessment of lesion morphology; thus, we used a scoring system rather than BI-RADS. Multiple breast lesions were divided into primary breast cancers as main lesions and additional suspicious lesions as daughter lesions.

Image analysis was conducted using a 1–10 scoring system. The scores of CE-MRI and DWI were used to determine lesion visibility and characteristics. The score ranged from 1 to 10, with corresponding descriptions as Table 1.

Table 1. Definition of group and score.

Group	Score	Definition
Unseen	1	absolutely not seen
	2	very subtle visibility
Partially or retrospectively seen	3	partial visibility where the lesion was only visible when its location was known (less than 50% visibility)
	4	partial visibility where the lesion was visible when its location was known (more than 50% visibility)
	5	signified complete visualization of the entire lesion when its location was known, retrospectively
Definitely seen	6	visualization of the lesion similar to moderate BPE on CE-MRI, albeit with numerous false-positive lesions
	7	complete confirmation of the lesion similar to moderate BPE on CE-MRI, along with a few false-positive lesions
	8	easily detected lesion, comparable to mild BPE on CE-MRI (main and daughter lesions)
	9	very easy identification of the lesion, resembling minimal BPE on CE-MRI
	10	very easy identification of the lesion, resembling minimal BPE on CE-MRI, with exceptionally clear main and daughter lesions

CE-MRI: contrast-enhanced magnetic resonance imaging; BPE: breast parenchymal enhancement.

The image gold standard for true lesions was established using conventional whole CE-MRI, mammography, and ultrasonography. The other radiologist, with five years of experience, analyzed sets of images and compared the results with the image gold standard.

Ductal carcinomas in situ (DCISs) and invasive cancers were counted as positive results. All other results of biopsy or excision analysis, including high-risk lesions such as atypical ductal hyperplasia, lobular carcinoma in situ, or papilloma, were considered as negative results.

We analyzed the correlation between BPE depending on MIP images of CE-MRI, BDS on MIP images of DWI, and clinical parameters such as age. Additionally, we analyzed the scores of primary breast cancer and examined additional suspicious lesions on both CE-MRI and DWI. Finally, the features of malignant breast lesions were also analyzed.

2.4. Histopathology Review

The biopsy or surgical specimen pathology reports were carefully examined to determine various tumor characteristics, such as size, depth, histologic type, grade, presence of lymph node metastasis, and immunohistochemical (IHC) subtypes. The IHC factors evaluated included estrogen receptor (ER), progesterone receptor (PR), human epidermal growth factor receptor 2 (HER2), Ki-67, epidermal growth factor receptor (EGFR), and CK5/6. IHC staining for ER, PR, HER2, Ki-67, and EGFR was conducted using specific primary antibodies on an automated Ventana BenchMark XT Slide Stainer (Ventana, Tucson, AZ, USA). The staining for CK5/6 was performed on the Dako Omnis (Dako, Carpinteria, CA, USA). For ER and PR positivity, a cut-off value of $\geq 1\%$ was used. HER2 expression intensity was semiquantitatively scored as 0, 1, 2, or 3, with a score of 3 indicating HER2 positivity, while scores of 0 or 1 indicated HER2 negativity. HER2 status for tumors with a score of 2 was determined using gene amplification [15]. Positive Ki-67 expression was defined as Ki-67 positivity in $\geq 14\%$ of cancer cell nuclei. EGFR and CK5/6 positivity were defined with a cut-off value of $\geq 1\%$. Based on the 2013 St. Gallen International Breast Cancer Conference recommendations, the IHC subtypes were classified as follows [16]: (1) Luminal A (ER or PR+, HER2-, and Ki-67low), (2) Luminal B (ER or PR+, HER2+, and/or Ki-67high), (3) HER2+ (ER-, PR-, and HER2+), (4) triple-negative basal-like (ER-, PR-, HER2-, EGFR, or CK5/6+), and (5) triple-negative non-basal-like (ER-, PR-, HER2-, EGFR-, CK5/6-).

2.5. Statistical Analysis

Summary statistics are presented as number (percentage) of categorical variables and as means (standard deviation (SD)) and medians (inter-quartile range (IQR)) in the case of continuous variables. Groups were compared using the chi square test or Fisher's exact test and Wilcoxon rank-sum tests for categorical and continuous variables, respectively. Comparison of CE-MRI and DWI was performed using the generalized estimating equation (GEE) or the cluster Wilcoxon rank-sum test, considering the same subjects as clusters, p value (CE-MRI vs. DWI in unseen category), and p value (CE-MRI vs. DWI in seen category). To investigate factors associated with higher score in DWI group than in CE-MRI, univariate and multivariable logistic regression analyses were performed. Variables were included in the multivariable model if their univariate significance was <0.05. All statistical analyses were performed using SAS version 9.4 (SAS Institute, Cary, NC, USA), with two-sided p-values < 0.05 considered statistically significant.

3. Results

The results of the study are summarized in Figure 1, presenting the flow diagram of the study population selection. A total of 1303 scans exhibited varying degrees of BPE, categorized as minimal (n = 422), mild (n = 410), moderate (n = 290), and marked (n = 181) (Figure 2). Within the marked BPE group, bilateral cases were analyzed separately (n = 3). Cases involving ipsilateral multiple masses with different pathologies were also analyzed separately (n = 2). In total, this study assessed 186 lesions in 181 patients with marked BPE on CE-MRI.

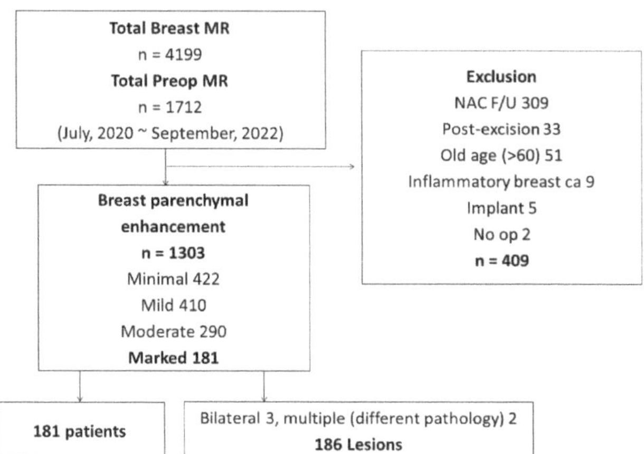

Figure 1. Flow diagram of study population selection.

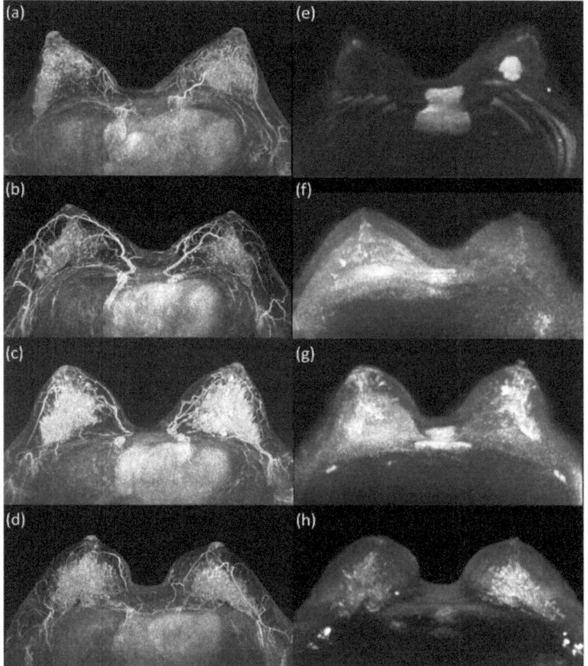

Figure 2. (**a–d**) Examples of marked breast parenchymal enhancement (BPE) on MIP of CE-MRI with (**e**) minimal, (**f**) mild, (**g**) moderate, and (**h**) marked background diffusion signal (BDS) on MIP of DWI.

The results of the study are presented in Table 2, providing important clinical, imaging, and pathologic characteristics of a total of 186 lesions included in the analysis. The mean age of the patients was 44.6 years (SD = 5.8), with a median of 46.0 years (IQR: 42.0–48.0). MRI revealed that the mean size of the lesions was 28.1 mm (SD = 19.3), with a median size of 21.5 mm (IQR: 15.0–36.0). Among the lesions, 66.1% were categorized as masses, 27.4% as non-masses, and 6.5% as both masses and non-masses. The mean ADC value was 0.9×10^{-3} mm^2/s (SD = 0.2), with a median of 0.9×10^{-3} mm^2/s (IQR: 0.8–1.0). The average CE-MRI score was 5.5 (SD = 2.0), with a median score of 6.0 (IQR: 4.0–7.0). The mean score in DWI was 6.2 (SD = 2.3), with a median score of 7.0 (IQR: 4.0–8.0). The combined score, considering both CE-MRI and DWI, had a mean of 6.6 (SD = 2.1) and a median of 7.0 (IQR: 5.0–8.0). Regarding pathologic characteristics, the majority of lesions were invasive ductal carcinoma (IDC) (67.7%), followed by ductal carcinoma in situ (DCIS) (23.7%). Immunohistochemical subtypes revealed 50.0% classified as Luminal A, 37.1% as Luminal B, 4.3% as HER2-positive, and 8.6% as triple-negative.

Table 2. Clinical, imaging, and pathologic characteristics of total lesions (N = 186 lesions of 181 patients).

Clinical Characteristics		N (%)
Age (years)	Mean $\pm$ SD	44.6 $\pm$ 5.8
	Median (IQR)	46.0 (42.0–48.0)
MRI imaging characteristics		
Size (mm)	Mean $\pm$ SD	28.1 $\pm$ 19.3
	Median (IQR)	21.5 (15.0–36.0)
Mass/non-mass	Mass	123 (66.1)
	Non-mass	51 (27.4)
	mass and non-mass	12 (6.5)
ADC value (10^{-3} mm^2/s)	Mean $\pm$ SD	0.9 $\pm$ 0.2
	Median (IQR)	0.9 (0.8–1.0)
CE-MRI score	Mean $\pm$ SD	5.5 $\pm$ 2.0
	Median (IQR)	6.0 (4.0–7.0)
DWI score	Mean $\pm$ SD	6.2 $\pm$ 2.3
	Median (IQR)	7.0 (4.0–8.0)
Combined score	Mean $\pm$ SD	6.6 $\pm$ 2.1
	Median (IQR)	7.0 (5.0–8.0)
Pathologic characteristics		
Pathology	DCIS	44 (23.7)
	IDC	126 (67.7)
	ILC	7 (3.8)
	Mucinous ca	6 (3.2)
	Papillary ca	1 (0.5)
	Tubular ca	2 (1.1)
Grade of invasive ca	Grade1	33 (23.2)
	Grade2	73 (51.4)
	Grade3	36 (25.4)
Grade of DCIS	Grade1	4 (9.1)
	Grade2	27 (61.4)
	Grade3	13 (29.5)
LN	Negative	133 (71.5)
	Positive	53 (28.5)

Table 2. *Cont.*

Clinical Characteristics		N (%)
ER	Negative	26 (14.0)
	Positive	160 (86.0)
PR	Negative	36 (19.4)
	Positive	150 (80.6)
HER2	Negative	153 (82.3)
	Positive	33 (17.7)
KI-67	$\leq$20%	108 (58.1)
	>20%	78 (41.9)
IHC type	Luminal A	93 (50.0)
	Luminal B	69 (37.1)
	Her2+	8 (4.3)
	Triple-	16 (8.6)

CE-MRI: contrast-enhanced magnetic resonance imaging; DWI: diffusion-weighted imaging; ADC: apparent diffusion coefficient; IHC: immunohistochemical; DCIS: ductal carcinoma in situ; LN: lymph node; ER: estrogen receptor; PR: progesterone receptor; HER2: human epidermal growth factor receptor 2; ca: carcinoma; SD: standard deviation; IQR: inter-quartile range.

Table 3 presents several clinically important findings regarding the detectability of the main lesion on MIPs of CE-MRI and MIPs of DWI. Among the total of 186 cases analyzed, the main lesion on MIPs of CE-MRI was partially or definitely seen in 165 cases (88.7%), while it was not seen in 21 cases (11.3%). However, the main lesion on MIPs of DWI was partially or definitely seen in 170 cases (91.4%), with only 16 cases (8.6%) showing no visibility (Figure 3). Statistical analysis revealed a significantly different detection of the main lesion on MIPs of CE-MRI based on size ($p = 0.002$), where the mean size of visible lesions was 29.4 mm compared with 18.4 mm for invisible lesions. However, the detectability based on size was not significantly different for the main lesion on MIPs of DWI ($p = 0.157$). No statistically significant differences were found in the detectability of the main lesion on MIPs of CE and MIPs of DWI based on the mass/non-mass categorization ($p = 0.092$ and $p = 0.146$, respectively). The ADC values showed no significant difference in detectability of the main lesion on MIPs of CE-MRI ($p = 0.235$), while a significant difference was observed for the main lesion on MIPs of DWI ($p = 0.017$). The combined scores of CE-MRI and DWI showed significant differences in detectability for both MIPs of CE-MRI and MIPs of DWI ($p < 0.001$). Notably, the detectability of the main lesion on MIPs of CE-MRI and DWI was associated with pathology, with statistically significant differences observed for both ($p = 0.012$ and $p = 0.216$, respectively). In summary, the study findings highlight the influence of size, ADC values, combined scores, and pathology on the detectability of the main lesion on MIPs of CE-MRI and DWI, providing valuable insights for the clinical assessment and interpretation of breast MRI results.

Univariable and multivariable logistic regression analyses were conducted to identify clinically important findings associated with a higher DWI score than CE-MRI score (Table 4). Among the variables examined, two variables showed significant association in the multivariable analysis. First, the presence of a marked BDS was significantly associated with a lower likelihood of a higher DWI score (odds ratio [OR] = 0.18, 95% confidence interval [CI] = 0.08–0.38, $p < 0.001$) (Figures 4 and 5). Second, the presence of a non-mass lesion was associated with a decreased likelihood of a higher DWI score compared with mass lesions (OR = 0.61, 95% CI = 0.28–1.29, $p = 0.196$) (Figure 6). Other variables, including size, ADC value, pathology type, tumor grade, lymph node status, hormone receptor status (ER and PR), HER2 status, KI-67 group, and immunophenotype, did not show a statistically significant association with higher DWI scores in the multivariable analysis ($p > 0.05$).

Table 3. Detectability of main breast lesions.

| | Total (N = 186) | | | | | |
| | MIP of CE-MRI | | | MIP of DWI | | |
	not seen (n = 21)	partially/definitely seen (n = 165)	*p* value	not seen (n = 16)	partially/definitely seen (n = 170)	*p* value
Size			0.002			0.157
Mean ± SD	18.4 ± 12.7	29.4 ± 19.6		22.4 ± 15.9	28.7 ± 19.5	
Median (IQR)	13.0 (10.0–25.0)	23.0 (16.0–37.0)		19.0 (9.5–32.0)	22.0 (15.0–36.0)	
Mass/non-mass			0.092			0.146
mass, mass and non-mass	12 (57.1)	123 (74.5)		9 (56.3)	126 (74.1)	
Non-mass	9 (42.9)	42 (25.5)		7 (43.8)	44 (25.9)	
ADC value			0.235			0.017
Mean ± SD	0.9 ± 0.2	0.9 ± 0.3		1.1 ± 0.3	0.9 ± 0.2	
Median (IQR)	0.9 (0.8–1.0)	0.9 (0.8–1.0)		1.0 (0.9–1.2)	0.9 (0.8–1.0)	
Score in CE-MRI			<0.001			<0.001
Mean ± SD	1.7 ± 0.5	5.9 ± 1.6		3.3 ± 2.0	5.7 ± 1.9	
Median (IQR)	2.0 (1.0–2.0)	6.0 (5.0–7.0)		2.5 (2.0–4.5)	6.0 (4.0–7.0)	
Score in DWI			0.002			<0.001
Mean ± SD	4.6 ± 2.7	6.5 ± 2.2		1.8 ± 0.4	6.7 ± 2.0	
Median (IQR)	4.0 (2.0–7.0)	7.0 (5.0–8.0)		2.0 (2.0–2.0)	7.0 (5.0–8.0)	
Combined score			<0.001			<0.001
Mean ± SD	4.6 ± 2.6	6.9 ± 1.8		3.3 ± 2.0	6.9 ± 1.8	
Median (IQR)	4.0 (2.0–7.0)	7.0 (6.0–8.0)		2.5 (2.0–4.5)	7.0 (6.0–8.0)	
Pathology			0.012			0.216
DCIS	10 (47.6)	34 (20.6)		6 (37.5)	38 (22.4)	
IDC + others	11 (52.4)	131 (79.4)		10 (62.5)	132 (77.6)	

CE-MRI: contrast-enhanced magnetic resonance imaging; DWI: diffusion-weighted imaging; MIP: maximal intensity projection; ADC: apparent diffusion coefficient; IDC: invasive ductal carcinoma; DCIS: ductal carcinoma in situ; Size = mm, ADC value = 10^{-3} mm^2/s; SD standard deviation; IQR: inter-quartile range.

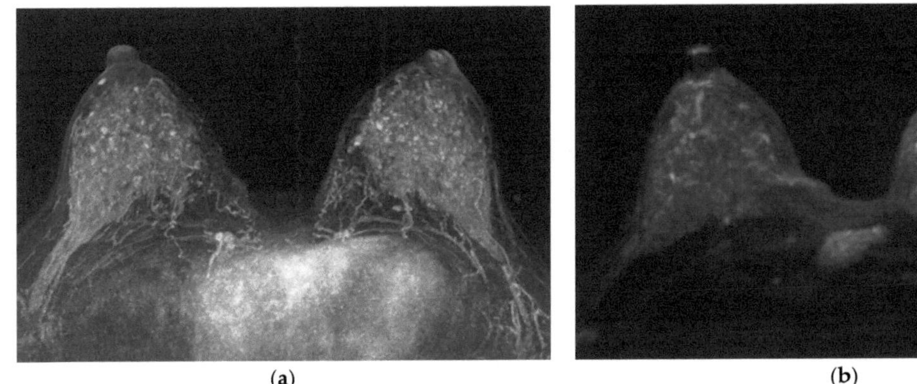

(a) **(b)**

Figure 3. Preoperative breast MRI of 39-year-old woman with invasive ductal carcinoma. (**a**) MIP of CE-MRI showed marked BPE and unseen suspicious breast lesion, (**b**) MIP of DWI showed mild BDS and a 1.3 cm mass in the 3 o'clock position of left breast. ADC value on ADC map was 0.73. In the scoring system, 1 on CE-MRI, 9 on DWI and combined. Triple-negative breast cancer was confirmed.

Table 4. Univariable and multivariable logistic regression for higher DWI scores.

	Higher DWI Scores					
	Yes (n= 85)	No (n = 101)	Univariate Odds Ratio (95% CI)	Univariate *p*	Multivariate Odds Ratio (95% CI)	Multivariate *p*
BDS						
Unmarked	72 (58.5)	51 (41.5)	reference		reference	
Marked	13 (20.6)	50 (79.4)	0.19 (0.09–0.38)	<0.001	0.18 (0.08–0.38)	<0.001
Size						
Mean ± SD	26.7 ± 16.5	29.3 ± 21.3	0.99 (0.98–1.01)	0.387		
Median (IQR)	22.0 (15.0–32.0)	21.0 (14.0–38.0)				
Size > 20 mm						
no	38 (44.2)	48 (55.8)	reference			
yes	47 (47.0)	53 (53.0)	1.12 (0.63–2.00)	0.705		
Mass/non-mass						
mass, mass and non-mass	68 (50.4)	67 (49.6)	reference		reference	
non-mass	17 (33.3)	34 (66.7)	0.50 (0.26–0.98)	0.043	0.61 (0.28–1.29)	0.196
ADC value						
Mean ± SD	0.8 ± 0.2	1.0 ± 0.3	0.08 (0.02–0.36)	0.001	0.13 (0.03–0.60)	0.009
Median (IQR)	0.8 (0.7–0.9)	0.9 (0.8–1.1)				
PATHOLOGY1				0.300		
DCIS	17 (38.6)	27 (61.4)	reference			
IDC	63 (50.0)	63 (50.0)	1.57 (0.78–3.16)	0.205		
ILC	1 (14.3)	6 (85.7)	0.36 (0.05–2.66)	0.318		
Mucinous ca	1 (16.7)	5 (83.3)	0.43 (0.06–3.30)	0.416		
Papillary ca	1 (100.0)	0 (0.0)	5.96 (0.05–768.9)	0.472		
Tubular ca	2 (100.0)	0 (0.0)	7.86 (0.18–340.1)	0.284		
PATHOLOGY2						
DCIS	17 (38.6)	27 (61.4)	reference			
IDC + others	68 (47.9)	74 (52.1)	1.44 (0.72–2.88)	0.296		
Grade in IDC (n = 142)				0.087		
Grade1	11 (33.3)	22 (66.7)	reference			
Grade2	35 (47.9)	38 (52.1)	1.80 (0.77–4.24)	0.176		
Grade3	22 (61.1)	14 (38.9)	3.04 (1.14–8.12)	0.027		
Grade in DCIS (n = 44)				0.778		
low	2 (50.0)	2 (50.0)	reference			
intermediate	11 (40.7)	16 (59.3)	0.70 (0.08–5.72)	0.737		
high	4 (30.8)	9 (69.2)	0.47 (0.05–4.63)	0.521		
LN						
Negative	58 (43.6)	75 (56.4)	reference			
Positive	27 (50.9)	26 (49.1)	1.34 (0.71–2.54)	0.370		

Table 4. *Cont.*

| | **Higher DWI Scores** | | | | | |
	Yes (n= 85)	**No (n = 101)**	**Univariate Odds Ratio (95% CI)**	**Univariate** p	**Multivariate Odds Ratio (95% CI)**	**Multivariate** p
ER						
Negative	18 (69.2)	8 (30.8)	reference		reference	
Positive	67 (41.9)	93 (58.1)	0.33 (0.14–0.80)	0.015	0.25 (0.06–1.06)	0.060
PR						
Negative	23 (63.9)	13 (36.1)	reference		reference	
Positive	62 (41.3)	88 (58.7)	0.41 (0.19–0.86)	0.019	1.34 (0.39–4.56)	0.644
HER2						
Negative	73 (47.7)	80 (52.3)	reference			
Positive	12 (36.4)	21 (63.6)	0.64 (0.29–1.38)	0.254		
KI-67						
<20%	49 (45.4)	59 (54.6)	reference			
≥20%	36 (46.2)	42 (53.8)	1.03 (0.58–1.85)	0.915		
IHC type				0.094		
Luminal A	41 (44.1)	52 (55.9)	reference			
Luminal B	27 (39.1)	42 (60.9)	0.82 (0.43–1.54)	0.535		
Her2+	5 (62.5)	3 (37.5)	1.99 (0.45–8.72)	0.363		
Triple-	12 (75.0)	4 (25.0)	3.51 (1.08–11.48)	0.037		

DWI: Diffusion-weighted imaging; BDS: background diffusion signal; ADC: apparent diffusion coefficient; IHC: immunohistochemical; IDC: invasive ductal carcinoma; ILC: invasive lobular carcinoma; DCIS: ductal carcinoma in situ; LN: lymph node; ER: estrogen receptor; PR: progesterone receptor; HER2: human epidermal growth factor receptor 2; Size = mm; ADC value = 10^{-3} mm^2/s.

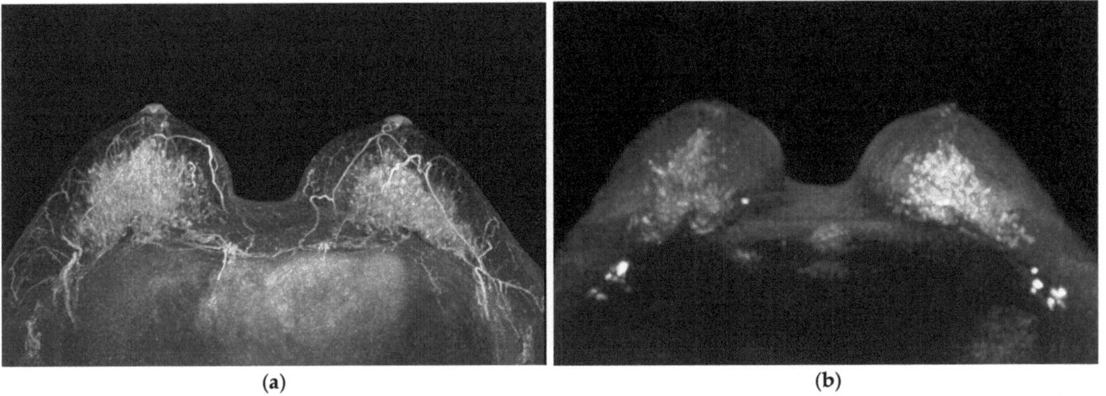

(a)   (b)

Figure 4. Preoperative breast MRIs of 43-year-old woman with breast malignant lesions. The presence of a marked BDS was significantly associated with a lower likelihood of a higher DWI score. (**a**) MIP of CE-MRI showed marked BPE and unseen suspicious breast lesion (score 1), (**b**) MIP of DWI showed marked BPE and unseen suspicious breast lesion (score 1). About 1.2 cm mucinous cancer was confirmed at the 5 o'clock position in left breast.

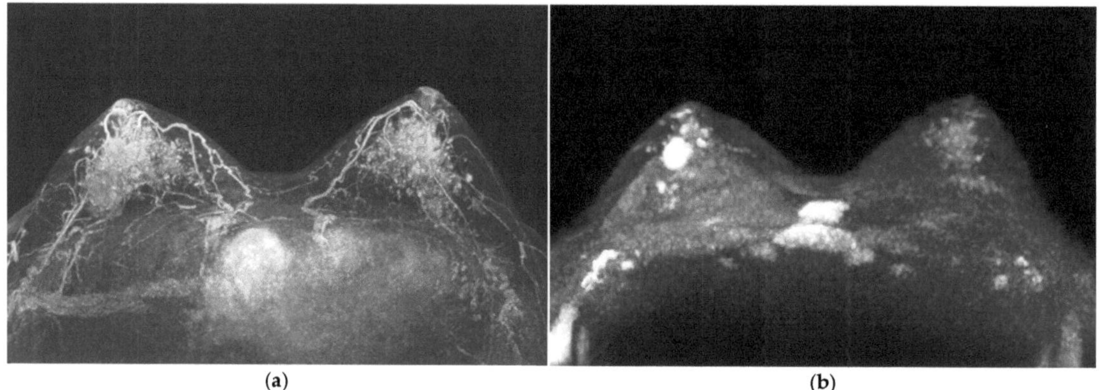

(a)   (b)

Figure 5. Preoperative breast MRIs of 49-year-old woman with breast malignant lesions. The presence of a marked BDS was significantly associated with a lower likelihood of a higher DWI score. (**a**) MIP of CE-MRI showed marked BPE and partially seen suspicious breast lesion (score 4), (**b**) MIP of DWI showed mild BDS and definitive suspicious breast lesion with false-positive lesions (score 7). About 1.5 cm-sized invasive ductal cancer was confirmed in the central portion in right breast.

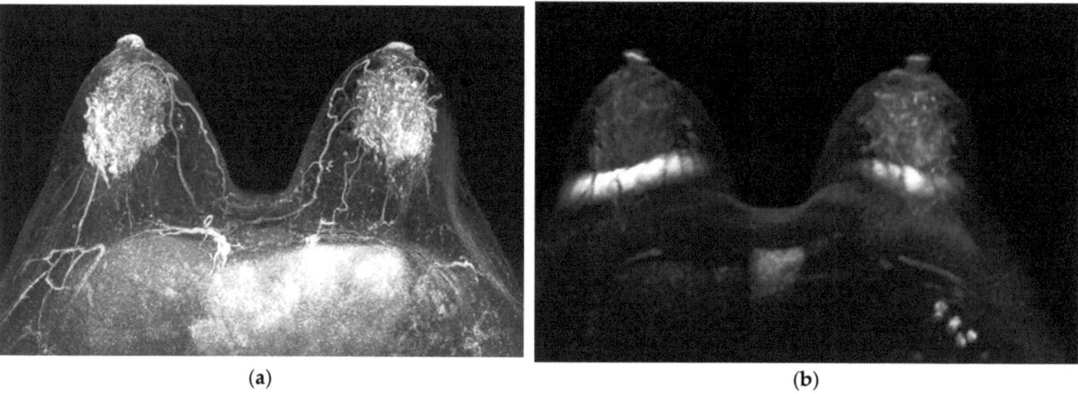

(a)   (b)

Figure 6. Preoperative breast MRI of a 52-year-old woman with ductal carcinoma in situ. The presence of a non-mass lesion was associated with a decreased likelihood of higher DWI score compared with the presence of mass lesions. (**a**) MIP of CE-MRI showed marked BPE and partially seen segmental non-mass enhancement lesion (score 4), (**b**) MIP of DWI showed mild BDS and unseen suspicious breast lesion (score 1). DCIS measuring about 4.5 cm was confirmed in the outer portion of the right breast.

4. Discussion

This study showed that the MIP of DWI significantly improved the detection of true suspicious lesions on MR imaging of young women with marked BPE.

Since its introduction in 1986, contrast-enhanced breast MRI has become the most sensitive method for detecting invasive breast cancer [17–19]. After the intravenous administration of a gadolinium-based contrast agent, BPE can lead to the enhancement of normal breast fibroglandular tissue. The extent of BPE can differ among women and even within the same individual, and it is believed to be associated with changes in the vascular supply and permeability of breast tissue, which are influenced by hormonal status [5]. However, BPE of normal breast parenchyma is a well-known and major clinical concern, significantly limiting breast tumor detection using CE-MRI. Telegrafo et al. and Demartini et al. reported that moderate and marked BPE reduces the sensitivity of CE-MRI imaging when compared with minimal and mild BPE [20,21]. Many earlier studies have shown that combining the DWI protocol results in a higher specificity compared with CE-MRI alone, by reducing false positives [7–10].

The evolving approach of using an AP for screening breast MRI offers several advantages, including shorter acquisition and interpretation times, reduced cost, and comparable diagnostic accuracy to the full protocol [13,14]. In a previous study, the AP used one pre-contrast and one post-contrast acquisition with MIP images, completing the MRI acquisition in just 3 min and the interpretation time in less than 30 s. Remarkably, the diagnostic performance was on par with the full protocol [14]. Additionally, recent apprehensions regarding the deposition of gadolinium-based contrast agents in neuronal tissues must not be dismissed [22]. Conversely, DWI is a valuable unenhanced technique that offers microstructural insights at the cellular level, enabling the detection of changes in tissue water related to modifications in tissues and intracellular structures [11]. Recent studies have demonstrated its potential for detecting and characterizing breast lesions, with technical advances enhancing its quality. In uncertain cases, the ADC on DWI can be utilized to reduce the need for biopsies [11,12,23,24]. A significant advantage is its high sensitivity for detecting breast cancer without the need for contrast material injection, as shown in a recent meta-analysis with an overall sensitivity of 84% and specificity of 79% [25]. Our study also detected a significant difference in the main lesion on MIPs of DWI based on ADC values ($p = 0.017$).

Kang et al. suggested that DWI MIPs could be a cost-effective alternative to the AP, leading to shorter acquisition and interpretation times, while avoiding the risk of gadolinium-based contrast agents. In their study, the AP's diagnostic performance, employing T1WI and rs-EPI DWI, closely resembled that of conventional CE-MRI, showcasing sensitivities ranging from 80.0% to 90.0%, specificities from 93.4% to 95.1%, PPV3s from 28.1% to 32.0%, and NPVs from 99.4% to 99.7%. The false-positive rates were minimal, ranging from 4.7% to 6.4% [26]. However, their study population differed from ours, as it involved postoperative breast MRI, which is generally easier due to treatment-related changes such as decreased lesion numbers after surgery and reduced BPE following radiation or anti-hormonal therapy [19].

Our study focused on preoperative CE-MRI, and the frequencies of various BPE categories were as follows: minimal (n = 422, 32.4%), mild (n = 410, 31.5%), moderate (n = 290, 22.3%), and marked (n = 181, 13.9%), with marked BPE showing the least prevalence. We specifically targeted young patients with breast cancer ($\leq$60 years old) with marked BPE. BPE is a valuable imaging marker for breast cancer risk assessment and can impact the interpretation of breast MRI [27,28]. And BPE is a known risk factor for breast cancer [5,29,30]. In our institution, the frequency of breast cancer in individuals over the age of 60 is significantly lower. The elderly population (>60) represents only a small proportion, and instances of marked BPE in this age group are exceedingly rare. Therefore, for this study, we defined the elderly age group as >60 and excluded it accordingly.

The MIP was central to our study, and its value has been demonstrated. The MIP images facilitated easy and rapid assessment and comparison. However, MIP images have limitations in evaluating shape, margin, and internal enhancement compared with the entire conventional CE-MRI images. Therefore, instead of using the BI-RADS, we employed a 1–10 scoring system, focusing on detectability. In our study, among 181 patients with marked BPE on the MIP of CE-MRI, the distribution of the BDS was as follows: minimal (n = 18, 9.7%), mild (n = 47, 25.3%), moderate (n = 58, 31.2%), and marked (n = 63, 33.9%), with the majority being unmarked (123, 66.1%) compared with marked (63, 33.9%). The observed results indicate a lack of correlation between BPE on CE-MRI and the BDS on DWI. While mammographic density and BPE are well-established risk factors for breast cancer, no significant correlation was found between them [5]. Consequently, we propose that mammographic density, BPE, and BDS are not correlated factors. Based on these findings, we confirmed that DWI is particularly helpful in breast cancer detection, especially in patients with marked BPE on CE-MRI but an unmarked BDS on DWI. Accordingly, the univariable and multivariable logistic regression analyses were conducted to identify clinically important findings associated with a higher score in DWI than in CE-MRI. Among the variables examined, two variables showed a significant association in the multivariable analysis. First, the presence of a marked BDS was significantly associated with a lower likelihood of a higher DWI. Second, the presence of a non-mass lesion was associated with a decreased likelihood of a higher DWI score compared with a mass lesion.

There were several limitations to our present study. First, we included a limited number of patients and adopted a retrospective study design. However, we did include a consecutive group of uniform patients who underwent preoperative breast MRI using a 3T MR scanner during the study period and definitive surgery. Second, we evaluated only the preoperative breast MRI, which may induce selection bias. This may affect image evaluations by radiologists. Third, subjective interpretations of MR imaging may also affect the image evaluation by radiologists.

Our primary focus in this study was to assess the detectability of breast cancer in young patients with marked BPE. We did not find any correlation between BPE on CE-MRI and a BDS on DWI. However, in cases where CE-MRI showed marked BPE but DWI did not show a marked BDS, additional analysis of DWI proved to be extremely helpful in breast cancer detection. Furthermore, we found that utilizing MIPs of DWI served as an effective tool for detecting breast malignancy.

5. Conclusions

In conclusion, the inclusion of MIPs of DWI in the preoperative evaluation of breast cancer patients, particularly young women with marked BPE, can be highly beneficial in improving the overall diagnostic process.

Author Contributions: G.-E.P. and N.-Y.J.: designed the study protocol, maintained medical/ethical contacts, organized clinical logistics, analyzed the data, and wrote the manuscript. B.-J.K. and S.-h.K.: analyzed the diagnostic images. G.-E.P.: analyzed the data and contributed to writing the manuscript. B.-J.K. and S.-h.K.: assisted with setting up MRI protocol and analysis of acquisition parameters related to artefact formation. G.-E.P., N.-Y.J., B.-J.K. and S.-h.K.: principal investigators and principal physicians of the contributing patients; they recruited and informed the patients. G.-E.P., N.-Y.J., B.-J.K. and S.-h.K. contributed to the implementation of the study protocol. All authors have read and agreed to the published version of the manuscript.

Funding: The study did not receive any funding.

Institutional Review Board Statement: The study was conducted in accordance with the Declaration of Helsinki and was approved by the Institutional Review Board for Clinical Research, Clinical Trial System of CMC (protocol code KC20OIDI0732).

Informed Consent Statement: Written informed consent was waived due to the retrospective study.

Data Availability Statement: The datasets used and/or analyzed during the current study are available from the corresponding author on reasonable request.

Acknowledgments: We acknowledge the Department of Biostatistics of the Catholic Research Coordinating Center for assistance with statistical analysis.

Conflicts of Interest: The authors declare no conflict of interest.

References

1. Mann, R.M.; Kuhl, C.K.; Moy, L. Contrast-enhanced MRI for breast cancer screening. *J. Magn. Reson. Imaging* **2019**, *50*, 377–390. [CrossRef]
2. Mann, R.M.; Cho, N.; Moy, L. Breast MRI: State of the art. *Radiology* **2019**, *292*, 520–536.
3. Korhonen, K.E.; Zuckerman, S.P.; Weinstein, S.P.; Tobey, J.; Birnbaum, J.A.; McDonald, E.S.; Conant, E.F. Breast MRI: False-Negative Results and Missed Opportunities. *RadioGraphics* **2021**, *41*, 645–664.
4. Song, D.; Kang, B.J.; Kim, S.H.; Lee, J.; Park, G.E. The Frequency and Causes of Not-Detected Breast Malignancy in Dynamic Contrast-Enhanced MRI. *Diagnostics* **2022**, *12*, 2575.

5. Lee, S.H.; Jang, M.-j.; Yoen, H.; Lee, Y.; Kim, Y.S.; Park, A.R.; Ha, S.M.; Kim, S.-Y.; Chang, J.M.; Cho, N. Background parenchymal enhancement at postoperative surveillance breast MRI: Association with future second breast cancer risk. *Radiology* **2023**, *306*, 90–99. [CrossRef]

6. Vreemann, S.; Dalmis, M.U.; Bult, P.; Karssemeijer, N.; Broeders, M.J.; Gubern-Mérida, A.; Mann, R.M. Amount of fibroglandular tissue FGT and background parenchymal enhancement BPE in relation to breast cancer risk and false positives in a breast MRI screening program: A retrospective cohort study. *Eur. Radiol.* **2019**, *29*, 4678–4690. [CrossRef]

7. Yabuuchi, H.; Matsuo, Y.; Okafuji, T.; Kamitani, T.; Soeda, H.; Setoguchi, T.; Sakai, S.; Hatakenaka, M.; Kubo, M.; Sadanaga, N. Enhanced mass on contrast-enhanced breast MR imaging: Lesion characterization using combination of dynamic contrast-enhanced and diffusion-weighted MR images. *J. Magn. Reson. Imaging Off. J. Int. Soc. Magn. Reson. Med.* **2008**, *28*, 1157–1165.

8. Pinker, K.; Bogner, W.; Baltzer, P.; Gruber, S.; Bickel, H.; Brueck, B.; Trattnig, S.; Weber, M.; Dubsky, P.; Bago-Horvath, Z. Improved diagnostic accuracy with multiparametric magnetic resonance imaging of the breast using dynamic contrast-enhanced magnetic resonance imaging, diffusion-weighted imaging, and 3-dimensional proton magnetic resonance spectroscopic imaging. *Investig. Radiol.* **2014**, *49*, 421–430. [CrossRef]

9. Baltzer, A.; Dietzel, M.; Kaiser, C.G.; Baltzer, P.A. Combined reading of contrast enhanced and diffusion weighted magnetic resonance imaging by using a simple sum score. *Eur. Radiol.* **2016**, *26*, 884–891.

10. Iima, M.; Honda, M.; Sigmund, E.E.; Ohno Kishimoto, A.; Kataoka, M.; Togashi, K. Diffusion MRI of the breast: Current status and future directions. *J. Magn. Reson. Imaging* **2020**, *52*, 70–90.

11. Kul, S.; Cansu, A.; Alhan, E.; Dinc, H.; Gunes, G.; Reis, A. Contribution of diffusion-weighted imaging to dynamic contrast-enhanced MRI in the characterization of breast tumors. *Am. J. Roentgenol.* **2011**, *196*, 210–217.

12. Partridge, S.C.; DeMartini, W.B.; Kurland, B.F.; Eby, P.R.; White, S.W.; Lehman, C.D. Differential diagnosis of mammographically and clinically occult breast lesions on diffusion-weighted MRI. *J. Magn. Reson. Imaging Off. J. Int. Soc. Magn. Reson. Med.* **2010**, *31*, 562–570.

13. Kim, S.H.; Shin, H.J.; Shin, K.C.; Chae, E.Y.; Choi, W.J.; Cha, J.H.; Kim, H.H. Diagnostic performance of fused diffusion-weighted imaging using t1-weighted imaging for axillary nodal staging in patients with early breast cancer. *Clin. Breast Cancer* **2017**, *17*, 154–163. [CrossRef]

14. Kuhl, C.K.; Schrading, S.; Strobel, K.; Schild, H.H.; Hilgers, R.-D.; Bieling, H.B. Abbreviated breast magnetic resonance imaging (MRI): First postcontrast subtracted images and maximum-intensity projection—A novel approach to breast cancer screening with MRI. *J. Clin. Oncol.* **2014**, *32*, 2304–2310. [CrossRef]

15. Jeong, Y.S.; Kang, J.; Lee, J.; Yoo, T.K.; Kim, S.H.; Lee, A. Analysis of the molecular subtypes of preoperative core needle biopsy and surgical specimens in invasive breast cancer. *J. Pathol. Transl. Med.* **2020**, *54*, 87–94.

16. Goldhirsch, A.; Winer, E.P.; Coates, A.S.; Gelber, R.D.; Piccart-Gebhart, M.; Thurlimann, B.; Senn, H.J.; Panel, M. Personalizing the treatment of women with early breast cancer: Highlights of the St Gallen International Expert Consensus on the Primary Therapy of Early Breast Cancer 2013. *Ann. Oncol.* **2013**, *24*, 2206–2223.

17. Shin, H.J.; Chae, E.Y.; Choi, W.J.; Ha, S.M.; Park, J.Y.; Shin, K.C.; Cha, J.H.; Kim, H.H. Diagnostic performance of fused diffusion-weighted imaging using unenhanced or postcontrast T1-weighted MR imaging in patients with breast cancer. *Medicine* **2016**, *95*, e3502.

18. Mann, R.M.; Kuhl, C.K.; Kinkel, K.; Boetes, C. Breast MRI: Guidelines from the European society of breast imaging. *Eur. Radiol.* **2008**, *18*, 1307–1318.

19. Lehman, C.D. *Screening MRI for Women at High Risk for Breast Cancer, Seminars in Ultrasound, CT and MRI*; Elsevier: Amsterdam, The Netherlands, 2006; pp. 333–338.

20. DeMartini, W.B.; Liu, F.; Peacock, S.; Eby, P.R.; Gutierrez, R.L.; Lehman, C.D. Background parenchymal enhancement on breast MRI: Impact on diagnostic performance. *Am. J. Roentgenol.* **2012**, *198*, W373–W380. [CrossRef]

21. Telegrafo, M.; Rella, L.; Ianora, A.S.; Angelelli, G.; Moschetta, M. Effect of background parenchymal enhancement on breast cancer detection with magnetic resonance imaging. *Diagn. Interv. Imaging* **2016**, *97*, 315–320.

22. McDonald, R.J.; McDonald, J.S.; Kallmes, D.F.; Jentoft, M.E.; Murray, D.L.; Thielen, K.R.; Williamson, E.E.; Eckel, L.J. Intracranial gadolinium deposition after contrast-enhanced MR imaging. *Radiology* **2015**, *275*, 772–782. [CrossRef]

23. Yabuuchi, H.; Matsuo, Y.; Sunami, S.; Kamitani, T.; Kawanami, S.; Setoguchi, T.; Sakai, S.; Hatakenaka, M.; Kubo, M.; Tokunaga, E. Detection of non-palpable breast cancer in asymptomatic women by using unenhanced diffusion-weighted and T2-weighted MR imaging: Comparison with mammography and dynamic contrast-enhanced MR imaging. *Eur. Radiol.* **2011**, *21*, 11–17.

24. Baltzer, P.A.; Benndorf, M.; Dietzel, M.; Gajda, M.; Camara, O.; Kaiser, W.A. Sensitivity and specificity of unenhanced MR mammography (DWI combined with T2-weighted TSE imaging, ueMRM) for the differentiation of mass lesions. *Eur. Radiol.* **2010**, *20*, 1101–1110. [CrossRef]

25. Chen, X.; Li, W.-l.; Zhang, Y.-l.; Wu, Q.; Guo, Y.-m.; Bai, Z.-L. Meta-analysis of quantitative diffusion-weighted MR imaging in the differential diagnosis of breast lesions. *BMC Cancer* **2010**, *10*, 693. [CrossRef]

26. Kang, J.W.; Shin, H.J.; Shin, K.C.; Chae, E.Y.; Choi, W.J.; Cha, J.H.; Kim, H.H. Unenhanced magnetic resonance screening using fused diffusion-weighted imaging and maximum-intensity projection in patients with a personal history of breast cancer: Role of fused DWI for postoperative screening. *Breast Cancer Res. Treat.* **2017**, *165*, 119–128.

27. Rella, R.; Contegiacomo, A.; Bufi, E.; Mercogliano, S.; Belli, P.; Manfredi, R. Background parenchymal enhancement and breast cancer: A review of the emerging evidences about its potential use as imaging biomarker. *Br. J. Radiol.* **2021**, *94*, 20200630.

28. Ray, K.M.; Kerlikowske, K.; Lobach, I.V.; Hofmann, M.B.; Greenwood, H.I.; Arasu, V.A.; Hylton, N.M.; Joe, B.N. Effect of background parenchymal enhancement on breast MR imaging interpretive performance in community-based practices. *Radiology* **2018**, *286*, 822–829.

29. Grimm, L.J.; Saha, A.; Ghate, S.V.; Kim, C.; Soo, M.S.; Yoon, S.C.; Mazurowski, M.A. Relationship between background parenchymal enhancement on high-risk screening MRI and future breast cancer risk. *Acad. Radiol.* **2019**, *26*, 69–75. [CrossRef]

30. Brenner, R. Background parenchymal enhancement at breast MR imaging and breast cancer risk. *Breast Dis. Year Book Q.* **2012**, *2*, 145–147. [CrossRef]

Review

Patients' Perceptions and Attitudes to the Use of Artificial Intelligence in Breast Cancer Diagnosis: A Narrative Review

Filippo Pesapane [1,*], Emilia Giambersio [2], Benedetta Capetti [3], Dario Monzani [3,4], Roberto Grasso [3,5], Luca Nicosia [1], Anna Rotili [1], Adriana Sorce [2], Lorenza Meneghetti [1], Serena Carriero [6], Sonia Santicchia [6], Gianpaolo Carrafiello [5,6], Gabriella Pravettoni [3,5] and Enrico Cassano [1]

[1] Breast Imaging Division, IEO European Institute of Oncology IRCCS, 20141 Milan, Italy; luca.nicosia@ieo.it (L.N.); anna.rotili@ieo.it (A.R.); lorenza.meneghetti@ieo.it (L.M.); enrico.cassano@ieo.it (E.C.)

[2] Postgraduation School in Radiodiagnostics, Università degli Studi di Milano, 20122 Milan, Italy; emilia.giambersio@unimi.it (E.G.); adriana.sorce@unimi.it (A.S.)

[3] Applied Research Division for Cognitive and Psychological Science, IEO European Institute of Oncology, IRCCS, 20141 Milan, Italy; benedetta.capetti@ieo.it (B.C.); dario.monzani@unipa.it (D.M.); roberto.grasso@ieo.it (R.G.); gabriella.pravettoni@ieo.it (G.P.)

[4] Department of Psychology, Educational Science and Human Movement (SPPEFF), University of Palermo, 90133 Palermo, Italy

[5] Department of Oncology and Hemato-Oncology, University of Milan, 20122 Milan, Italy; gianpaolo.carrafiello@unimi.it

[6] Foundation IRCCS Cà Granda-Ospedale Maggiore Policlinico, 20122 Milan, Italy; serena.carriero@policlinico.mi.it (S.C.); sonia.santicchia@policlinico.mi.it (S.S.)

* Correspondence: filippo.pesapane@ieo.it; Tel.: +39-02-574891

Citation: Pesapane, F.; Giambersio, E.; Capetti, B.; Monzani, D.; Grasso, R.; Nicosia, L.; Rotili, A.; Sorce, A.; Meneghetti, L.; Carriero, S.; et al. Patients' Perceptions and Attitudes to the Use of Artificial Intelligence in Breast Cancer Diagnosis: A Narrative Review. *Life* **2024**, *14*, 454. https://doi.org/10.3390/life14040454

Academic Editors: Riccardo Autelli, Taobo Hu, Mengping Long and Lei Wang

Received: 17 February 2024
Revised: 26 March 2024
Accepted: 27 March 2024
Published: 29 March 2024

Abstract: Breast cancer remains the most prevalent cancer among women worldwide, necessitating advancements in diagnostic methods. The integration of artificial intelligence (AI) into mammography has shown promise in enhancing diagnostic accuracy. However, understanding patient perspectives, particularly considering the psychological impact of breast cancer diagnoses, is crucial. This narrative review synthesizes literature from 2000 to 2023 to examine breast cancer patients' attitudes towards AI in breast imaging, focusing on trust, acceptance, and demographic influences on these views. Methodologically, we employed a systematic literature search across databases such as PubMed, Embase, Medline, and Scopus, selecting studies that provided insights into patients' perceptions of AI in diagnostics. Our review included a sample of seven key studies after rigorous screening, reflecting varied patient trust and acceptance levels towards AI. Overall, we found a clear preference among patients for AI to augment rather than replace the diagnostic process, emphasizing the necessity of radiologists' expertise in conjunction with AI to enhance decision-making accuracy. This paper highlights the importance of aligning AI implementation in clinical settings with patient needs and expectations, emphasizing the need for human interaction in healthcare. Our findings advocate for a model where AI augments the diagnostic process, underlining the necessity for educational efforts to mitigate concerns and enhance patient trust in AI-enhanced diagnostics.

Keywords: artificial intelligence; breast cancer; screening; diagnosis; psychological burden; population; survey; policy; healthcare

1. Introduction

Breast cancer has long stood as one of the most prevalent forms of neoplastic diseases affecting women globally. Its prominence among the most common forms of cancer has persisted over decades, shaping healthcare strategies and research endeavors worldwide. The year 2020, in particular, marked a significant landmark, with an estimated 19.3 million new cancer cases reported and nearly 10 million cancer-related deaths recorded worldwide. Among these statistics, breast cancer emerged as the most common, with approximately 2.2 million new cases diagnosed and close to 685,000 deaths attributed to the disease, thus being more ordinarily diagnosed even than lung cancer [1]. Breast cancer in the male population represents a rare entity with an estimated incidence of 1.2 per 100,000 in the US. Specific risk factors such as gynecomastia, BRCA mutations, Klinefelter syndrome, previous radiation exposure to the chest, and high estrogen levels are tightly linked to these diagnoses; therefore, routine screening mammography is not recommended for asymptomatic men [2].

The evolution of artificial intelligence (AI) in medical imaging and diagnostics has ushered in an era of precision medicine, significantly impacting breast cancer detection and management. AI algorithms, particularly in mammography [3], have demonstrated potential in enhancing diagnostic accuracy, reducing false positives and negatives, aiding risk stratification and prognostication [4], and sensibly reducing the time to examine images, which are very useful in breast cancer screening [5,6].

However, the successful implementation of AI in clinical practice hinges not only on its technical efficacy but also on patients' acceptance and attitudes towards this technology. Understanding patients' perceptions is crucial in the context of breast cancer diagnosis, where the psychological burden of screening and diagnosis is substantial [7]. Patient attitudes towards AI in healthcare can influence their willingness to engage with AI-assisted diagnostic procedures and can impact their trust in the outcomes of such diagnostics.

Patients' attitudes towards AI in medicine are a topic of growing interest. Attitudes of patients towards AI differ, as some are optimistic about its potential to enhance healthcare while others harbor concerns, especially about possible misdiagnosis and privacy breaches [8]. Additionally, research has shown that patients generally prefer human doctors to AI-powered machines in diagnosis, screening, and treatment [9]. Overall, these findings underscore the importance of understanding and addressing patient attitudes towards AI in medicine.

The primary aim of this narrative review is to elucidate patient perceptions regarding the potential role of deep-learning algorithms in the detection of breast cancer. Specifically, we sought to understand patient apprehensions about the use of AI software in routine radiological practice. Central to our analysis are questions about patient trust, such as whether patients show more confidence in the clinical judgment of radiologists compared to AI predictions, or if there is a noticeable shift towards relying on algorithmic analysis. Additionally, we aimed to determine the extent of AI involvement that patients find acceptable or preferable in their diagnostic journey, thus providing insights that could guide the integration of AI in medical practice in a manner that is sensitive to patient needs and concerns.

We performed this narrative literature review utilizing databases such as PubMed, Embase, Medline, and Scopus, spanning publications from January 2000 to December 2023. Our search strategy involved a carefully constructed string of key terms to ensure a thorough exploration of the relevant literature. The search string employed was: ("artificial intelligence" OR "AI") AND ("breast cancer" OR "mammography") AND ("patient perspective" OR "patient opinion" OR "quality of life" OR "QoL") AND ("screening" OR "diagnosis" OR "radiology").

Through this meticulous approach, our initial search yielded a total of 49 results. Subsequently, we subjected these findings to a double reading assessment of the entire papers, excluding 42 papers and selecting 7 studies that contribute to shed light on the intricate relationship between AI, breast cancer diagnosis, and the patient experience, allowing for a deeper exploration of this critical field [9–15].

2. Receiving a Diagnosis of Breast Cancer

2.1. The Physical and Psychological Aftermath

Receiving a breast cancer diagnosis marks the onset of a challenging journey, encompassing not only the physical battles against the disease but also confronting its psychological repercussions. [16] Breast cancer exhibits a notable frequency of coexisting conditions, including psychological discomfort [17–19] issues related to anxiety and mood [20,21], feelings of depression [22], and enduring fatigue coupled with reduced social engagements, emerging as prevalent reactions to the diagnosis and therapeutic interventions associated with breast cancer [23]. In particular, it has been observed that the psychological impact of the illness is significant, especially in the transition to motherhood for women of childbearing age [19]. For these women, fears and concerns associated with a cancer diagnosis are primarily linked to the disease and its potential effects on pregnancy and the child's health [19]. Furthermore, individuals diagnosed with primary breast cancer remain susceptible to enduring psychological challenges over an extended duration [24,25], underscoring the substantial influence of this health condition on the overall well-being of affected individuals.

The improvements in early cancer detection and efficacy of innovative treatments developed in recent years have supported a prolonged lifespan of cancer patients, generating, however, the onset of long-term psychological and physical consequences and altered quality of life [26]. During treatment, in fact, whether involving minor or major procedures, patients may grapple with temporary or permanent alterations to their bodies, giving rise to significant psychological challenge [27–30]. The removal of breasts, the development of swollen arms due to lymphedema, chemotherapy-induced baldness, pharmacologically triggered menopause, heightened skin sensitivity from radiation, and the use of prosthetics can impact the self-perception, body image, sexual function, and overall emotional well-being of women with breast cancer [16,31,32].

Moreover, 90% of breast cancer survivors experience sequalae following treatments, including a decline in physical strength of their upper body, and chronic neuropathic pain or nonpainful sensations in the amputated breast following surgery [33]. Accordingly, the integration of supportive measures, tailored to address both the physical and emotional strains, is essential in fostering resilience and recovery.

2.2. Psychological Burden of Carrying a BRCA Genetic Mutation

The discovery of a BRCA mutation carries with it not just a heightened risk for breast cancer but also a profound psychological burden, stemming from the anticipation of cancer and its implications on familial and personal health [34]. BRCA1 and BRCA2 stand out as the predominant genes associated with this specific cancer type compared to others [35]. It is reported that 55–72% of women who inherit a harmful BRCA1 variant and 45–69% of women who inherit a harmful BRCA2 mutation will develop breast cancer by 70–80 years of age, while the chances of developing breast cancer among the general population at some point in their lives is about 13% [36].

Several factors contribute to the decision-making process regarding preventive strategies among BRCA carriers. Some factors are linked to information processing [37] while others are associated with psychosocial variables such as risk perception, cancer-related worry, levels of emotion dysregulation, family history, and having young children [38,39].

The potential psychological responses and related considerations when a mutation is detected in an individual could include elevated levels of distress, anxiety, and depression [40]. These psychological manifestations can be attributed to an increased risk of future illnesses and implications not only for the health of the tested individuals but also for their entire family [41]. In addition, increased psychological distress may be triggered when genetic testing is conducted during a woman's fertile age, highlighting the necessity also for fertility counseling [42]. Addressing these needs comprehensively can alleviate the psychological impact and empower women carrying a genetic mutation to make informed choices about their health [43,44].

2.3. Artificial Intelligence and Breast Cancer: Patients' Perspectives

The integration of AI in medical diagnostics, particularly in breast cancer detection, is a rapidly evolving field. This evolution has prompted a need to understand the patient's perspective on AI's role in their healthcare. To address this gap, our research focuses on analyzing existing literature that explores patient attitudes towards AI in breast cancer diagnosis.

The main take-home messages of our review are summarized in Table 1.

Table 1. A concise overview of the major findings and implications from our review.

Aspect	Take-Home Messages
AI's Potential in Diagnosis	AI enhances diagnostic accuracy and efficiency in breast cancer screening.
Patient Concerns	Varied concerns about AI's trustworthiness, personal interaction, and accountability.
Role of Radiologists	Patients prefer AI as a complement to radiologists, not a replacement.
Demographic Variations	Perceptions of AI vary by demographic; tailored patient education is crucial.
Legal and Ethical Considerations	Need for explainable AI and governance frameworks to address legal/ethical issues.
Future Focus	Harmonize AI with patient needs, ensuring it supports human elements of healthcare.

In 2022, Borondy Kitts A.B. [41] reported that patient engagement in radiology AI revolves around two key areas: data sharing for AI development and AI's use in patient care. Patients generally support data sharing if it benefits others or research but have concerns about privacy risks and trust issues. In terms of AI in medical care, patients are open to AI assisting radiologists but lack trust in unsupervised AI. They worry about liability, loss of human connection, and bias in AI algorithms. Building trust in AI requires transparency, security, and privacy measures. According to the author, radiologists can prepare patients by implementing data-sharing agreements for algorithm development and having discussions about AI use in their care. This presents an opportunity for radiologists to maintain strong patient relationships as AI becomes more integrated into healthcare.

In 2021, Ongena et al. [39] published the results of a survey administered to women undergoing screening mammography in the Netherlands; specifically, they investigated four precise themes regarding AI in radiology: trust and accountability (trust in AI in taking over diagnostic interpretation tasks of the radiologist, both with regard to accuracy, communication, and confidentiality), personal interaction (preference of personal interaction over AI-based communication), efficiency (belief in whether AI could improve diagnostic workflow), and a newly developed scale measuring the attitude towards AI in general medicine. They also took into consideration social status and the level of education of the patients. Their results showed that their population does not trust AI enough for its use in standalone interpretation of screening mammograms. Respondents were slightly more optimistic about the use of AI as a tool that could help select patients that require a second reader or not. However, a considerable proportion (41%) still opposed the idea of using AI as a tool to select patients for second reading. Seventeen percent of women explicitly objected against using AI as an actual second reader. Therefore, the combination of a radiologist as a first reader and an AI system as a second reader seems to be the most feasible approach to the population at present, as suggested also by the above-mentioned recent clinical trials [5,6].

In 2020, Lennox-Chhugani et al. [42] administered a similar survey that investigated topics concerning attitudes of women to the use of AI in the breast screening process in four National Health Service trusts providing acute care in the East Midlands of England.

The study revealed that women generally have a limited understanding of the current mammogram reading process, with only a minority recognizing the involvement of two human readers in blind readings. However, sentiment analysis of free-text responses indicated that a significant proportion of women expressed positivity towards the use of AI in breast screening, with the largest percentage holding positive views. Additionally, thematic analysis highlighted perceived benefits of AI in breast screening, including increased efficiency, improved reliability, and greater safety. Many women expressed the belief that AI integration in breast screening is inevitable and beneficial for the future. Interestingly, women of screening age showed a higher inclination towards positive views on AI in breast screening compared to younger women, despite being less likely to use AI in everyday health advice or hold positive views of its impact in society.

In 2020, Adams et al. [43] investigated similar topics by a roundtable discussion where radiologists engaged with patients and invited them to share their opinions and concerns about the use of AI in radiology. They noted that the four themes that recurred the most during their conversation were the following: fear of the unknown, trust, human connection, and cultural acceptability. On the other hand, patients agreed that AI could have a positive impact on the workflow of radiologists by improving access and reducing waiting times, reducing time to diagnosis, and even increasing diagnostic accuracy.

In 2021, Pesapane et al. [9] conducted studies, including a survey, to investigate the attitudes and perceptions of patients towards the use of AI in mammography. Researchers administered an anonymous questionnaire to participants in a breast cancer screening program, focusing on their opinions about the introduction of AI in mammography. This questionnaire was developed in collaboration with psycho-oncologists and subsequently validated. The findings revealed that a significant portion of the sample (88%) held a positive view of AI's role in mammography screening, recognizing its potential utility and security. Notably, 94% of respondents believed that radiologists should always provide their interpretation of mammograms. Furthermore, 90% opined that AI could assist in identifying cases warranting further investigation. A substantial majority (77%) concurred that AI should be employed at least as a secondary reader. A critical insight emerged regarding the attribution of responsibility for potential AI errors. About 52% of the participants believed that both software developers and radiologists share the responsibility for any mistakes made by AI systems. The survey also uncovered intriguing variations in opinions across different demographics. Women from diverse age groups and educational backgrounds exhibited distinct perspectives on AI's potential use and involvement in medicine, highlighting the importance of considering demographic factors when assessing patient attitudes towards AI in healthcare. Women with a higher education level (e.g., high school diploma or university degree) were positively associated with optimistic thinking on the use of AI, although some concern was also observed among the more educated. Particularly, authors reported a lower perceived accuracy in medical AI knowledge as educational level increased. This subjective evaluation of personal knowledge about medical AI was explained by the "Dunning–Kruger" [45] effect, which describes how people with limited skills or knowledge in an area of expertise tend to overestimate their own knowledge or competence in that domain. Also, according to this survey, women held both software and radiologists accountable for errors.

The matter of accountability of errors when implementing AI is extremely controversial. Standardized AI governance frameworks and proper AI regulation and legislation are still loosely defined and largely underdeveloped [46]. The attribution of responsibility by patients is related to their understanding of the AI apparatus and workflow, which is not always fully explainable due to the "black box" nature of its networks [47,48]. The concept of "explainable AI" has recently been developed to unravel the inexplicable algorithms of AI in order to address and resolve ethical and legal issues, also concerning fault and accountability, and to make its use in clinical practice more acceptable and understandable by patients [49].

Additionally, in 2022, Bunnel and Rowe [44] investigated the effects that the implementation of AI in breast imaging could have on the relationship between the radiologist and the patient and their ways of communication. The results of their analysis showed that patients perceive and appreciate the competency of the radiologist by mutual effective communication and human interpretation of AI-generated diagnoses. According to patients, radiologists are able to administer adequate care when their competency and expertise are unaffected by AI integration, and they effectively identify potential AI errors.

The key findings of these investigations and surveys are summarized in Table 2.

Table 2. A synopsis of the key findings in the research and surveys conducted by the authors referenced in our review.

Study	Key Findings
Borondy Kitts (2022) [41]	- Patients support data sharing for AI development but have concerns about privacy and trust. - They are open to AI assisting radiologists but lack trust in unsupervised AI. - Building trust in AI requires transparency and privacy measures.
Ongena et al. (2021) [39]	- Population lacks trust in AI for standalone interpretation of mammograms. - Slightly more optimistic about AI assisting in patient selection for further review. - Prefer the combination of a radiologist as the first reader and AI as the second reader
Lennox-Chhugani et al. (2020) [42]	- Women express positivity towards AI in breast screening, citing increased efficiency and reliability. - Many believe AI integration in breast screening is inevitable and beneficial for the future
Adams et al. (2020) [43]	- Patients express fear of the unknown and concerns about trust and human connection regarding AI in radiology. - They believe AI could positively impact radiologists' workflow
Pesapane et al. (2021) [9]	- Majority of participants hold a positive view of AI's role in mammography screening. - Most believe radiologists should always provide their interpretation of mammograms. - Patients hold both software developers and radiologists accountable for AI errors.
Bunnel and Rowe (2022) [44]	- Patients appreciate effective communication and human interpretation of AI-generated diagnoses by radiologists. - Radiologists are perceived as competent when their expertise is unaffected by AI integration.

Interestingly, a survey concerning similar topics was conducted in the UK by de Vries et al. [50] that, on the other hand, was aimed at the evaluation of the opinions of screening readers on the use and future applications of AI in mammography. Accredited breast cancer screening readers were asked to respond and give their opinions on four different scenarios for future, possible utilization of AI: a "partial replacement scenario" with a specialist and an AI algorithm examining the mammograms where, in case of disagreement, a different specialist would make the final decision; a "total replacement scenario" with AI algorithms examining the mammograms without input from radiologists, thus making the final decision; a "triage scenario" with AI algorithms initially examining the mammograms where, if suspicious findings are detected, a specialist would be required to review the image; a "companion scenario" with mammograms continuing to be examined by specialists as is the current practice, with on-demand access to an AI algorithm to help them make their decisions.

The data obtained from the survey evidenced that breast screening readers in the UK favor the introduction of AI, with over 63% of participants having a positive or strongly positive view of AI use in screening. Respondents overall preferred partial replacement (AI replaces one human reader) over other AI implementation scenarios. They objected to the total replacement scenario, while views on the triage and companion scenarios were mixed.

Some comments added by the responding radiologists also suggested other possible uses of AI in the screening setting, such as maximizing image quality, interpreting breast density, and then assessing risk and possible masking from breast density and fat—the parenchyma ratio—so that the algorithm can suggest whether or not to perform tomosynthesis. Approximately half of the respondents thought first readers (52%) and second readers (51%) should have access to the AI opinion. Most respondents (68%) thought that third readers or an arbitration panel should have access to the AI opinion.

In summary, the collective findings extracted from the referenced articles and surveys conducted among patients and breast radiologists alike score a harmonious consensus. Both groups emphasize the insufficiency of exclusively relying on AI for mammogram assessment, preferring instead its partial integration into the decision-making process. AI should function as an auxiliary tool, potentially assuming the role of a second or third reader in conjunction with a human radiologist. Such a collaborative approach not only optimizes diagnostic accuracy but also values the continued significance of human expertise and judgment in breast cancer detection and diagnosis.

Finally, we must acknowledge that the studies considered in this review predominantly originate from populations residing in medium- to high-income countries. These countries are typically at the forefront of adopting and integrating AI applications into various domains, including healthcare. Moreover, their populations tend to be more aware of technological advancements and medical innovations [9]. However, it is crucial to acknowledge that the impact of AI in healthcare, particularly in fields such as breast cancer screening, is not limited to affluent nations, as the implementation of AI holds significant promise in addressing healthcare disparities, especially in low-income countries [51–53]. As low-income countries often face challenges in establishing and maintaining comprehensive screening programs due to limited resources, infrastructure, and healthcare accessibility [54], AI has the potential to be a transformative tool, offering more accessible and cost-effective solutions for breast cancer screening [53]. It could, in fact, help provide easy access to better healthcare, such as screening mammography for vulnerable and at-risk women through algorithm-assisted, telemedicine-based platforms [55].

However, a paradox emerges when considering the global application of AI in healthcare. While AI has the potential to reduce healthcare disparities by improving access to screening and diagnostic services in low-income countries, the current concentration of AI-related research and development in high-income regions poses a challenge. The majority of AI algorithms are developed and validated on datasets primarily derived from affluent populations, which may not adequately represent the diversity of health conditions, demographics, and healthcare systems in low-income countries [7].

Furthermore, the adoption of AI in low-income regions can be hindered by several factors, including limited access to high-quality medical data for algorithm training, inadequate infrastructure, and a lack of awareness and acceptance among healthcare providers and communities. These challenges may inadvertently exacerbate healthcare disparities, as the benefits of AI may not be equally accessible to all populations [52].

To address these issues, initiatives should focus on developing AI solutions that are adaptable to resource-constrained settings, promoting data sharing and collaboration, and fostering education and awareness about the potential benefits of AI in healthcare. Efforts to expand the reach of AI in healthcare should be guided by a commitment to inclusivity, equitable access, and a thorough understanding of the specific challenges faced by underserved populations [56]. Only then can AI fulfill its promise of reducing disparities rather than accentuating them.

3. Conclusions

While recognizing AI's potential for enhanced diagnostic accuracy and efficiency in mammography, patients express varied concerns about trust, personal interaction, and accountability, highlighting the need for a balanced approach in clinical practice.

Demographic differences in perceptions and concerns underline the importance of tailored patient education about medical AI. Legal and ethical considerations, particularly regarding error accountability and AI's "black box" nature, necessitate the resolute development of an ever-increasing explainable AI [57] as well as standardized ethics and governance frameworks [58] capable of ensuring the ethical sustainability of AI and of maintaining and strengthening patient trust [59].

Nevertheless, a fundamental ethical requirement, reported by participants themselves, remains that of considering AI always as an empowering and enabling tool, which should never replace human evaluation of images altogether or hinder the direct interaction and communication between the radiologist and the patient.

In conclusion, the integration of AI presents substantial advancements in breast cancer screening. However, its effective clinical implementation necessitates addressing patient concerns and preserving the crucial role of radiologists in providing empathetic patient care. Moving forward, it is essential to prioritize efforts aimed at aligning AI technology with patient preferences and requirements, ensuring that AI complements rather than replaces the human aspects of healthcare delivery.

Author Contributions: Conception and design: F.P., A.R., L.N. and E.G. Administrative support: E.C., R.G., G.P. and G.C. Provision of study materials or patients: All Authors, Collection and assembly of data: F.P., E.G., B.C., A.R. and D.M. Data analysis and interpretation: All Authors. Manuscript writing: F.P., E.G., B.C., R.G., L.N., A.R., G.C. and E.C. All authors have read and agreed to the published version of the manuscript.

Funding: This research received no external funding.

Data Availability Statement: No new data were created or analyzed in this study. Data sharing is not applicable to this article.

Acknowledgments: This work was partially supported by the Italian Ministry of Health Ricerca Corrente 5 × 1000.

Conflicts of Interest: The authors declare no conflict of interest.

References

1. Sung, H.; Ferlay, J.; Siegel, R.L.; Laversanne, M.; Soerjomataram, I.; Jemal, A.; Bray, F.; Bsc, M.F.B.; Me, J.F.; Soerjomataram, M.I.; et al. Global Cancer Statistics 2020: GLOBOCAN Estimates of Incidence and Mortality Worldwide for 36 Cancers in 185 Countries. *CA Cancer J. Clin.* **2021**, *71*, 209–249. [CrossRef] [PubMed]

2. Khan, N.A.J.; Tirona, M. An updated review of epidemiology, risk factors, and management of male breast cancer. *Med. Oncol.* **2021**, *38*, 39. [CrossRef] [PubMed]

3. Pesapane, F.; Trentin, C.; Ferrari, F.; Signorelli, G.; Tantrige, P.; Montesano, M.; Cicala, C.; Virgoli, R.; D'acquisto, S.; Nicosia, L.; et al. Deep learning performance for detection and classification of microcalcifications on mammography. *Eur. Radiol. Exp.* **2023**, *7*, 69. [CrossRef]

4. Pesapane, F.; Battaglia, O.; Pellegrino, G.; Mangione, E.; Petitto, S.; Manna, E.D.F.; Cazzaniga, L.; Nicosia, L.; Lazzeroni, M.; Corso, G.; et al. Advances in breast cancer risk modeling: Integrating clinics, imaging, pathology and artificial intelligence for personalized risk assessment. *Future Oncol.* **2023**, *19*, 2547–2564. [CrossRef]

5. Lång, K.; Josefsson, V.; Larsson, A.-M.; Larsson, S.; Högberg, C.; Sartor, H.; Hofvind, S.; Andersson, I.; Rosso, A. Artificial intelligence-supported screen reading versus standard double reading in the Mammography Screening with Artificial Intelligence trial (MASAI): A clinical safety analysis of a randomised, controlled, non-inferiority, single-blinded, screening accuracy study. *Lancet Oncol.* **2023**, *24*, 936–944. [CrossRef] [PubMed]
6. Dembrower, K.; Crippa, A.; Colón, E.; Eklund, M.; Strand, F. Artificial intelligence for breast cancer detection in screening mammography in Sweden: A prospective, population-based, paired-reader, non-inferiority study. *Lancet Digit. Health* **2023**, *5*, e703–e711. [CrossRef] [PubMed]
7. Pesapane, F.; Cassano, E. Enhancing Breast Imaging Practices: Addressing False-Positive Findings, Personalization, and Equitable Access. *Radiology* **2023**, *309*, e232189. [CrossRef] [PubMed]
8. Khullar, D.; Casalino, L.P.; Qian, Y.; Lu, Y.; Krumholz, H.M.; Aneja, S. Perspectives of Patients about Artificial Intelligence in Health Care. *JAMA Netw. Open* **2022**, *5*, e2210309. [CrossRef] [PubMed]
9. Pesapane, F.; Rotili, A.; Valconi, E.; Agazzi, G.M.; Montesano, M.; Penco, S.; Nicosia, L.; Bozzini, A.; Meneghetti, L.; Latronico, A.; et al. Women's perceptions and attitudes to the use of AI in breast cancer screening: A survey in a cancer referral centre. *Br. J. Radiol.* **2023**, *96*, 20220569. [CrossRef] [PubMed]
10. Campbell-Enns, H.; Woodgate, R. The psychosocial experiences of women with breast cancer across the lifespan: A systematic review protocol. *JBI Database Syst. Rev. Implement. Rep.* **2015**, *13*, 112–121. [CrossRef] [PubMed]
11. Arnaboldi, P.; Riva, S.; Crico, C.; Pravettoni, G. A systematic literature review exploring the prevalence of post-traumatic stress disorder and the role played by stress and traumatic stress in breast cancer diagnosis and trajectory. *Breast Cancer Targets Ther.* **2017**, *9*, 473–485. [CrossRef] [PubMed]
12. Arnaboldi, P.; Lucchiari, C.; Santoro, L.; Sangalli, C.; Luini, A.; Pravettoni, G. PTSD symptoms as a consequence of breast cancer diagnosis: Clinical implications. *SpringerPlus* **2014**, *3*, 392. [CrossRef] [PubMed]
13. Faccio, F.; Mascheroni, E.; Ionio, C.; Pravettoni, G.; Peccatori, F.A.; Pisoni, C.; Cassani, C.; Zambelli, S.; Zilioli, A.; Nastasi, G.; et al. Motherhood during or after breast cancer diagnosis: A qualitative study. *Eur. J. Cancer Care* **2020**, *29*, e13214. [CrossRef] [PubMed]
14. Gallagher, J.; Parle, M.; Cairns, D. Appraisal and psychological distress six months after diagnosis of breast cancer. *Br. J. Health Psychol.* **2002**, *7*, 365–376. [CrossRef] [PubMed]
15. Kissane, D.W.; Ildn, J.; Bloch, S.; Vitetta, L.; Clarke, D.M.; Smith, G.C.; McKenzie, D.P. Psychological morbidity and quality of life in Australian women with early-stage breast cancer: A cross-sectional survey. *Med. J. Aust.* **1998**, *169*, 192–196. [CrossRef] [PubMed]
16. Fann, J.R.; Thomas-Rich, A.M.; Katon, W.J.; Cowley, D.; Pepping, M.; McGregor, B.A.; Gralow, J. Major depression after breast cancer: A review of epidemiology and treatment. *Gen. Hosp. Psychiatry* **2008**, *30*, 112–126. [CrossRef] [PubMed]
17. Denieffe, S.; Gooney, M. A meta-synthesis of women's symptoms experience and breast cancer. *Eur. J. Cancer Care* **2011**, *20*, 424–435. [CrossRef] [PubMed]
18. Cohee, A.A.; Adams, R.N.; Fife, B.L.; Von Ah, D.M.; Monahan, P.O.; Zoppi, K.A.; Cella, D.; Champion, V.L. Relationship Between Depressive Symptoms and Social Cognitive Processing in Partners of Long-Term Breast Cancer Survivors. *Oncol. Nurs. Forum* **2017**, *44*, 44–51. [CrossRef] [PubMed]
19. Schmidt, M.E.; Wiskemann, J.; Steindorf, K. Quality of life, problems, and needs of disease-free breast cancer survivors 5 years after diagnosis. *Qual. Life Res.* **2018**, *27*, 2077–2086. [CrossRef] [PubMed]
20. Dinapoli, L.; Colloca, G.; Di Capua, B.; Valentini, V. Psychological Aspects to Consider in Breast Cancer Diagnosis and Treatment. *Curr. Oncol. Rep.* **2021**, *23*, 38. [CrossRef] [PubMed]
21. Izci, F.; Ilgun, A.S.; Findikli, E.; Ozmen, V. Psychiatric Symptoms and Psychosocial Problems in Patients with Breast Cancer. *J. Breast Health* **2016**, *12*, 94–101. [CrossRef] [PubMed]
22. Sebri, V.; Mazzoni, D.; Triberti, S.; Pravettoni, G. The Impact of Unsupportive Social Support on the Injured Self in Breast Cancer Patients. *Front. Psychol.* **2021**, *12*, 722211. [CrossRef] [PubMed]
23. Durosini, I.; Triberti, S.; Savioni, L.; Sebri, V.; Pravettoni, G. The Role of Emotion-Related Abilities in the Quality of Life of Breast Cancer Survivors: A Systematic Review. *Int. J. Environ. Res. Public Health* **2022**, *19*, 12704. [CrossRef] [PubMed]
24. Sebri, V.; Durosini, I.; Mazzoni, D.; Pravettoni, G. The Body after Cancer: A Qualitative Study on Breast Cancer Survivors' Body Representation. *Int. J. Environ. Res. Public Health* **2022**, *19*, 12515. [CrossRef] [PubMed]
25. Lopes, J.V.; Bergerot, C.D.; Barbosa, L.R.; Calux, N.M.d.C.T.; Elias, S.; Ashing, K.T.; Domenico, E.B.L.D. Impact of breast cancer and quality of life of women survivors. *Rev. Bras. Enferm.* **2018**, *71*, 2916–2921. [CrossRef] [PubMed]
26. Jing, L.; Zhang, C.; Li, W.; Jin, F.; Wang, A. Incidence and severity of sexual dysfunction among women with breast cancer: A meta-analysis based on female sexual function index. *Support. Care Cancer* **2019**, *27*, 1171–1180. [CrossRef] [PubMed]
27. Lovelace, D.L.; McDaniel, L.R.; Golden, D. Long-Term Effects of Breast Cancer Surgery, Treatment, and Survivor Care. *J. Midwifery Womens Health* **2019**, *64*, 713–724. [CrossRef] [PubMed]
28. Bjørnslett, M.; Dahl, A.A.; Sørebø, Ø.; Dørum, A. Psychological distress related to BRCA testing in ovarian cancer patients. *Fam. Cancer* **2015**, *14*, 495–504. [CrossRef]
29. Borreani, C.; Manoukian, S.; Bianchi, E.; Brunelli, C.; Peissel, B.; Caruso, A.; Morasso, G.; Pierotti, M. The psychological impact of breast and ovarian cancer preventive options in *BRCA1* and *BRCA2* mutation carriers. *Clin. Genet.* **2014**, *85*, 7–15. [CrossRef] [PubMed]

30. Kuchenbaecker, K.B.; Hopper, J.L.; Barnes, D.R.; Phillips, K.A.; Mooij, T.M.; Roos-Blom, M.J.; Jervis, S.; Van Leeuwen, F.E.; Milne, R.L.; Andrieu, N.; et al. Risks of Breast, Ovarian, and Contralateral Breast Cancer for BRCA1 and BRCA2 Mutation Carriers. *JAMA* **2017**, *317*, 2402. [CrossRef] [PubMed]

31. Landsbergen, K.M.; Brunner, H.G.; Manders, P.; Hoogerbrugge, N.; Prins, J.B. Educational-support groups for BRCA mutation carriers satisfy need for information but do not affect emotional distress. *Genet. Couns.* **2010**, *21*, 423–437. [PubMed]

32. Jamal, L.; Schupmann, W.; Berkman, B.E. An ethical framework for genetic counseling in the genomic era. *J. Genet. Couns.* **2020**, *29*, 718–727. [CrossRef] [PubMed]

33. van Driel, C.M.G.; Oosterwijk, J.C.; Meijers-Heijboer, E.J.; van Asperen, C.J.; Zeijlmans van Emmichoven, I.A.; de Vries, J.; Mourits, M.; Henneman, L.; Timmermans, D.; de Bock, G. Psychological factors associated with the intention to choose for risk-reducing mastectomy in family cancer clinic attendees. *Breast* **2016**, *30*, 66–72. [CrossRef] [PubMed]

34. Guarino, A.; Polini, C.; Forte, G.; Favieri, F.; Boncompagni, I.; Casagrande, M. The Effectiveness of Psychological Treatments in Women with Breast Cancer: A Systematic Review and Meta-Analysis. *J. Clin. Med.* **2020**, *9*, 209. [CrossRef] [PubMed]

35. Meiser, B. Psychological impact of genetic testing for cancer susceptibility: An update of the literature. *Psycho-Oncology* **2005**, *14*, 1060–1074. [CrossRef] [PubMed]

36. Knabben, L.; Siegenthaler, F.; Imboden, S.; Mueller, M.D. Fertility in BRCA mutation carriers: Counseling BRCA-mutated patients on reproductive issues. *Horm. Mol. Biol. Clin. Investig.* **2022**, *43*, 171–177. [CrossRef] [PubMed]

37. Dibble, K.E.; Donorfio, L.K.M.; Britner, P.A.; Bellizzi, K.M. Stress, anxiety, and health-related quality of life in BRCA1/2-positive women with and without cancer: A comparison of four US female samples. *Gynecol. Oncol. Rep.* **2022**, *42*, 101033. [CrossRef] [PubMed]

38. Scotto, L.; Pizzoli, S.F.M.; Marzorati, C.; Mazzocco, K.; Pravettoni, G. The impact of prophylactic mastectomy on sexual well-being: A systematic review. *Sex. Med. Rev.* **2024**, *2024*, qead054. [CrossRef]

39. Ongena, Y.P.; Yakar, D.; Haan, M.; Kwee, T.C. Artificial Intelligence in Screening Mammography: A Population Survey of Women's Preferences. *J. Am. Coll. Radiol.* **2021**, *18*, 79–86. [CrossRef] [PubMed]

40. Derevianko, A.; Pizzoli, S.F.M.; Pesapane, F.; Rotili, A.; Monzani, D.; Grasso, R.; Cassano, E.; Pravettoni, G. The Use of Artificial Intelligence (AI) in the Radiology Field: What Is the State of Doctor–Patient Communication in Cancer Diagnosis? *Cancers* **2023**, *15*, 470. [CrossRef]

41. Borondy Kitts, A. Patient Perspectives on Artificial Intelligence in Radiology. *J. Am. Coll. Radiol.* **2023**, *20*, 863–867. [CrossRef] [PubMed]

42. Lennox-Chhugani, N.; Chen, Y.; Pearson, V.; Trzcinski, B.; James, J. Women's attitudes to the use of AI image readers: A case study from a national breast screening programme. *BMJ Health Care Inform.* **2021**, *28*, e100293. [CrossRef] [PubMed]

43. Adams, S.J.; Tang, R.; Babyn, P. Patient Perspectives and Priorities Regarding Artificial Intelligence in Radiology: Opportunities for Patient-Centered Radiology. *J. Am. Coll. Radiol.* **2020**, *17*, 1034–1036. [CrossRef] [PubMed]

44. Bunnell, A.; Rowe, S. The Effect of AI-Enhanced Breast Imaging on the Caring Radiologist-Patient Relationship. *Pac. Symp. Biocomput.* **2023**, *28*, 472–483. [PubMed]

45. Kruger, J.; Dunning, D. Unskilled and unaware of it: How difficulties in recognizing one's own incompetence lead to inflated self-assessments. *J. Pers. Soc. Psychol.* **1999**, *77*, 1121–1134. [CrossRef]

46. Stogiannos, N.; Malik, R.; Kumar, A.; Barnes, A.; Pogose, M.; Harvey, H.; McEntee, M.F.; Malamateniou, C. Black box no more: A scoping review of AI governance frameworks to guide procurement and adoption of AI in medical imaging and radiotherapy in the UK. *Br. J. Radiol.* **2023**, *96*, 20221157. [CrossRef] [PubMed]

47. Jeyaraman, M.; Balaji, S.; Jeyaraman, N.; Yadav, S. Unraveling the Ethical Enigma: Artificial Intelligence in Healthcare. *Cureus* **2023**, *15*, e43262. [CrossRef] [PubMed]

48. Hulsen, T. Explainable Artificial Intelligence (XAI): Concepts and Challenges in Healthcare. *AI* **2023**, *4*, 652–666. [CrossRef]

49. Ali, S.; Abuhmed, T.; El-Sappagh, S.; Muhammad, K.; Alonso-Moral, J.M.; Confalonieri, R.; Guidotti, R.; Del Ser, J.; Díaz-Rodríguez, N.; Herrera, F. Explainable Artificial Intelligence (XAI): What we know and what is left to attain Trustworthy Artificial Intelligence. *Inf. Fusion* **2023**, *99*, 101805. [CrossRef]

50. de Vries, C.F.; Colosimo, S.J.; Boyle, M.; Lip, G.; Anderson, L.A.; Staff, R.T.; Harrison, D.; Black, C.; Murray, A.; Wilde, K.; et al. AI in breast screening mammography: Breast screening readers' perspectives. *Insights Imaging* **2022**, *13*, 186. [CrossRef]

51. Handtke, O.; Schilgen, B.; Mösko, M. Culturally competent healthcare—A scoping review of strategies implemented in healthcare organizations and a model of culturally competent healthcare provision. *PLoS ONE* **2019**, *14*, e0219971. [CrossRef] [PubMed]

52. Perry, H.; Eisenberg, R.L.; Swedeen, S.T.; Snell, A.M.; Siewert, B.; Kruskal, J.B. Improving Imaging Care for Diverse, Marginalized, and Vulnerable Patient Populations. *RadioGraphics* **2018**, *38*, 1833–1844. [CrossRef] [PubMed]

53. Pesapane, F.; Tantrige, P.; Rotili, A.; Nicosia, L.; Penco, S.; Bozzini, A.C.; Raimondi, S.; Corso, G.; Grasso, R.; Pravettoni, G.; et al. Disparities in Breast Cancer Diagnostics: How Radiologists Can Level the Inequalities. *Cancers* **2023**, *16*, 130. [CrossRef] [PubMed]

54. López-Gómez, M.; Malmierca, E.; de Górgolas, M.; Casado, E. Cancer in developing countries: The next most preventable pandemic. The global problem of cancer. *Crit. Rev. Oncol. Hematol.* **2013**, *88*, 117–122. [CrossRef] [PubMed]

55. Malhotra, K.; Wong, B.N.X.; Lee, S.; Franco, H.; Singh, C.; Cabrera Silva, L.A.; Iraqi, H.; Sinha, A.; Burger, S.; Breedt, D.S.; et al. Role of Artificial Intelligence in Global Surgery: A Review of Opportunities and Challenges. *Cureus* **2023**, *15*, e43192. [CrossRef] [PubMed]

56. Patel, M.M.; Parikh, J.R. Education of Radiologists in Healthcare Disparities. *Clin. Imaging* **2022**, *81*, 98–102. [CrossRef] [PubMed]

57. Floridi, L.; Cowls, J. A Unified Framework of Five Principles for AI in Society. In *Machine Learning and the City*; Wiley: Hoboken, NJ, USA, 2022; pp. 535–545.
58. Morley, J.; Machado, C.C.V.; Burr, C.; Cowls, J.; Joshi, I.; Taddeo, M.; Floridi, L. The ethics of AI in health care: A mapping review. *Soc. Sci. Med.* **2020**, *260*, 113172. [CrossRef] [PubMed]
59. Mökander, J.; Floridi, L. Ethics-Based Auditing to Develop Trustworthy AI. *Minds Mach.* **2021**, *31*, 323–327. [CrossRef]

Review

Progress and Challenges of Immunotherapy Predictive Biomarkers for Triple Negative Breast Cancer in the Era of Single-Cell Multi-Omics

Jiangnan Yu, Zhikun Guo and Lei Wang *

International Cancer Center, Shenzhen University Medical School, Shenzhen 518054, China; jiangnanyu@szu.edu.cn (J.Y.); 2100243054@email.szu.edu.cn (Z.G.)
* Correspondence: lei.wang@szu.edu.cn

Abstract: Triple-negative breast cancer (TNBC) is a highly aggressive subtype of breast cancer with a poor prognosis. Despite conventional treatments, including surgery, radiation, and chemotherapy, the overall response rate to PD-1/PD-L1 immune checkpoint inhibitors remains low, with limited predictive significance from current biomarkers such as PD-L1 expression, tumor-infiltrating lymphocytes (TILs), and tumor mutational burden (TMB). To address this challenge, recent advancements in single-cell sequencing techniques have enabled deeper exploration of the highly complex and heterogeneous TNBC tumor microenvironment at the single-cell level, revealing promising TNBC predictive biomarkers for immune checkpoint inhibitors. In this review, we discuss the background, motivation, methodology, results, findings, and conclusion of multi-omics analyses that have led to the identification of these emerging biomarkers. Our review suggests that single-cell multi-omics analysis holds great promise for the identification of more effective biomarkers and personalized treatment strategies for TNBC patients.

Keywords: TNBC; predictive biomarker; immunotherapy; multi-omics

Citation: Yu, J.; Guo, Z.; Wang, L. Progress and Challenges of Immunotherapy Predictive Biomarkers for Triple Negative Breast Cancer in the Era of Single-Cell Multi-Omics. *Life* **2023**, *13*, 1189. https://doi.org/10.3390/life13051189

Academic Editor: Shinichiro Kashiwagi

Received: 13 March 2023
Revised: 24 April 2023
Accepted: 9 May 2023
Published: 16 May 2023

1. Introduction

Breast cancer is a frequent malignant disease in women worldwide and is categorized into three major subtypes based on the molecular level: human epidermal growth factor receptor 2 (HER2), estrogen receptor (ER), and progesterone receptor (PR) [1]. TNBC is clinically negative for expression of the HER2, ER, and PR, which is more likely to recur than the other two subtypes (absence of HER2 or absence of ER and PR) [2]. TNBC is often characterized by a high histological grade, strong invasiveness, and high rate of metastasis. TNBC has a poor prognosis due to its aggressive clinical characteristics and lack of response to receptor-targeted therapy [3].

Therefore, there is an urgent need for more effective treatment for TNBC. TNBC accounts for 20% of breast cancer [4]. In recent years, remarkable progress has been made in exploiting the intrinsic mechanism of the host immune system to eliminate cancer cells. The advancement in immunotherapy provides a potential novel therapeutic approach for managing this devastating subtype of breast cancer. It is anticipated that immunotherapy intervention will elicit a specific response that targets and eradicates tumor cells while preserving normal cells. Diverse immunotherapy techniques have been developed and investigated, including the use of neutralizing or inhibitory antibodies to block immune checkpoints, induction of cytotoxic T lymphocytes (CTLs), adoptive cell transfer-based therapy, and modulation of the tumor microenvironment to enhance CTL activity [4].

TNBC patients may be given neoadjuvant treatment (chemotherapy before resection) in early stage tumors, which could shrink tumor size and protect normal breast tissue [5]. Immunotherapies also appear to be durable in metastatic TNBC, which suggests that immunotherapies may bring better treatment strategies to responding patients. Immune checkpoint antagonists targeting cytotoxic T lymphocyte-associated antigen-4 (CTLA-4), programmed cell death-1 (PD-1) and programmed death ligand-1 (PD-L1) have completely changed cancer treatment, induced lasting objective reactions, and sometimes translated into overall survival (OS) benefits of multiple cancer types including breast cancer [6] (Figure 1).

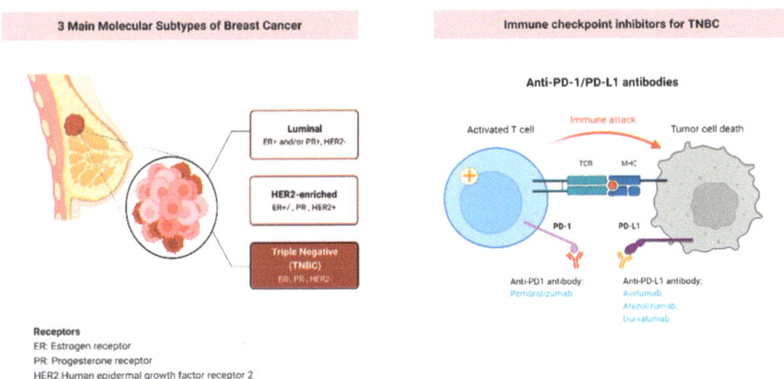

Figure 1. Immune checkpoint inhibitors for patients with TNBC.

This review contributes to the recent exploration of the highly complex and heterogeneous TNBC tumor microenvironment at the single-cell level. We summarize the major contributions of single-cell multi-omics in TNBC research, including the identification of novel immune cell subpopulations and cellular interactions, the characterization of dynamic changes in tumor heterogeneity and clonal evolution during treatment, and the discovery of potential therapeutic targets and biomarkers for immune checkpoint inhibitors.

2. PD-1/PD-L1 Checkpoint Inhibitor in Triple Negative Breast Cancer

PD-1 and PD-L1 are important immunotherapy targets in TNBC treatment. PD-1 receptors are upregulated on activated T cells and bind to the related ligand, PD-L1. Through the interaction with PD-L1 on the surface of tumor cells and immune cells, the PD-1 signal antagonizes T cell activation during the immune response stage [7]. Some immune checkpoint inhibitors (ICIs) targeting PD-1 and PD-L1 have shown favorable treatment effects in TNBC patients.

Pembrolizumab—Humanized monoclonal antibodies that target PD-1 (pembrolizumab) improve event-free survival (EFS) in TNBC [8]. Patients with stage II-III TNBC usually receive neoadjuvant chemotherapy before surgery. Nevertheless, about 30% of patients will experience disease progression within five years after typical treatment, which indicates the need for more effective upfront treatment in TNBC [9]. Currently, data from the phase III KeYNOTe-522 trial shows that in this case, neoadjuvant pembrolizumab plus chemotherapy followed by adjuvant pembrolizumab has advantages over neoadjuvant chemotherapy (NCT03036488) [10]. Pembrolizumab has been tested in several clinical trials, demonstrating its safety and clinical activity across a range of tumor types [11,12]. These data led to FDA approval of pembrolizumab in combination with chemotherapy as of July 2021.

Avelumab, atezolizumab, durvalumab—ICIs, monoclonal antibodies against PD-L1 (avelumab, atezolizumab, durvalumab), have generated durable responses across many tumor types including TNBC [13]. Although avelumab and atezolizumab are already applied to ICI monotherapy in metastatic triple-negative breast cancer (mTNBC), low response rates have been observed in pretreated metastatic disease: in the phase Ib JAVELIN trail (NCT01772004), the overall response rate (ORR) of avelumab in 58 heavily pretreated patients was 5.2% [14], while the phase I trial of atezolizumab (NCTO1375842) resulted in an ORR of 10% in 115 pre-treatment patients, with no response observed in the PD-L1 negative subgroup [15]. The GeparNuevo trial (NCT02685059) demonstrated that durvalumab improved pathologic complete response (pCR) rates when durvalumab was started two weeks before chemotherapy, which was a subgroup analysis underpowered for significance testing [16].

3. Current Predictive Biomarkers for PD1/PD-L1 Checkpoint Inhibitors

In patients with advanced-stage TNBC, monotherapy with PD-1 or PD-L1 antibodies has limited efficacy and might only benefit a small portion of patients [17]; chemoimmunotherapy approaches have improved tumor progression-free and overall survival, but these trials have yet to undergo detailed biomarker analysis [5]. Overall, immunotherapy still faces some difficulties: therapeutic resistance, unclear mechanisms, and poor response (<20%), which indicates that more efficient biomarkers are needed to identify TNBC patients who can benefit from immunotherapies in prediction and prognosis.

Because of the low response rate of immunotherapy, established and developing prognostic and predictive biomarkers are important to clinical therapies guide. Some known biomarkers of breast cancer, such as PDL1, TILs, and TMB, are helpful in the management of breast cancer.

3.1. Intratumoral PD-L1 Expression and Tumor Infiltrating Lymphocytes

The assessment of TILs and tumor PD-L1 expression has been proposed as potential predictors of clinical outcome in breast cancer. However, the reliability of these biomarkers in predicting the response to immunotherapy in early stage TNBC remains uncertain, as response to checkpoint inhibitors has been observed in tumors lacking PD-L1 expression [18]. There were 20% percent tumor cells that are PD-L1 positive in TNBC, and PD-L1 present in 20% of TNBC samples [19]. The inhibitory interactions between tumor-infiltrating immune cells and PD-1$^+$ T cells associated with poor prognostic features [19]. PD-L1 can be measured and quantified on tumor or immune cells. Tumor PD-L1 negative patients can also benefit from ICIS because other immune cells can express PD-L1, and ICIs are activated the whole immune system. Nevertheless, recent clinical trials, such as KEYNOTE-119, have demonstrated that PD-L1 positivity alone may not be a sufficient biomarker to select patients who will benefit from pembrolizumab monotherapy in the metastatic setting [20,21].

In addition, TILs have been shown to be promising microenvironment biomarkers with independent predictive value for the clinical benefits of ICI. In the metastatic setting, CD8$^+$ T cell infiltration, in particular, has been predictive of overall survival benefit with atezolizumab in IMpassion130 [22]. TILs seem to be slightly associated with PD-L1, but it has independent predictive value for the clinical benefits of ICI [23,24]. In early TNBC, an increase in TILs has been associated with improved disease-free survival, overall survival, and pathological complete response rate following neoadjuvant chemotherapy [25,26]. In metastatic TNBC, higher TIL levels have also been associated with improved prognosis. Despite its potential as a low-cost biomarker with additive predictive value to PD-L1 expression, no TILs test has yet entered routine clinical practice [21,22]. However, no TILs test has entered routine clinical practice, future research should further explore the potential of TILs as a predictive biomarker in ICI therapy, particularly in combination with PD-L1 expression.

3.2. Tumor Mutational Burden

Tumor mutational burden (TMB) is a metric used to measure the number of somatic mutations per megabase (mut/Mb) of DNA, typically determined through whole exome or gene panel sequencing. A recent analysis by Isaacs et al. found that breast cancer has a relatively low TMB of 2.63 muts/mb, with only 5% of tumors classified as hypermutated (>10 mut/MB) [20]. Breast cancer tumors with high TMB appear to be more sensitive to checkpoint inhibitors; However, there was no difference in OS among patients with high TMB breast cancer who received immunotherapy [27]. In 2020, the FDA approved pembrolizumab for the treatment of high TMB ($\geq$10 mut/Mb) non-resectable or metastatic solid tumors that have progressed after previous treatment or have no alternative treatment options, making it a potential treatment option for patients with high TMB TNBC [18].

4. Predictive Biomarkers Revealed by Single-Cell Multi-Omics

4.1. T Cell Expansion and Differentiation

Although ICIs combined with neoadjuvant chemotherapy improves pCR and EFS in TNBC [28], only a subset of tumors responds to neoadjuvant ICI. To understand the response of which underlying mechanisms and associated markers determine neoadjuvant ICI treatment response, Bassez and Vos et al. conducted a single-cell multi-omics analysis of pre-treatment and on-treatment biopsies from treatment-naive patients receiving anti-PD1 ($n = 29$) or neoadjuvant chemotherapy before anti-PD1 ($n = 11$) therapy [29]. They found that one third of the tumors contained PD1-expressing T cells that clonally expanded after anti-PD1 therapy, regardless of tumor subtype, while some gene sets were positively or negatively correlated with T cell expansion following anti-PD1 treatment [29]. Clonal expansion of T cells underlies response to ICI therapy for several cancer types, such as melanoma or lung cancer, and the single-cell characterization of pre- and on-treatment biopsies of breast cancer are absent in the previous research [30–32].

By contrasting patients' immune microenvironment alterations with and without emerging clonal expansion before and after treatment utilizing single-cell multi-omics, Bassez and Vos et al. revealed the regulation of differentiation of multiple immune cells in response to immunotherapy, and the possible mechanisms: the CD4$^+$ and CD8$^+$ T cell subtypes are the main targeted cells of anti-PD-1 therapy; the extent of differentiation and clonal proliferation of the lineage corresponding to CD8$^+$ experienced T cells (CD8$^+_{\text{TEX}}$), type-1 helper (T_{H1}) and follicular helper (T_{FH}) cells can be used to predict response to anti-PD-1 therapy, and it is likely that anti-PD-1 therapy will further enhance the differentiation of these cells [29].

Furthermore, PD-L1-expressing macrophages such as CCR2$^+$ or MMP9$^+$ and multiple dendritic cell subtypes were positively associated with T cell expansion and treatment response [33,34], whereas the proportion of CX3CR1$^+$ macrophages was negatively associated with clonal proliferation of T cells [35–37]. This study further found that the predominant cell type expressing PD-L1 in breast cancer is not tumor cells but macrophages and dendritic cells, whereas high expression of PD-L1 on macrophages and dendritic cells was predictive of immunotherapy response. In addition, macrophage phenotypes expressing PD-L1, including CCR2$^+$ and MMP9$^+$ macrophages, correlated positively with T cell expansion, which shows ICIs response. Inhibitory macrophages (CX3CR1$^+$, C3$^+$) were inversely correlated with T cell expansion, which shows limited ICIs response [29].

Therefore, T_{EX} cell abundance, T cell clonality, and richness were regarded as highly predictive markers of T cell expansion. Immune checkpoint markers or CD4$^+$ T cell activation gene markers are also highly predictive, whether in the initial BC treatment or after neoadjuvant chemotherapy, anti-PD1 is given [29]. Interestingly, the expression of these markers in TNBC is more obvious than that in ER$^+$ BC, which may explain why ICI has provided the greatest benefit in TNBC treatment so far [29].

Virassamy et al. revealed that tumor CD8$^+$ T cells with tissue-resident memory phenotypes mediate local immunity and immune checkpoint reaction of breast cancer. This study explores the role of tissue-resident memory T (TRM) cells in breast cancer and their contribution to anti-tumor immunity and immune checkpoint blockade efficacy. The study found that intratumoral CD8$^+$ T cells with a TRM-like phenotype display significantly enhanced cytotoxic capacity and provide local immune protection against tumor rechallenge. Treatment with anti-PD-1 and anti-CTLA-4 therapy resulted in the expansion of these intratumoral populations [38]. The study established two intratumoral sub-populations: one more enriched in markers of terminal exhaustion (TEX-like) and the other with a bona fide resident phenotype (TRM-like) [38]. A TRM gene signature extracted from tumor-free tissue was significantly associated with improved clinical outcomes in TNBC patients treated with checkpoint inhibitors.

First, the team assumed that immune checkpoint inhibitor therapy targeted local tumor microenvironment, and CD8$^+$ TRM cells might be crucial to its therapeutic effect on cancer; The phenotypic characteristics and cytokine requirements for the production and maintenance of different intratumoral CD8$^+$ T cell subsets in cancer were established. They further confirmed the molecular differences of these intratumoral populations and proved that the CD69$^+$CD103$^+$ subgroup showed an increased expression of TEX-related genes (such as Tox and Eomes) [39] and showed a transcriptional spectrum similar to that of terminally depleted T cells in the context of chronic lymphocytic choriomeningitis virus (LCMV) [40]. The CD69$^+$CD103$^+$ subgroup showed enhanced anti-tumor function in mediating tumor lysis [38]. These T cell subsets were further verified by unbiased clustering of CD8$^+$ single cell transcriptome data analyzed before ICB treatment. The two clusters distinguished by the expression of Itgae and Tox have significant transcriptional similarity with the large number of RNA sequences of the classified CD69$^+$CD103$^+$ and CD103$^-$ subgroups, respectively.

Phenotypic and transcriptional studies have established two intratumoral subpopulations: one is richer in terminal failure markers (TEX-like), and the other has a real resident phenotype (tissue resident memory T (TRM-like)) [38]. The treatment of anti-PD-1 and anti-CTLA-4 led to the expansion of these tumor populations, and the TRM-like subgroup showed significantly enhanced cytotoxicity. TRM-like CD8$^+$ T cells can also provide local immune protection against tumor challenge, and TRM gene markers extracted from tumor-free tissues are significantly related to the improvement of clinical prognosis of TNBC patients treated with checkpoint inhibitors [38].

It is reported that CD8$^+$ tumor-infiltrating lymphocytes with TRM cell phenotypes are related to the good prognosis of TNBC patients. However, the relative contribution of CD8$^+$ TRM cells to breast cancer anti-tumor immunity and immune checkpoint blocking efficacy is still unknown [38]. Overall, the study highlights the importance of TRM-like CD8$^+$ T cells in breast cancer anti-tumor responses and ongoing protective immunity.

4.2. CXCL13$^+$ CD4$^+$ and CD8$^+$ T Cells

In TNBC, the combining chemotherapy paclitaxel with PD-L1 checkpoint inhibitors atezolizumab did not benefit all patients [41,42]; to illuminate the different immune responses in TNBC patients (22 patients with TNBC pre- and on-treated with paclitaxel or its combination with atezolizumab), Zhang et al. leveraged single-cell multi-omics to investigate the dynamic map of tumor microenvironment and immune cells derived from peripheral blood. The tumor tissue and peripheral blood immune cells from patients with TNBC who received two treatment schemes were analyzed at the single cell level, and the tumor microenvironment and peripheral blood immune characteristics of response patients and non-response patients were compared. The dynamic changes of immune cells under different treatment strategies and the mechanism of anti-PD-L1 immunotherapy combined with paclitaxel chemotherapy in TNBC have been revealed as well as the selection of biomarkers. Zhang et al. found that the tumor microenvironment in response to patients was enriched with two groups of T cells with high expression of CXCL13 (CD8-CXCL13 and CD4-CXCL13) [43], and also highly expressed T cytotoxicity and exhaustion-related genes [44].

To investigate the connection systematically between the composition and proportion changes of different immune cells and the treatment effect, the research developed two indices: predictive index (Pi) and therapeutic index (Ti) [43]. Through the analysis of Pi and Ti, the researchers found that CD8-CXCL13 and CD4-CXCL13 at higher baseline levels can predict better immune treatment response, in addition, the proportion of these two groups of CXCL13$^+$ T cells in response patients increased significantly after combined treatment [43].

In addition, researchers found that two groups of pro-inflammatory macrophages with high expression of CXCL9 and CXCL10 were enriched in the tumor microenvironment of response patients, and there was a significant positive correlation between these two groups of pro-inflammatory macrophages and CXCL13$^+$ T cells [45,46]. CXCL9 and CXCL10 can participate in the recruitment of T cells [45], and the characteristic genes of proinflammatory macrophages are regulated by IFNG and TNF signals, indicating that there is a positive feedback signal between CXCL13$^+$ T cells that play a killing function and proinflammatory macrophages that express CXCL9 and CXCL10 [43]. On the contrary, CXCL13$^+$ T cells were hardly detected in the tumor microenvironment of non-responsive patients, but a large number of macrophages with immunosuppressive function were enriched [43]. It is noteworthy that the researchers found that the peripheral blood mononuclear cells of response patients showed pro-inflammatory characteristics, while the peripheral blood mononuclear cells of non-response patients showed anti-inflammatory characteristics, suggesting that the peripheral blood can reflect the tumor microenvironment characteristics to a certain extent [43].

4.3. Tumor-Resilient T Cell Assessed by Tres Model

In recent years, cancer immunotherapy represented by anti-PD-1/PD-L1, anti-CTLA4, and CAR-T has made considerable progress [47]. However, the effect of various immunotherapies on solid tumors is not satisfactory [47,48]. For T cells, a solid tumor is a battlefield with a suppressive environment [49]. The tumor will establish a microenvironment full of various immunosuppressive factors to suppress and differentiate the invading T-cell soldiers [49,50]. Despite being armed with various anti-cancer mechanisms, most T cells cannot persist in such a harsh environment [50].

They have developed a computational model called Tres (tumor-resilient T cell, https://resilience.ccr.cancer.gov/ accessed on 12 March 2023) by analyzing single-cell T-cell transcriptomes from ICI-treated melanoma or lung tumors to find the characteristics of T cells that are still active under the suppression of various inhibitors in solid tumors and predict the efficiency of T cells in immunotherapy [51–53]. Tres also identifies FIBP as a new checkpoint for T-cell immunometabolism and a possible new target for immunotherapy [54].

Tres is a computational model that uses single-cell transcriptome data to identify the characteristics of T cells resistant to immunosuppressive types, and Tres was also trained by TNBC-published single-cell transcriptomic profiles of T cells from responders and non-responders to ICIs [38]. It presented better predictive signature correlates in responders than non-responders in pre- and post-treatment of ICIs [49]. The application of single-cell sequencing technology in tumor research has produced a large number of single-cell gene expression profiles, depicting various states of T-cell subsets from tumors [53]. Tres, a computational model assesses the cytokines perceived by each T cell in the tumor environment; for example, TGFβ and PGE2 are common immunosuppressive factors [55], and TRAIL is the trigger of T cell in cell death [50]. If the downstream pathway of these cytokines is activated, it indicates that the T cell is in an unfavorable environment. At the same time, the health of T cells can be measured by the cell cycle and the expression of DNA replication pathway genes [55]. The activity of these pathways in suppressed or dying cells is often low. Based on the variables calculated and evaluated above, Tres looks for which T cells are under the pressure of various inhibitors, still remain healthy. These T cells are defined as tumor-resilient T cells (Tres) [54]. These Tres features demonstrate important clinical applications.

Based on the simplest correlation coefficient calculation, if these T cell samples are positively correlated with the characteristics of the tumor-resilient T cells model, the corresponding immunotherapy will achieve good results. If there is a negative correlation, the corresponding immunotherapy effect will be unfavorable [54]. It is particularly pointed out that the Tres model is almost correct in predicting the accuracy of patients with poor efficacy in cell therapy using only pre-manufacturing samples [54].

Zhang et al. analyzed the genetic characteristics of Tres. In 168 tumors and single-cell expression data from 19 kinds of cancer, the high expression of the FIBP gene in T cells almost indicates the low match of the Tres model, which means that T cells with high expression of the FIBP gene are not regarded as a tumor-resilient T cell [54]. Many solid tumors have high cholesterol concentrations [56]. Although an appropriate amount of cholesterol will guarantee the activity of T cells, excessive cholesterol concentration will greatly reduce the tumor-killing ability of T cells and lead to T cell exhaustion [56].

Therefore, the Tres model uses single-cell data to identify immunotherapy response biomarkers and predict cell therapy response from pre-manufacture samples, which provide an important research and development tool for cancer immunotherapy guidelines.

4.4. CD8$^+$ T Cell-Intrinsic IL-6

Although immune checkpoint inhibitors (ICI) have been identified as effective cancer therapies, overcoming drug resistance remains a key challenge. Huseni et al. determined that interleukin 6 (IL-6) is associated with poor reaction to atezolizumab (anti-PD-L1) in large clinical trials of advanced renal cancer, breast cancer, and bladder cancer [57]. The pleiotropic cytokine IL-6 is associated with tumor progression and is supposed to affect anti-tumor immunity through a variety of mechanisms [58–60]. Plasma IL-6 has a negative effect on the survival rate of melanoma patients treated with ICI [61,62], and IL-6 appears to be a potential driver of ICI resistance [63–65].

In this study, Huseni et al. found that high levels of IL-6 are a characteristic of atezolizumab-resistant disease in patients with advanced cancer [57]. IL-6 inhibits the effector differentiation of CD8$^+$ T cells (also known as cytotoxic T lymphocytes or CTLs), and high plasma IL-6 is associated with lower expression of effector genes in CTLs of cancer patients [57]. IL-6 impairs anti-PD-L1 efficacy by restricting the anti-tumor functions of cytotoxic T cells and IL-6-STAT3 signaling inhibits classical cytotoxic differentiation of CTLs in vitro [57]. In preclinical tumor models, blocking IL6R or gene ablation of intrinsic IL-6 signaling in CTLs, in combination with anti-PD-L1 therapy, enhances the anti-tumor CTL response, and improves tumor control [57].

In the PCD4989g clinical trial, patients with mTNBC treated with atezolizumab13, or in the IMvigor210 and IMvigor211 trials [66–68], patients with metastatic urothelial bladder cancer (UC), compared with the healthy control group, had elevated plasma IL-6, and was associated with low OS in multivariate survival analysis. According to the single-cell RNA sequencing, the circulating CTL of cancer patients with high plasma IL-6 levels showed a suppressed functional feature, and IL-6-STAT3 signal transduction inhibited the classical cytotoxic differentiation of CTL in vitro [57]. Therefore, based on clinical and experimental evidence, drugs targeting the IL-6 signal are reasonable partners for cancer patients and ICIs in combination treatment [57].

4.5. TME Phenotypes Do Not Respond to Checkpoint Inhibitors

In an effort to understand the lack of response to immune checkpoint inhibitors (ICI), Hammerl et al. conducted a study analyzing 681 triple-negative breast cancers (TNBCs) for spatial immune cell contextures in relation to clinical outcomes and pathways of T cell evasion [69–71]. Through this analysis, the authors identified three main spatial phenotypes: inflamed, excluded, and ignored, and recognized their association with clinical outcomes in TNBC and other cancer types [67]. The inflamed phenotype, characterized by the presence of intratumoral lymphocytes, is related to anti-PD-1 response, while the excluded and ignored phenotypes, characterized by lymphocytes restricted to the invasive margin or a lack of lymphocytes, respectively, are related to anti-PD-1 resistance [67].

Combined with multiple immunofluorescence and sequencing technology, Hammerl et al. revealed: immune excluded phenotypes (related to anti-PD-1 resistance), showed collagen-10 deposition, enhanced glycolysis, and TGFβ/VEGF pathway activation; immune ignored phenotypes (related to anti-PD1 resistance), showing high-density CD163$^+$ myeloid cells or activating WNT/PPARγ pathways; inflamed phenotype, which was associated with anti-PD-1 response, exhibited necrosis, high-density CLEC9A$^+$ dendritic cells, high TCR clonality, and enhanced expression of T cell co-inhibitory receptors [72]. These results suggest that the spatial immunophenotypes of primary TNBC have unique immune-determinants, as well as tumor microenvironment (TME) and immune response-mediated T cell escape pathway.

The TONIC test found that the proportion of inflammatory phenotype increased after cisplatin and doxorubicin induction treatment, which indicated that the spatial phenotype was plastic [73], while the cold TNBC (i.e., excluded and ignored) phenotypes can be remodeled, suggesting the possibility of treatment benefit for these two types of patients. Therefore, the immune excluded type and the ignored immunophenotypes in TNBC and metastatic TNBC validated by the gene classifier accurately do not respond to anti-PD1 treatment, which can be considered as a variant of cold tumor [72]. The spatial phenotype classifier demonstrated good predictive value, which could potentially improve the efficacy of anti-PD-1 treatment.

5. Conclusions

In the era of single-cell multi-omics, there has been significant progress and challenges in predicting immunotherapy biomarkers for triple-negative breast cancer. The use of single-cell multi-omics techniques has enabled the identification of novel biomarkers and molecular pathways that play a critical role in the response to immunotherapy. Additionally, these techniques have provided a deeper understanding of the complex immune landscape of triple-negative breast cancer and the heterogeneity of individual tumor cells, which has helped to refine biomarker discovery and validation. However, there are still many challenges that need to be addressed and other kinds of biomarkers are needed. One challenge is the lack of standardization in data analysis and interpretation across different studies, which can lead to inconsistencies in biomarker identification and validation. Another challenge is the limited sample size and heterogeneity of patient cohorts, which can affect the accuracy and reproducibility of biomarker discovery.

Furthermore, the integration of different multi-omics datasets is still in its early stages and requires more advanced computational methods and analytical tools. Despite these challenges, the use of single-cell multi-omics techniques offers great potential for identifying predictive biomarkers for immunotherapy in triple-negative breast cancer and for developing more personalized and effective treatment strategies.

Immunotherapies show the prospect of breast cancer treatment and the potential of activating the immune system to eliminate cancer cells. Inhibitors of PD-1/PD-L1 checkpoints can induce a long-lasting clinical response in some breast cancer patients with metastatic TNBC. Although some TNBC patients show PD-L1 negative expression in tumors, they can still benefit from ICIs. Intratumoral PD-L1 expression is highly heterogenous and PD-L1 expression on either cancer cells or immune cells can be changing dynamically. More importantly, we believe that the clinical efficacy of ICIs treatment requires stimulating the systemic anti-tumor immunity of TNBC patients and it is now reported that tumor-specific T cell activation in the tumor-draining lymph node can be targeted by ICI treatment [74].

In addition, the combination of checkpoint inhibitors and chemo/targeted therapies in neoadjuvant first-line treatments has already demonstrated clinical benefit and potential. However, the major challenge is that the current biomarkers, such as intratumoral PD-L1 expression, TILs, and TMB, have limited predictability and reliability to select patients with TNBC. Therefore, there is an urgent need to develop novel predictive biomarkers by using deep multi-omics analysis at single-cell level. The involvement of multi-omics technology in TNBC predictive biomarker research has made promising discoveries so far (summarized in Figure 2). Future research, especially deep immune profiling of paired tumor, lymph node, and blood samples of pre- vs. post treatment are needed to further explore biomarkers during the ICI-induced systemic immune changes.

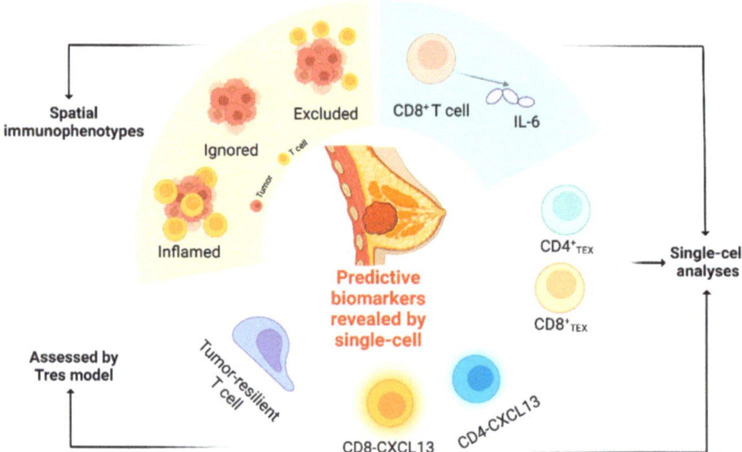

Figure 2. Summary of emerging anti-PD1/PD-L1 predictive biomarker revealed by scRNA multi-omics. Figures created with BioRender.com.

Funding: This research was funded by China National Natural Science Foundation (82173095), Shenzhen Natural Science Fund (JCYJ20210324094203010 and Stable Support Plan Program 202208100940170001).

Institutional Review Board Statement: Not applicable.

Informed Consent Statement: Not applicable.

Data Availability Statement: Not applicable.

Conflicts of Interest: The authors declare no conflict of interest.

References

1. Harbeck, N.; Penault-Llorca, F.; Cortes, J.; Gnant, M.; Houssami, N.; Poortmans, P.; Ruddy, K.; Tsang, J.; Cardoso, F. Breast cancer. *Nat. Rev. Dis. Primers* **2019**, *5*, 66. [CrossRef] [PubMed]
2. Waks, A.G.; Winer, E.P. Breast Cancer Treatment: A Review. *JAMA* **2019**, *321*, 288–300. [CrossRef] [PubMed]
3. Liu, Z.; Li, M.; Jiang, Z.; Wang, X. A Comprehensive Immunologic Portrait of Triple-Negative Breast Cancer. *Transl. Oncol.* **2018**, *11*, 311–329. [CrossRef] [PubMed]
4. Jia, H.; Truica, C.I.; Wang, B.; Wang, Y.; Ren, X.; Harvey, H.A.; Song, J.; Yang, J.-M. Immunotherapy for triple-negative breast cancer: Existing challenges and exciting prospects. *Drug Resist. Updat. Rev. Comment. Antimicrob. Anticancer. Chemother.* **2017**, *32*, 1–15. [CrossRef]
5. Savas, P.; Loi, S. Expanding the Role for Immunotherapy in Triple-Negative Breast Cancer. *Cancer Cell* **2020**, *37*, 623–624. [CrossRef]
6. Emens, L.A. Breast Cancer Immunotherapy: Facts and Hopes. *Clin. Cancer Res. Off. J. Am. Assoc. Cancer Res.* **2018**, *24*, 511–520. [CrossRef]
7. Pardoll, D.M. The blockade of immune checkpoints in cancer immunotherapy. *Nat. Rev. Cancer* **2012**, *12*, 252–264. [CrossRef]
8. Sidaway, P. Pembrolizumab improves EFS in TNBC. *Nat. Rev. Clin. Oncol.* **2022**, *19*, 220. [CrossRef]
9. Cortazar, P.; Zhang, L.; Untch, M.; Mehta, K.; Costantino, J.P.; Wolmark, N.; Bonnefoi, H.; Cameron, D.; Gianni, L.; Valagussa, P.; et al. Pathological complete response and long-term clinical benefit in breast cancer: The CTNeoBC pooled analysis. *Lancet* **2014**, *384*, 164–172. [CrossRef]
10. Schmid, P.; Cortes, J.; Dent, R.; Pusztai, L.; McArthur, H.; Kümmel, S.; Bergh, J.; Denkert, C.; Park, Y.H.; Hui, R.; et al. Event-free Survival with Pembrolizumab in Early Triple-Negative Breast Cancer. *N. Engl. J. Med.* **2022**, *386*, 556–567. [CrossRef]
11. Lipson, E.J.; Forde, P.M.; Hammers, H.-J.; Emens, L.A.; Taube, J.M.; Topalian, S.L. Antagonists of PD-1 and PD-L1 in Cancer Treatment. *Semin. Oncol.* **2015**, *42*, 587–600. [CrossRef]
12. Emens, L.A.; Ascierto, P.A.; Darcy, P.K.; Demaria, S.; Eggermont, A.M.M.; Redmond, W.L.; Seliger, B.; Marincola, F.M. Cancer immunotherapy: Opportunities and challenges in the rapidly evolving clinical landscape. *Eur. J. Cancer* **2017**, *81*, 116–129. [CrossRef] [PubMed]

13. Keenan, T.E.; Tolaney, S.M. Role of Immunotherapy in Triple-Negative Breast Cancer. *J. Natl. Compr. Cancer Netw. JNCCN* **2020**, *18*, 479–489. [CrossRef] [PubMed]

14. Dirix, L.Y.; Takacs, I.; Jerusalem, G.; Nikolinakos, P.; Arkenau, H.T.; Forero-Torres, A.; Boccia, R.; Lippman, M.E.; Somer, R.; Smakal, M.; et al. Avelumab, an anti-PD-L1 antibody, in patients with locally advanced or metastatic breast cancer: A phase 1b JAVELIN Solid Tumor study. *Breast Cancer Res. Treat.* **2018**, *167*, 671–686. [CrossRef]

15. Emens, L.A.; Cruz, C.; Eder, J.P.; Braiteh, F.; Chung, C.; Tolaney, S.M.; Kuter, I.; Nanda, R.; Cassier, P.A.; Delord, J.-P.; et al. Long-term Clinical Outcomes and Biomarker Analyses of Atezolizumab Therapy for Patients with Metastatic Triple-Negative Breast Cancer: A Phase 1 Study. *JAMA Oncol.* **2019**, *5*, 74–82. [CrossRef]

16. Loibl, S.; Untch, M.; Burchardi, N.; Huober, J.; Sinn, B.V.; Blohmer, J.-U.; Grischke, E.-M.; Furlanetto, J.; Tesch, H.; Hanusch, C.; et al. A randomised phase II study investigating durvalumab in addition to an anthracycline taxane-based neoadjuvant therapy in early triple-negative breast cancer: Clinical results and biomarker analysis of GeparNuevo study. *Ann. Oncol. Off. J. Eur. Soc. Med. Oncol.* **2019**, *30*, 1279–1288. [CrossRef] [PubMed]

17. Bianchini, G.; De Angelis, C.; Licata, L.; Gianni, L. Treatment landscape of triple-negative breast cancer—expanded options, evolving needs. *Nat. Rev. Clin. Oncol.* **2022**, *19*, 91–113. [CrossRef]

18. Sukumar, J.; Gast, K.; Quiroga, D.; Lustberg, M.; Williams, N. Triple-negative breast cancer: Promising prognostic biomarkers currently in development. *Expert Rev. Anticancer. Ther.* **2021**, *21*, 135–148. [CrossRef]

19. Sabatier, R.; Finetti, P.; Mamessier, E.; Adelaide, J.; Chaffanet, M.; Ali, H.R.; Viens, P.; Caldas, C.; Birnbaum, D.; Bertucci, F. Prognostic and predictive value of PDL1 expression in breast cancer. *Oncotarget* **2015**, *6*, 5449–5464. [CrossRef]

20. Isaacs, J.; Anders, C.; McArthur, H.; Force, J. Biomarkers of Immune Checkpoint Blockade Response in Triple-Negative Breast Cancer. *Curr. Treat. Options Oncol.* **2021**, *22*, 38. [CrossRef]

21. Cortés, J.; Lipatov, O.N.; Im, S.-A.; Gonçalves, A.; Lee, K.; Schmid, P.; Tamura, K.; Testa, L.; Witzel, I.; Ohtani, S.; et al. KEYNOTE-119: Phase III study of pembrolizumab (pembro) versus single-agent chemotherapy (chemo) for metastatic triple negative breast cancer (mTNBC). *Ann. Oncol.* **2019**, *30*, v859–v860. [CrossRef]

22. Emens, L.; Loi, S.; Rugo, H.; Schneeweiss, A.; Diéras, V.; Iwata, H.; Barrios, C.; Nechaeva, M.; Molinero, L.; Ngu-yen Duc, A.; et al. Abstract GS1-04: IMpassion130: Efficacy in immune biomarker subgroups from the global, randomized, double-blind, placebo-controlled, phase III study of atezolizumab + nab-paclitaxel in patients with treatment-naïve, locally advanced or metastatic triple-negative breast cancer. *Cancer Res.* **2019**, *79*, GS1-04-GS01-04. [CrossRef]

23. Loi, S.; Winer, E.; Lipatov, O.; Im, S.-A.; Goncalves, A.; Cortes, J.; Lee, K.S.; Schmid, P.; Testa, L.; Witzel, I.; et al. Abstract PD5-03: Relationship between tumor-infiltrating lymphocytes (TILs) and outcomes in the KEYNOTE-119 study of pembrolizumab vs chemotherapy for previously treated metastatic triple-negative breast cancer (mTNBC). *Cancer Res* **2020**, *80*, PD5-03–PD05-03. [CrossRef]

24. Loi, S.; Adams, S.; Schmid, P.; Cortés, J.; Cescon, D.W.; Winer, E.P.; Toppmeyer, D.L.; Rugo, H.S.; De Laurentiis, M.; Nanda, R.; et al. Relationship between tumor infiltrating lymphocyte (TIL) levels and response to pembrolizumab (pembro) in metastatic triple-negative breast cancer (mTNBC): Results from KEYNOTE-086. *Ann. Oncol.* **2017**, *28*, v608. [CrossRef]

25. Denkert, C.; Von Minckwitz, G.; Darb-Esfahani, S.; Lederer, B.; Heppner, B.I.; Weber, K.E.; Budczies, J.; Huober, J.; Klauschen, F.; Furlanetto, J.; et al. Tumour-infiltrating lymphocytes and prognosis in different subtypes of breast cancer: A pooled analysis of 3771 patients treated with neoadjuvant therapy. *Lancet Oncol.* **2018**, *19*, 40–50. [CrossRef]

26. Adams, S.; Gray, R.J.; Demaria, S.; Goldstein, L.; Perez, E.A.; Shulman, L.N.; Martino, S.; Wang, M.; Jones, V.E.; Saphner, T.J.; et al. Prognostic Value of Tumor-Infiltrating Lymphocytes in Triple-Negative Breast Cancers from Two Phase III Randomized Adjuvant Breast Cancer Trials: ECOG 2197 and ECOG 1199. *J. Clin. Oncol. Off. J. Am. Soc. Clin. Oncol.* **2014**, *32*, 2959–2966. [CrossRef] [PubMed]

27. Samstein, R.M.; Lee, C.-H.; Shoushtari, A.N.; Hellmann, M.D.; Shen, R.; Janjigian, Y.Y.; Barron, D.A.; Zehir, A.; Jordan, E.J.; Omuro, A.; et al. Tumor mutational load predicts survival after immunotherapy across multiple cancer types. *Nat. Genet.* **2019**, *51*, 202–206. [CrossRef] [PubMed]

28. Schmid, P.; Cortes, J.; Pusztai, L.; McArthur, H.; Kümmel, S.; Bergh, J.; Denkert, C.; Park, Y.H.; Hui, R.; Harbeck, N.; et al. Pembrolizumab for Early Triple-Negative Breast Cancer. *N. Engl. J. Med.* **2020**, *382*, 810–821. [CrossRef] [PubMed]

29. Bassez, A.; Vos, H.; Van Dyck, L.; Floris, G.; Arijs, I.; Desmedt, C.; Boeckx, B.; Vanden Bempt, M.; Nevelsteen, I.; Lambein, K.; et al. A single-cell map of intratumoral changes during anti-PD1 treatment of patients with breast cancer. *Nat. Med.* **2021**, *27*, 820–832. [CrossRef]

30. Amaria, R.N.; Reddy, S.M.; Tawbi, H.A.; Davies, M.A.; Ross, M.I.; Glitza, I.C.; Cormier, J.N.; Lewis, C.; Hwu, W.-J.; Hanna, E.; et al. Neoadjuvant immune checkpoint blockade in high-risk resectable melanoma. *Nat. Med.* **2018**, *24*, 1649–1654. [CrossRef]

31. Tumeh, P.C.; Harview, C.L.; Yearley, J.H.; Shintaku, I.P.; Taylor, E.J.; Robert, L.; Chmielowski, B.; Spasic, M.; Henry, G.; Ciobanu, V.; et al. PD-1 blockade induces responses by inhibiting adaptive immune resistance. *Nature* **2014**, *515*, 568–571. [CrossRef]

32. Forde, P.M.; Chaft, J.E.; Smith, K.N.; Anagnostou, V.; Cottrell, T.R.; Hellmann, M.D.; Zahurak, M.; Yang, S.C.; Jones, D.R.; Broderick, S.; et al. Neoadjuvant PD-1 Blockade in Resectable Lung Cancer. *N. Engl. J. Med.* **2018**, *378*, 1976–1986. [CrossRef] [PubMed]

33. Hart, K.M.; Bak, S.P.; Alonso, A.; Berwin, B. Phenotypic and Functional Delineation of Murine CX3CR1+ Monocyte-Derived Cells in Ovarian Cancer. *Neoplasia* **2009**, *11*, 564–573. [CrossRef]

34. Zheng, J.; Yang, M.; Shao, J.; Miao, Y.; Han, J.; Du, J. Chemokine receptor CX3CR1 contributes to macrophage survival in tumor metastasis. *Mol. Cancer* **2013**, *12*, 141. [CrossRef] [PubMed]
35. Gubin, M.M.; Esaulova, E.; Ward, J.P.; Malkova, O.N.; Runci, D.; Wong, P.; Noguchi, T.; Arthur, C.D.; Meng, W.; Alspach, E.; et al. High-Dimensional Analysis Delineates Myeloid and Lymphoid Compartment Remodeling during Successful Immune-Checkpoint Cancer Therapy. *Cell* **2018**, *175*, 1014–1030.e1019. [CrossRef] [PubMed]
36. Zha, H.; Wang, X.; Zhu, Y.; Chen, D.; Han, X.; Yang, F.; Gao, J.; Hu, C.; Shu, C.; Feng, Y.; et al. Intracellular Activation of Complement C3 Leads to PD-L1 Antibody Treatment Resistance by Modulating Tumor-Associated Macrophages. *Cancer Immunol. Res.* **2019**, *7*, 193–207. [CrossRef]
37. Kwak, J.W.; Laskowski, J.; Li, H.Y.; McSharry, M.V.; Sippel, T.R.; Bullock, B.L.; Johnson, A.M.; Poczobutt, J.M.; Neuwelt, A.J.; Malkoski, S.P.; et al. Complement Activation via a C3a Receptor Pathway Alters CD4(+) T Lymphocytes and Mediates Lung Cancer Progression. *Cancer Res.* **2018**, *78*, 143–156. [CrossRef]
38. Virassamy, B.; Caramia, F.; Savas, P.; Sant, S.; Wang, J.; Christo, S.N.; Byrne, A.; Clarke, K.; Brown, E.; Teo, Z.L.; et al. Intratumoral CD8+ T cells with a tissue-resident memory phenotype mediate local immunity and immune checkpoint responses in breast cancer. *Cancer Cell* **2023**, *41*, 585–601.e8. [CrossRef]
39. Seo, W.; Jerin, C.; Nishikawa, H. Transcriptional regulatory network for the establishment of CD8+ T cell exhaustion. *Exp. Mol. Med.* **2021**, *53*, 202–209. [CrossRef]
40. Man, K.; Gabriel, S.S.; Liao, Y.; Gloury, R.; Preston, S.; Henstridge, D.C.; Pellegrini, M.; Zehn, D.; Berberich-Siebelt, F.; Febbraio, M.A.; et al. Transcription Factor IRF4 Promotes CD8$^+$ T Cell Exhaustion and Limits the Development of Memory-like T Cells during Chronic Infection. *Immunity* **2017**, *47*, 1129–1141.e1125. [CrossRef]
41. Schmid, P.; Adams, S.; Rugo, H.S.; Schneeweiss, A.; Barrios, C.H.; Iwata, H.; Diéras, V.; Hegg, R.; Im, S.-A.; Shaw Wright, G.; et al. Atezolizumab and Nab-Paclitaxel in Advanced Triple-Negative Breast Cancer. *N. Engl. J. Med.* **2018**, *379*, 2108–2121. [CrossRef]
42. Siu, L.L.; Ivy, S.P.; Dixon, E.L.; Gravell, A.E.; Reeves, S.A.; Rosner, G.L. Challenges and Opportunities in Adapting Clinical Trial Design for Immunotherapies. *Clin. Cancer Res. Off. J. Ameri-Can Assoc. Cancer Res.* **2017**, *23*, 4950–4958. [CrossRef]
43. Zhang, Y.; Chen, H.; Mo, H.; Hu, X.; Gao, R.; Zhao, Y.; Liu, B.; Niu, L.; Sun, X.; Yu, X.; et al. Single-cell analyses reveal key immune cell subsets associated with response to PD-L1 blockade in triple-negative breast cancer. *Cancer Cell* **2021**, *39*, 1578–1593.e1578. [CrossRef] [PubMed]
44. Yost, K.E.; Satpathy, A.T.; Wells, D.K.; Qi, Y.; Wang, C.; Kageyama, R.; McNamara, K.L.; Granja, J.M.; Sarin, K.Y.; Brown, R.A.; et al. Clonal replacement of tumor-specific T cells following PD-1 blockade. *Nat. Med.* **2019**, *25*, 1251–1259. [CrossRef] [PubMed]
45. House, I.G.; Savas, P.; Lai, J.; Chen, A.X.Y.; Oliver, A.J.; Teo, Z.L.; Todd, K.L.; Henderson, M.A.; Giuffrida, L.; Petley, E.V.; et al. Macrophage-Derived CXCL9 and CXCL10 Are Required for Antitumor Immune Responses Following Immune Checkpoint Blockade. *Clin. Cancer Res. Off. J. Am. Assoc.-Ation Cancer Res.* **2020**, *26*, 487–504. [CrossRef]
46. The Cancer Genome Atlas (TCGA) Research Network. Comprehensive molecular portraits of human breast tumours. *Nature* **2012**, *490*, 61–70. [CrossRef]
47. Ribas, A.; Wolchok, J.D. Cancer immunotherapy using checkpoint blockade. *Science* **2018**, *359*, 1350–1355. [CrossRef]
48. Rosenberg, S.A.; Yang, J.C.; Sherry, R.M.; Kammula, U.S.; Hughes, M.S.; Phan, G.Q.; Citrin, D.E.; Restifo, N.P.; Robbins, P.F.; Wunderlich, J.R.; et al. Durable Complete Responses in Heavily Pretreated Patients with Metastatic Melanoma Using T-Cell Transfer Immunotherapy. *Clin. Cancer Res. Off. J. Am. Assoc. Cancer Res.* **2011**, *17*, 4550–4557. [CrossRef]
49. Sharma, P.; Hu-Lieskovan, S.; Wargo, J.A.; Ribas, A. Primary, Adaptive, and Acquired Resistance to Cancer Immunotherapy. *Cell* **2017**, *168*, 707–723. [CrossRef] [PubMed]
50. Tschumi, B.O.; Dumauthioz, N.; Marti, B.; Zhang, L.; Lanitis, E.; Irving, M.; Schneider, P.; Mach, J.P.; Coukos, G.; Romero, P.; et al. CART cells are prone to Fas- and DR5-mediated cell death. *J. Immunother. Cancer* **2018**, *6*, 71. [CrossRef] [PubMed]
51. Sade-Feldman, M.; Yizhak, K.; Bjorgaard, S.L.; Ray, J.P.; de Boer, C.G.; Jenkins, R.W.; Lieb, D.J.; Chen, J.H.; Frederick, D.T.; Barzily-Rokni, M.; et al. Defining T Cell States Associated with Response to Checkpoint Immunotherapy in Melanoma. *Cell* **2019**, *176*, 404. [CrossRef] [PubMed]
52. Caushi, J.X.; Zhang, J.; Ji, Z.; Vaghasia, A.; Zhang, B.; Hsiue, E.H.; Mog, B.J.; Hou, W.; Justesen, S.; Blosser, R.; et al. Transcriptional programs of neoantigen-specific TIL in anti-PD-1-treated lung cancers. *Nature* **2021**, *596*, 126–132. [CrossRef]
53. Suvà, M.L.; Tirosh, I. Single-Cell RNA Sequencing in Cancer: Lessons Learned and Emerging Challenges. *Mol. Cell* **2019**, *75*, 7–12. [CrossRef]
54. Zhang, Y.; Vu, T.; Palmer, D.C.; Kishton, R.J.; Gong, L.; Huang, J.; Nguyen, T.; Chen, Z.; Smith, C.; Livák, F.; et al. A T cell resilience model associated with response to immunotherapy in multiple tumor types. *Nat. Med.* **2022**, *28*, 1421–1431. [CrossRef]
55. Kanehisa, M.; Sato, Y.; Kawashima, M.; Furumichi, M.; Tanabe, M. KEGG as a reference resource for gene and protein annotation. *Nucleic Acids Res.* **2016**, *44*, D457–D462. [CrossRef]
56. Luo, J.; Yang, H.; Song, B.-L. Mechanisms and regulation of cholesterol homeostasis. *Nat. Rev. Mol. Cell Biol.* **2020**, *21*, 225–245. [CrossRef]
57. Huseni, M.A.; Wang, L.; Klementowicz, J.E.; Yuen, K.; Breart, B.; Orr, C.; Liu, L.-F.; Li, Y.; Gupta, V.; Li, C.; et al. CD8(+) T cell-intrinsic IL-6 signaling promotes resistance to anti-PD-L1 immunotherapy. *Cell Rep. Med.* **2023**, *4*, 100878. [CrossRef]
58. Hunter, C.A.; Jones, S.A. IL-6 as a keystone cytokine in health and disease. *Nat. Immunol.* **2015**, *16*, 448–457. [CrossRef] [PubMed]
59. Schaper, F.; Rose-John, S. Interleukin-6: Biology, signaling and strategies of blockade. *Cytokine Growth Factor Rev.* **2015**, *26*, 475–487. [CrossRef] [PubMed]

60. Jones, S.A.; Jenkins, B.J. Recent insights into targeting the IL-6 cytokine family in inflammatory diseases and cancer. *Nat. Rev. Immunol.* **2018**, *18*, 773–789. [CrossRef]

61. Tsukamoto, H.; Fujieda, K.; Miyashita, A.; Fukushima, S.; Ikeda, T.; Kubo, Y.; Senju, S.; Ihn, H.; Nishimura, Y.; Oshiumi, H. Combined Blockade of IL6 and PD-1/PD-L1 Signaling Abrogates Mutual Regulation of Their Immunosuppressive Effects in the Tumor Microenvironment. *Cancer Res.* **2018**, *78*, 5011–5022. [CrossRef]

62. Laino, A.S.; Woods, D.; Vassallo, M.; Qian, X.; Tang, H.; Wind-Rotolo, M.; Weber, J. Serum interleukin-6 and C-reactive protein are associated with survival in melanoma patients receiving immune checkpoint inhibition. *J. Immunother. Cancer* **2020**, *8*, e000842. [CrossRef]

63. Liu, H.; Shen, J.; Lu, K. IL-6 and PD-L1 blockade combination inhibits hepatocellular carcinoma cancer development in mouse model. *Biochem. Biophys. Res. Commun.* **2017**, *486*, 239–244. [CrossRef] [PubMed]

64. Tsukamoto, H.; Fujieda, K.; Hirayama, M.; Ikeda, T.; Yuno, A.; Matsumura, K.; Fukuma, D.; Araki, K.; Mizuta, H.; Nakayama, H.; et al. Soluble IL6R Expressed by Myeloid Cells Reduces Tumor-Specific Th1 Differentiation and Drives Tumor Progression. *Cancer Res.* **2017**, *77*, 2279–2291. [CrossRef] [PubMed]

65. Mace, T.A.; Shakya, R.; Pitarresi, J.R.; Swanson, B.; McQuinn, C.W.; Loftus, S.; Nordquist, E.; Cruz-Monserrate, Z.; Yu, L.; Young, G.; et al. IL-6 and PD-L1 antibody blockade combination therapy reduces tumour progression in murine models of pancreatic cancer. *Gut* **2018**, *67*, 320–332. [CrossRef]

66. Balar, A.V.; Galsky, M.D.; Rosenberg, J.E.; Powles, T.; Petrylak, D.P.; Bellmunt, J.; Loriot, Y.; Necchi, A.; Hoffman-Censits, J.; Perez-Gracia, J.L.; et al. Atezolizumab as first-line treatment in cisplatin-ineligible patients with locally advanced and metastatic urothelial carcinoma: A single-arm, multicentre, phase 2 trial. *Lancet* **2017**, *389*, 67–76. [CrossRef] [PubMed]

67. Rosenberg, J.E.; Hoffman-Censits, J.; Powles, T.; van der Heijden, M.S.; Balar, A.V.; Necchi, A.; Dawson, N.; O'Donnell, P.H.; Balmanoukian, A.; Loriot, Y.; et al. Atezolizumab in patients with locally advanced and metastatic urothelial carcinoma who have progressed following treatment with platinum-based chemotherapy: A single-arm, multicentre, phase 2 trial. *Lancet* **2016**, *387*, 1909–1920. [CrossRef]

68. Powles, T.; Durán, I.; Van Der Heijden, M.S.; Loriot, Y.; Vogelzang, N.J.; De Giorgi, U.; Oudard, S.; Retz, M.M.; Castellano, D.; Bamias, A.; et al. Atezolizumab versus chemotherapy in patients with platinum-treated locally advanced or metastatic urothelial carcinoma (IMvigor211): A multicentre, open-label, phase 3 randomised controlled trial. *Lancet* **2018**, *391*, 748–757. [CrossRef]

69. Gruosso, T.; Gigoux, M.; Manem, V.S.K.; Bertos, N.; Zuo, D.; Perlitch, I.; Saleh, S.M.I.; Zhao, H.; Souleimanova, M.; Johnson, R.M.; et al. Spatially distinct tumor immune microenvironments stratify triple-negative breast cancers. *J. Clin. Investig.* **2019**, *129*, 1785–1800. [CrossRef]

70. Galon, J.; Costes, A.; Sanchez-Cabo, F.; Kirilovsky, A.; Mlecnik, B.; Lagorce-Pagès, C.; Tosolini, M.; Camus, M.; Berger, A.; Wind, P.; et al. Type, Density, and Location of Immune Cells Within Human Colorectal Tumors Predict Clinical Outcome. *Science* **2006**, *313*, 1960–1964. [CrossRef]

71. Galon, J.; Bruni, D. Approaches to treat immune hot, altered and cold tumours with combination immunotherapies. *Nat. Rev. Drug Discov.* **2019**, *18*, 197–218. [CrossRef] [PubMed]

72. Hammerl, D.; Martens, J.W.M.; Timmermans, M.; Smid, M.; Trapman-Jansen, A.M.; Foekens, R.; Isaeva, O.I.; Voorwerk, L.; Balcioglu, H.E.; Wijers, R.; et al. Spatial immunophenotypes predict response to anti-PD1 treatment and capture distinct paths of T cell evasion in triple negative breast cancer. *Nat. Commun.* **2021**, *12*, 5668. [CrossRef] [PubMed]

73. Voorwerk, L.; Slagter, M.; Horlings, H.M.; Sikorska, K.; Van De Vijver, K.K.; De Maaker, M.; Nederlof, I.; Kluin, R.J.C.; Warren, S.; Ong, S.; et al. Immune induction strategies in metastatic triple-negative breast cancer to enhance the sensitivity to PD-1 blockade: The TONIC trial. *Nat. Med.* **2019**, *25*, 920–928. [CrossRef]

74. van Krimpen, A.; Gerretsen, V.I.V.; Mulder, E.; van Gulijk, M.; van den Bosch, T.P.; von der Thüsen, J.; Grünhagen, D.J.; Verhoef, C.; Mustafa, D.; Aerts, J.G.; et al. Immune suppression in the tumor-draining lymph node corresponds with distant disease recurrence in patients with melanoma. *Cancer Cell* **2022**, *40*, 798–799. [CrossRef] [PubMed]

Article

Eosinophilic Dermatoses: Cause of Non-Infectious Erythema after Volume Replacement with Diced Acellular Dermal Matrix in Breast Cancer?

Jean Schneider [1], Seung Taek Lim [2], Yeong Yi An [3] and Young Jin Suh [2,*]

1 School of Medicine, Texas Tech University Health Science Center, Lubbock, TX 79430, USA; jsschneider4055@gmail.com
2 Division of Breast and Thyroid Surgical Oncology, Department of Surgery, The Catholic University of Korea St. Vincent's Hospital, Suwon 16247, Republic of Korea; in-somnia@hanmail.net
3 Department of Radiology, The Catholic University of Korea St. Vincent's Hospital, Suwon 16247, Republic of Korea; didi97@catholic.ac.kr
* Correspondence: yjsuh@catholic.ac.kr; Tel.: +82-31-881-8690; Fax: +82-31-247-5347

Abstract: Introduction: Non-infectious erythema, or Red Breast Syndrome (RBS), has been observed on the skin where acellular dermal matrix was implanted, although the exact cause is yet to be determined. Patients and Methods: A total of 214 female patients underwent breast-conserving surgery (BCS) and volume replacement using diced acellular dermal matrix (dADM) for breast cancer between December 2017 and December 2018. After collecting and evaluating relevant clinical data, inflammation markers, along with NK cell status presented by IFN-γ secretion assay, were measured using ELISA. Results: Nineteen patients (8.88%) presented with RBS after BCS and dADM use. A significant increase of platelet-to-lymphocyte ratio was noted in the non-RBS group ($p = 0.02$). Compared to the RBS group ($p = 0.042$), the WBC level of the non-RBS group showed significant decrease over time. Eosinophil counts increased significantly at follow-up but went up higher in the RBS group. Multivariate analysis showed preoperative chemotherapy significantly increased the hazard of RBS (OR 3.274, $p = 0.047$ and OR 17.098, $p < 0.001$, respectively). Discussion: Though no causal relationship between RBS and immune status was proven, the results suggest an association between preoperative chemotherapy and RBS in addition to the possible role of eosinophilia in leading to eosinophilic dermatoses, which warrants further exploration and elucidation.

Keywords: erythema; breast cancer; acellular dermal matrix; red breast syndrome; eosinophilic dermatoses

Citation: Schneider, J.; Lim, S.T.; Yi An, Y.; Suh, Y.J. Eosinophilic Dermatoses: Cause of Non-Infectious Erythema after Volume Replacement with Diced Acellular Dermal Matrix in Breast Cancer? *Life* **2024**, *14*, 608. https://doi.org/10.3390/life14050608

Academic Editors: Taobo Hu, Mengping Long, Lei Wang, Riccardo Autelli and Alejandro Martin Sanchez

Received: 28 January 2024
Revised: 21 April 2024
Accepted: 30 April 2024
Published: 9 May 2024

1. Introduction

The introduction of oncoplastic breast surgery (OPBS) has made it possible for patients to avoid cosmetic defects resulting from the removal of breast tissue as part of breast cancer treatment [1].

Previously, we reported initial cases of breast cancer patients who had undergone breast conserving surgery followed by inserting acellular dermal matrix into the cavity after removing index tumor [2]. Compared to using an implant or mastopexy with complex design, a more cost effective and convenient reconstruction (with better results that spare preoperative natural mammary ptosis with time and symmetry) is to use a diced acellular dermal matrix (dADM) of human origin which is implanted into the empty cavity immediately after the removal of the breast cancer [3–6]. There have been reports of non-infectious erythema described as "red breast syndrome" (RBS) in plastic surgery where implants are placed after a mastectomy [7–13], in which all breast parenchyma is removed, but there have been no studies on the incidence of RBS and the factors associated with it after breast-conserving surgery for breast cancer and placement of a new form of ADM, dADM, for volume replacement in the dead space where the cancer was removed.

It cannot be claimed that dADM is the best material for volume replacement, but it may have significant advantages among the materials currently available. Therefore, it is very important to differentiate RBS from the infection, and it should be common practice to give adequate treatment aside from empirical antibiotics therapy, which is not helpful to alleviate non-infectious erythema.

The exact incidence of 'Red Breast Syndrome (RBS)' is unknown, but some estimations between 5~10% are reported [7,13], while the infection rate related to breast conserving surgery (BCS) with surrounding tissue or artificial material such as mesh is reported to be about 11% [14]. However, non-infectious erythema or RBS, which been noted to manifest unexpectedly after ADM use, involves local heat and redness on the skin over the dADM implantation site and is refractory to antibiotic therapy [7,8]. Various reports have suggested that this is due to endotoxins [9], the use of preservatives [7], neovascularization [10], delayed hypersensitivity reaction [8], or graft vs. host reaction [11]. But because dADM has been processed and irradiated in order to theoretically eradicate antigenicity, we suspect that the patient's immunity may play a role when non-infectious erythema occurs in these RBS cases [12].

This study aims to identify the cause of this potentially bothersome problem by elucidating a possible relation between dADM and breast tissue in terms of immunological status at the time of diagnosis and follow-up.

2. Patients and Methods

The study protocol was reviewed and approved by the institutional review board of the Catholic University of Korea (VC17OESI0168, VC20RESI0225). This clinical trial is registered as KCT0003886 on the site of Clinical Research Information Service of Korea that is participating as one of the primary registries in the WHO International Clinical Trials Registry Platform. The study involved a total of 214 breast cancer patients treated between December 2017 and December 2018. All patients had resectable female breast cancer without suspicious skin invasion, regardless of the size or location of the index tumor. Exclusion criteria were inflammatory or infectious disease within a month before the surgery, autoimmune disease, and blood clotting disorders. Written informed consent was obtained for the use of dADM to fill the defect after removal of the breast cancer and the storage of clinical information in the database, its use for the purpose of this study only. All patients in the study underwent standard breast-conserving surgery. Volume replacement was performed by filling with dADMs inserted into the empty space left after the resection was completed, and the skin was sutured [3]. Diced acellular dermal matrix is a unique form of ADM was used to fill the dead space after tumor removal in the breast to reconstruct during breast-conserving surgery for cancer; the details of its production and sterilization process have been reported elsewhere [12]. All procedures were performed in accordance with the ethical standards of the institutional and/or national research committee and with the Helsinki Declaration of 1964 and its later amendments.

We collected electronic clinical data from patients that had dADM (Megaderm®, L&C Bio, Seongnam, Republic of Korea), inflammation markers such as neutrophil-to-lymphocyte ratio (N/L), platelet-to-neutrophil ratio (P/N) and platelet-to-lymphocyte ratio (P/L), and natural killer (NK) cell status using a quantitative sandwich ELISA (Enzyme-linked Immunosorbent Assay) kit to measure the released interferon-γ (IFN-γ) from natural killer (NK) cells to quantify NK cell activity. NK cell status was previously measured and collected using radioactive material such as ^{51}Cr [15], but this approach is not used due to a longer time for the assay and greater funding requirements NK-IFN-γ secretion assay to determine NK cell status was performed by ELISA using NK Vue-Kit (NKMAX, Seongnam, Republic of Korea). Fresh whole blood (1 mL) was obtained using tubes containing Promoca (NKMAX, Seongnam, Republic of Korea). Promoca is a stimulatory cytokine that can specifically stimulate NK cells. The main cell population secreting IFN-γ after stimulating whole blood with Promoca was NK cells. After incubation at 37 °C for 20–24 h, the samples were centrifuged at 11,500× g for 1 min, and the supernatant was transferred to a 1.5 mL microtube, which was then stored at $-20\ °C$ until of IFN-γ levels reached the recommended amount according to the manufacturer's instructions. Briefly, 50 µL of six standards, controls, and samples were incubated in an antihuman IFN γ-coated plate at room temperature for 2 h and washed with washing buffer. IFN-γ conjugate was added and further incubated at room temperature for 1 h. After washing and incubation with 100 µL of the substrate at room temperature for 30 min in the dark, the absorbance value was measured at 450 nm. Concentrations of IFN-γ were determined with a calibration curve. The measuring range was 40~2000 pg/mL and the total imprecision for two levels of controls was less than the 15% coefficient of variations. We differentiated wound infection from RBS, since RBS is defined as a type of erythema without identifiable pathogens. Based on previous reports of RBS, we used blood samples taken immediately after diagnosis as the baseline and compared results with blood samples taken at 6 months postoperatively, after all chemotherapy or radiation treatments that could affect postoperative blood tests had been completed. However, in cases where RBS occurred, blood was drawn to differentiate from infection, and this blood was used for further analysis to better understand the circumstances under which RBS occurred.

3. Statistical Evaluation

Categorical variables were reported as the number and percentages, and continuous variables were reported with mean $\pm$ standard deviation. The normality of distribution of continuous variables was tested by the Shapiro–Wilk or Kolmogorov–Smirnov test, and variance equality was assessed by Levene's test. The comparison of continuous variables between groups was assessed using the student's t-test or Mann–Whitney U-test. The chi square of Fisher's exact test was used in categorical variables to assess the relationship between groups. Statistical analysis was performed using SPSS for Windows version 17.0 and a p value < 0.05 was accepted as statistically significant.

4. Results

This study is comprised of 214 female breast cancer patients treated between December 2017 and December 2018 (Table 1). Nineteen patients (8.88%) developed RBS within 6 months after breast-conserving surgery with dADM for reconstruction by the time of completion of systemic chemotherapy and/or external radiation treatment. Five out of 19 patients with RBS removed the dADM. Among 19 patients with RBS, 12 patients were premenopausal and seven were postmenopausal women. Two out of seven postmenopausal women with RBS had hypertension before surgery, while the others had no comorbid disease. In the RBS group, the index tumor was located in the upper outer quadrant in eight patients, in the upper inner quadrant in six, in the subareolar location in two, in the lower outer quadrant in two, and in the and lower inner quadrant in one case. Except for incision type and chemotherapy profile, there were no significant differences between the RBS and non-RBS groups, including age, menopausal status, BSA (body surface area), BMI (body mass index), TNM, breast volume, postoperative hormonal treatment status, tumor location, molecular subtype, and comorbid diseases [16].

Seven patients in the RBS group received neoadjuvant chemotherapy and had curative-intent operation. The number of patients who had adjuvant chemotherapy in the RBS group was 15. One patient did not receive systemic chemotherapy perioperatively. Of the nineteen patients, 17 patients developed RBS during systemic chemotherapy. The chemotherapeutic regimens were docetaxel–trastuzumab–pertuzumab (DHP) (one case), docetaxel–anthracycline–cyclophosphamide (TAC) (eight cases), and docetaxel–cyclophosphamide (TC) (nine cases). Two patients, who received three cycles of preoperative palliative chemotherapy with a triweekly TC regimen because of suspected bone metastasis, developed RBS on completion of three cycles postoperatively (approximately 8 weeks postoperatively); one patient, who was in clinically more than partial remission after triweekly neoadjuvant DHP regimen, developed RBS 6 weeks postoperatively on completion of radiation therapy, which was started immediately after surgery; and the other six patients, who received three cycles of triweekly neoadjuvant TC regimen, developed RBS 2–3 weeks after the initiation of radiation therapy following the completion of three cycles of postoperative triweekly adjuvant TC regimen. Among the eight patients who received the postoperative triweekly adjuvant TAC regimen, three developed RBS after three cycles, two after four cycles, one after five cycles, and the remaining two patients developed RBS two weeks after starting radiation therapy. One patient, who received an adjuvant triweekly TC regimen after surgery, developed RBS 3 weeks after completing four cycles and just before starting radiation therapy.

Table 1. Patients' characteristics.

			RBS (n = 19)	Non-RBS (n = 195)	p Value
Age			47.9474 ± 9.38364	51.8718 ± 8.83053	0.067
Menopausal status	Premenopausal		12 (63.2%)	121 (62.1%)	0.924
	Postmenopausal		7 (36.8%)	74 (37.9%)	
Tumor location	UOQ		8 (42.1%)	72 (36.9%)	
	UIQ		6 (31.6%)	63 (32.3%)	
	LOQ		2 (10.5%)	21 (10.8%)	0.994
	LIQ		1 (5.3%)	16 (8.2%)	
	SA		2 (10.5%)	23 (11.8%)	
Incision	Circumareolar		2 (10.5%)	28 (14.4%)	
	Periareolar		13 (68.4%)	153 (78.5%)	
	Peri-breast		0 (0%)	9 (4.6%)	0.016
	Radial		4 (21.1%)	5 (2.6%)	
Breast volume (cc)			959.1053 ± 355.751	1052.9795 ± 411.67598	0.339
Body surface area (m^2)			1.6253 ± 0.12677	1.6405 ± 0.32539	0.840
Body mass index (Kg/m^2)			25.5968 ± 3.40849	24.1507 ± 4.95705	0.216
Comorbid disease	DM	No	19 (100%)	183 (93.8%)	0.606
		Yes	0 (0%)	12 (6.2%)	
	Hypertension	No	17 (89.5%)	173 (88.7%)	0.921
		Yes	2 (10.5%)	22 (11.3%)	
	Other	No	19 (100%)	192 (98.5%)	0.756
		Yes	0 (0%)	3 (1.5%)	
Molecular subtype	LUM A		9 (47.4%)	130 (66.7%)	
	LUM B		4 (21.1%)	24 (12.3%)	
	HER+		4 (21.1%)	19 (9.7%)	0.181
	Triple negative		2 (10.5%)	22 (11.3%)	
TNM	0		2 (10.5%)	32 (16.4%)	
	I		5 (26.3%)	84 (43.1%)	
	II		7 (36.8%)	62 (31.8%)	0.109
	III		4 (21.1%)	13 (6.7%)	
	IV		1 (5.3%)	4 (2.1%)	
Chemotherapy	Neoadjuvant	No	12 (63.2%)	171 (87.7%)	0.01
		Yes	7 (36.8%)	24 (12.3%)	
	Adjuvant	No	4 (21.1%)	67 (34.4%)	0.240
		Yes	15 (78.9%)	128 (65.6%)	
	Neoadjuvant→Adjuvant	No	13 (68.4%)	172 (88.2%)	0.028
		Yes	6 (31.6%)	23 (11.8%)	
	Palliative	No	17 (89.5%)	188 (96.4%)	<0.001
		Yes	2 (10.5%)	7 (3.6%)	
Hormonal therapy	No		6 (31.6%)	40 (20.5%)	0.254
	Yes		13 (68.4%)	155 (79.5%)	

No other malignancies including hematologic abnormalities were reported thus far. There were no cases of RBS more than 6 months after breast-conserving surgery.

NK activity represented by IFN-γ was not statistically different at diagnosis or at follow-up (6 months after operation) between the RBS and non-RBS groups (Table 2). Within the RBS group, there was no significant increase or decrease of NK or inflammation markers (N/L, P/N, and P/L). On the contrary, the P/L ratio significantly increased at follow-up compared to the initial value in the non-RBS group ($p = 0.02$). Otherwise, there were no significant changes from initial to follow-up, including NK value change, in the non-RBS group.

Table 2. NK cell activity and inflammation markers.

	Initial			Follow Up		
	RBS ($n = 19$)	Non-RBS ($n = 195$)	p Value	RBS ($n = 19$)	Non-RBS ($n = 195$)	p Value
NK	1168.326316 ± 728.3503391	913.436096 ± 718.6330553	0.143	868.878947 ± 691.2452641	949.281482 ± 728.2787949	0.646
N/L	1.737826 ± 0.8557119	1.952532 ± 0.9806427	0.358	1.937236 ± 1.0851416	2.157756 ± 1.2841531	0.470
P/N	108.801468 ± 76.5852187	105.300069 ± 160.8408602	0.925	98.267985 ± 42.4433964	87.012974 ± 35.0998656	0.192
P/L	157.625805 ± 81.4910094	152.905757 ± 56.0619944	0.738	159.601787 ± 58.3481064	168.268940 ± 72.2833186	0.613
	RBS ($n = 19$)			Non-RBS ($n = 195$)		
	Initial	Follow Up	p Value	Initial	Follow Up	p Value
NK	1168.326316 ± 728.3503391	868.878947 ± 691.2452641	0.202	913.436096 ± 718.6330553	953.691935 ± 727.7360589	0.591
N/L	1.737826 ± 0.8557119	1.937236 ± 1.0851416	0.533	1.952532 ± 0.9806427	2.157756 ± 1.2841531	0.077
P/N	108.801468 ± 76.5852187	98.267985 ± 42.4433964	0.603	105.300069 ± 160.8408602	87.012974 ± 35.0998656	0.122
P/L	157.625805 ± 81.4910094	159.601787 ± 58.3481064	0.932	152.905757 ± 56.0619944	168.268940 ± 72.2833186	0.02

NK = natural killer cell activity. N/L = neutrophil to lymphocyte ratio. P/N = platelet to neutrophil ratio. P/L = platelet to lymphocyte ratio.

By differential count of CBC (Table 3), we found that all the values at initial and follow-up failed to show significant differences between the RBS and non-RBS groups. However, the levels of hemoglobin, hematocrit, platelets, and white blood cells decreased significantly at follow-up compared to the initial levels in the non-RBS group ($p < 0.001$, $p = 0.001$, $p < 0.001$, $p = 0.042$, respectively). All the decreased values remained within the reference range, indicating no clinically significant change. In the RBS group, platelet values decreased significantly at follow-up, but still remained within the reference range ($p = 0.023$). By the fraction of white blood cells, all cells except eosinophils showed no significant changes from the initial value to the follow-up. Eosinophil counts at follow-up were increased in the RBS group ($p = 0.04$) and the non-RBS group showed significantly increased levels of eosinophils ($p < 0.001$) but the fold of increase compared to the initial value was much higher in the RBS group compared to the non-RBS group. This suggests there may a role of eosinophils in the development of RBS without definite infection, which is supported by the non-significant changes of segmented neutrophils from the initial to the follow-up in both groups.

Between the RBS and non-RBS groups, the incision type seemed meaningfully different on univariate analysis ($p = 0.015$) (Table 4).

Radial incisions were more common in the RBS group, in contrast to all cases of peri-breast (inframammary) incisions that were utilized in the non-RBS group (Table 1) (Figure 1). Even though radial incision seems likely to pose much a higher odds ratio, leading to RBS on univariate analysis, there was no definite risk leading to RBS after radical incisions were examined on multivariate analysis (Table 5). Peri-breast incisions were made alongside the lowermost line of the breast including the inframammary line, according to the location of the index tumor, to reach through at the nearest point (Figure 1). Sentinel lymph node biopsy or axillary lymph node dissection was also carried out. Since the percentage of patients that received preoperative chemotherapy, either neoadjuvant or palliative, was relatively higher in the radial incision patients (two out of nine in the neoadjuvant group versus four out of nine in the palliative group), radial incision was a significant factor leading to RBS on univariate analysis. Even though patients with palliative chemotherapy seemed to be common in the RBS group, only one patient underwent preoperative palliative chemotherapy. After univariate and multivariate analyses, neoadjuvant and palliative chemotherapy preoperatively proved to be significant factors in provoking RBS, with an odds ratio of 3.274 (neoadjuvant, $p = 0.047$) and 17.098 (palliative, $p < 0.001$), respectively. Although these results have significance, the case numbers are not large enough to get an accurate statistical power. It should be evaluated in larger scale studies to get the definitive clinical meaning.

Table 3. Differential count of CBC.

	Initial			Follow Up		
	RBS ($n = 19$)	**Non-RBS ($n = 195$)**	***p* Value**	**RBS ($n = 19$)**	**Non-RBS ($n = 195$)**	***p* Value**
Hemoglobin (g/dL)	13.005263 ± 1.3554309	12.989231 ± 1.2428504	0.958	12.3842 ± 1.27160	12.4728 ± 1.38955	0.790
Hematocrit (%)	39.163158 ± 3.7985685	39.016769 ± 3.2851918	0.855	38.1053 ± 3.25891	37.8800 ± 3.17772	0.769
Platelet ($\times 10^3/\mu$L)	$273.947368 \pm 61.0341195$	$274.943590 \pm 73.1470902$	0.954	227.6316 ± 59.00114	231.5128 ± 61.79700	0.793
White blood cell ($\times 10^3/\mu$L)	5.613684 ± 1.5691548	6.033436 ± 1.6691522	0.294	4.8674 ± 1.33476	5.3612 ± 4.27633	0.618
Segmented neutrophil ($\times 10/\mu$L)	54.731579 ± 11.6569032	56.653846 ± 11.3055193	0.481	54.0211 ± 12.71345	57.8374 ± 9.62992	0.217
Lymphocyte ($\times 10/\mu$L)	35.694737 ± 10.0949643	33.380513 ± 9.8606492	0.331	32.8789 ± 10.05566	33.0754 ± 27.85745	0.976
Monocyte ($\times 10/\mu$L)	7.399474 ± 5.0950635	7.170256 ± 5.3150959	0.857	8.4053 ± 3.14245	8.0774 ± 4.67696	0.765
Eosinophil ($\times 10/\mu$L)	1.431579 ± 0.9189684	2.148205 ± 2.2124144	0.164	4.2895 ± 5.57752	3.3749 ± 3.71695	0.331
Basophil ($\times 10/\mu$L)	0.6089 ± 0.77994	0.4995 ± 0.44203	0.554	0.4053 ± 0.25050	0.4436 ± 0.38462	0.671

	RBS ($n = 19$)			Non-RBS ($n = 195$)		
	Initial	**Follow Up**	***p* Value**	**Initial**	**Follow Up**	***p* Value**
Hemoglobin (g/dL)	13.005263 ± 1.3554309	12.384211 ± 1.2715971	0.154	12.989231 ± 1.2428504	12.472821 ± 1.3895522	<0.001
Hematocrit (%)	39.163158 ± 3.7985685	38.105263 ± 3.2589149	0.353	39.016769 ± 3.2851918	37.880000 ± 3.1777188	0.001
Platelet ($\times 10^3/\mu$L)	$273.947368 \pm 61.0341195$	$227.631579 \pm 59.0011398$	0.023	$274.943590 \pm 73.1470902$	$231.512821 \pm 61.7970038$	<0.001
White blood Cell ($\times 10^3/\mu$L)	5.613684 ± 1.5691548	4.867368 ± 1.3347611	0.123	6.033436 ± 1.6691522	5.361174 ± 4.2763289	0.042
Segmented Neutrophil ($\times 10/\mu$L)	54.731579 ± 11.6569032	54.021053 ± 12.7134478	0.859	56.653846 ± 11.3055193	57.837436 ± 9.6299246	0.266
Lymphocyte ($\times 10/\mu$L)	35.694737 ± 10.0949643	32.878947 ± 10.0556551	0.395	33.380513 ± 9.8606492	33.075385 ± 27.8574466	0.885
Monocyte ($\times 10/\mu$L)	7.399474 ± 5.0950635	8.405263 ± 3.1424466	0.469	7.170256 ± 5.3150959	8.077436 ± 4.6769564	0.074
Eosinophil ($\times 10/\mu$L)	1.431579 ± 0.9189684	4.289474 ± 5.5775238	0.04	2.148205 ± 2.2124144	3.374923 ± 3.7169474	<0.001
Basophil ($\times 10/\mu$L)	0.608947 ± 0.7799351	0.405263 ± 0.2504966	0.290	0.499486 ± 0.4420275	0.443590 ± 0.3846238	0.184

Table 4. Univariate analysis.

		OR	95% CI	*p*-Value
Age		1.051	0.996–1.108	0.069
Menopause	Premenopausal	1		
	Postmenopausal	0.954	0.359–2.531	0.924
Tumor location	OUQ	1		
	UIQ	0.857	0.282–2.604	0.786
	LOQ	0.857	0.169–4.348	0.852
	LIQ	0.563	0.066–4.821	0.6
	SA	0.783	0.155–3.951	0.767
Incision type	Circumareolar	1		
	Peri-areolar	1.190	0.254–5.561	0.825
	Inframammary			
	Peri-breast	0	0	0.999
	Radial	11.200	1.600–78.400	0.015
Breast volume		0.999	0.998–1.001	0.338
Body surface area (m^2)		0.830	0.135–5.102	0.84
Body mass index (Kg/m^2)		1.071	0.964–1.190	0.203
Diabetes	No	1		
	Yes	0	0	0.999
Hypertension	No	1		HBP
	Yes	1.081	0.234–4.996	0.921
Other	No	1		
	Yes	0	0	0.999
Molecular subtype	Luminal A	1		
	Luminal B	2.293	0.655–8.031	0.194
	HER2	3.018	0.845–10.771	0.089
	Triple negative	1.303	0.264–6.438	0.745
TNM	0	1		
	I	0.952	0.176–5.159	0.955
	II	1.806	0.355–9.205	0.477
	III	4.923	0.801–30.253	0.085
	IV	4.000	0.292–54.715	0.299
Neoadjuvant CTx.	No	1		
	Yes	4.156	1.491–11.588	0.006
Adjuvant CTx.	No	1		Adjuvant CTx.
	Yes	1.963	0.627–6.149	0.247
Neo + Adjuvant CTx.	No	1		
	Yes	3.452	1.195–9.969	0.022
Palliative CTx.	No	1		
	Yes	19.532	5.986–63.733	<0.001
Hormonal Tx.	No	1		
	Yes	0.559	0.200–1.563	0.268

CTx = chemotherapy. Tx = therapy.

(**a**) For lower outer tumor

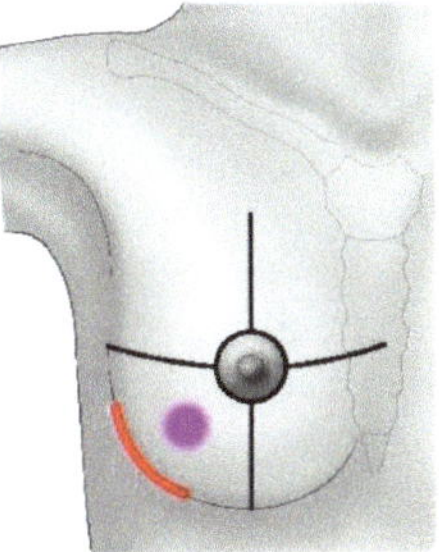

(**b**) For mid-lower or subareolar tumor

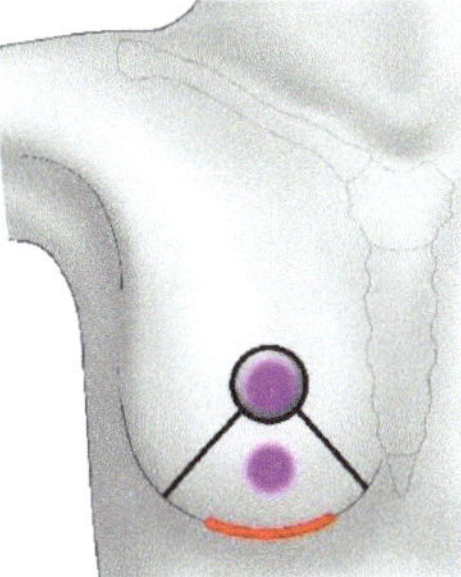

(**c**) For lower-inner or upper-inner tumor

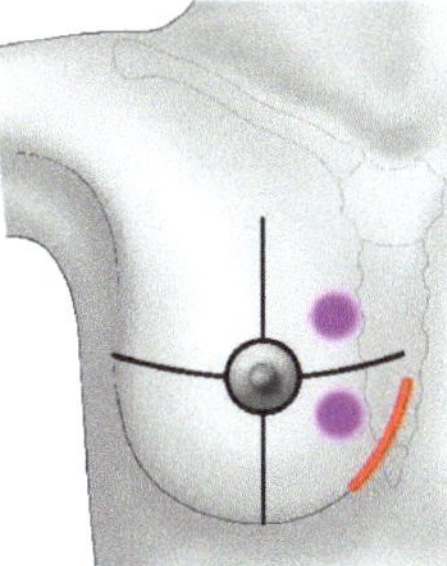

Figure 1. Peri-breast (Inframammary) incisions.

Table 5. Multivariate analysis.

		OR	95% CI	*p* Value
Incision type	Circumareolar	1		
	Periareolar	1.330	0.238–7.430	0.745
	Peri-breast	0	0	0.999
	Radial	5.125	0.497–52.887	0.17
Neoadjuvant chemotherapy	No	1		
	Yes	3.274	1.018–10.526	0.047
Palliative chemotherapy	No	1		
	Yes	17.098	5.060–57.767	<0.001

5. Discussion

Over the past few decades, there has been a significant shift in the surgical treatment of breast cancer from mastectomy to breast-conserving surgery, and this shift has been accelerated by the rapid introduction of the concept of oncoplastic surgery. In breast-conserving surgery, the cavity left after cancer removal is filled by volume replacement using adjacent breast parenchyma or adipofascial tissue, or if this is not possible, various materials have been used to fill the empty space at the index tumor site. In the latter case, there was no material that consistently showed a satisfactory outcome, but we devised a volume replacement method using diced acellular dermal matrix. Non-infectious erythema is not seen in cases where ADMs are not used and has been reported by many plastic surgeons who use ADM and have reported a number of possible mechanisms. However, no characteristic histologic findings have been described in these reports, and none of the proposed mechanisms have been proven to be causal.

We were unable to prove a direct relationship between NK activity at diagnosis and RBS after breast-conserving surgery and volume replacement with dADM for breast cancer [17–19]. Though there was no significant chronological change since diagnosis, NK activity reflected by IFN-γ secretion seemed to be decreased in the RBS group at follow-up, contrary to the increased IFN-γ secretion in the non-RBS group. This should not be interpreted definitively using only the data collected thus far. It may be due to the reflection of local tissue reaction during development of RBS, but there are likely to be other factors to be considered. Average values of the N/L, P/N, and P/L ratio at the time of diagnosis and follow-up, as well as chronological changes within each group, are shown in Table 2. These values are commonly used to indirectly evaluate patients' immunological status or prognostic/predictive value in various solid tumors including breast cancer [20,21]. The initial and follow-up ratio of N/L, P/N, and P/L between the RBS and non-RBS groups was not statistically different, except that the platelet/lymphocyte ratio was significantly increased at follow-up compared to the initial ratio in the non-RBS group ($p = 0.02$). Considering the fact that many more patients with preoperative systemic chemotherapy were included in the RBS group, it may harbor some clinical meaning. But statistical evaluation was not possible due to the small number of such patients [22,23].

In this study, there were no significant differences in eosinophil counts between the RBS and non-RBS groups at the time of diagnosis or follow-up. However, the RBS and non-RBS groups showed significantly increased eosinophils at follow-up, compared to respective initial values. In the context of eosinophilia, the increment of eosinophil counts at follow-up was 2.99 times in the RBS group and 1.57 times in the non-RBS group, compared to initial values, respectively. Eosinophils act to defend against infectious stimuli, especially parasites, and play key roles in various immune-mediated skin and constitutional diseases such as allergic inflammation. During this defense mechanism, eosinophils release mediators that act in immune regulations as well as mediate skin symptoms [24]. Eosinophilic skin diseases show eosinophilic infiltration that may or may not accompany eosinophilia. Idiopathic eosinophilic dermatoses are known to be accompanied by eosinophilic infiltration that can affect certain tissue layers or adnexal structures of dermis, subcutaneous fat or other structures [25].

Initially, antibiotics were given to alleviate suspected infection; however, this had a minimal effect when there was no isolable pathogen from culture by aspiration under the erythema. Once we were unable to identify any specific pathogens causing the erythema, we began to remove dADM to minimize further subcutaneous fat necrosis that might lead to devastating skin necrosis. Most of the RBS cases that required surgical management showed profuse subcutaneous fat necrosis at re-exploration. Once the dADM was removed, the skin erythema began to disappear within a couple of weeks in most cases. Even though it is difficult to prove that there is a direct causal relationship with all these findings, we suspect that dADM may activate the immune system and affect specific skin layers, such as dermis or subcutaneous fat, to induce allergic reaction and cause eosinophilic dermatitis. Therefore, we suspect that the removal of dADM can lead to fast recovery from RBS leaving no sequelae behind, because it should be the triggering factor of the non-infectious erythema. Taken together with the increment of eosinophil counts in the non-RBS group, dADM may induce the host immune system by recruiting eosinophils over the area where dADMs are stacked. Nevertheless, most patients undergoing breast-conserving surgery and volume replacement with dADM can recover from this rare eosinophilic dermatitis without the need of removing the dADM [26]. Preoperative chemotherapy and granulocyte stimulating factors used to facilitate recovery from neutropenia after chemotherapy may affect the host immune system, especially during the course of systemic treatment [27,28]. The analysis of many more cases of RBS should be required to determine an objective causal relationship between eosinophils and RBS after breast-conserving surgery and volume replacement with dADM. Lastly, no case of any other malignancies, including hematologic diseases such as anaplastic large cell lymphoma, has been seen after certain synthetic breast implant insertion for reconstruction, for more than 3 years since the first case of dADM reconstruction with breast-conserving surgery for breast cancer [29,30].

6. Conclusions

We believe that, in cases of suspected red breast syndrome, it is important to first differentiate whether there is an infection or not; if RBS is confirmed, unnecessary antibiotics should be avoided, and short-term steroid use can effectively relieve symptoms. In this study, we proposed eosinophilia as a possible reason for the development of RBS after volume displacement with dADM after conventional breast-conserving surgery. To our knowledge, this is the first report of de novo RBS in patients undergoing partial mastectomy for resectable breast cancer and the implantation of a dADM as volume displacement.

Author Contributions: Conceptualization, J.S. and Y.J.S.; methodology, S.T.L.; software, S.T.L.; validation, J.S., and Y.J.S.; formal analysis, J.S.; investigation, Y.J.S.; resources, Y.J.S.; data curation, J.S.; writing—original draft preparation, J.S.; writing—review and editing, Y.Y.A.; visualization, Y.Y.A.; supervision, Y.J.S.; project administration, S.T.L. All authors have read and agreed to the published version of the manuscript.

Funding: This research received no external funding.

Institutional Review Board Statement: The study was conducted in accordance with the Declaration of Helsinki, and approved by the Institutional Review Board of the Catholic University of Korea St. Vincent's Hospital (protocol code VC20RESI0225 and date of approval 29 Oct 2020).

Informed Consent Statement: Informed consent was obtained from all subjects involved in the study.

Data Availability Statement: Data supporting reported results can be found on electronic medical record database only, any kind of act related to share, gain, create related to that database is completely prohibited due to privacy or ethical restrictions according to the law and regulations of the Republic of Korea and institutional review board of the Catholic University of Korea St. Vincent's hospital.

Acknowledgments: We acknowledge valuable support to gather information from the database done by nurse practitioners; Hye Sug Han, Ju Hee Shin, Su Yeon Jung, Da Yeon Jung and Sang Mi Park.

Conflicts of Interest: The authors declare no conflicts of interest.

References

1. Rainsbury, R.M. Surgery insight: Oncoplastic breast-conserving reconstruction—Indications, benefits, choices and outcomes. *Nat. Clin. Pract. Oncol.* **2007**, *4*, 657–664. [CrossRef] [PubMed]
2. Vashi, C. Clinical outcomes for breast cancer patients undergoing mastectomy and reconstruction with use of DermACELL, a sterile, room temperature acellular dermal matrix. *Plast. Surg. Int.* **2014**, *2014*, 704323. [CrossRef] [PubMed]
3. Gwak, H.; Jeon, Y.W.; Lim, S.T.; Park, S.Y.; Suh, Y.J. Volume replacement with diced acellular dermal matrix in oncoplastic breast-conserving surgery: A prospective single-center experience. *World J. Surg. Oncol.* **2020**, *18*, 60–66. [CrossRef] [PubMed]
4. Asaad, M.; Selber, J.C.; Adelman, D.M.; Baumann, D.P.; Hassid, V.J.; Crosby, M.A.; Liu, J.; Butler, C.E.; Clemens, M.W. Allograft vs xenograft bioprosthetic mesh in tissue expander breast reconstruction: A blinded prospective randomized controlled trial. *Aesthet. Surg. J.* **2021**, *41*, NP1931–NP1939. [CrossRef]
5. Giordano, S.; Garvey, P.B.; Clemens, M.W.; Baumann, D.P.; Selber, J.C.; Rice, D.C.; Butler, C.E. Synthetic mesh versus acellular dermal matrix for oncologic chest wall reconstruction: A comparative analysis. *Ann. Surg. Oncol.* **2020**, *27*, 3009–3017. [CrossRef]
6. Baek, W.Y.; Byun, I.H.; Kim, Y.S.; Lew, D.H.; Jeong, J.; Roh, T.S. Patient satisfaction with implant based breast reconstruction associated with implant volume and mastectomy specimen weight ratio. *J. Br. Cancer* **2017**, *20*, 98–103. [CrossRef] [PubMed]

7. Nahabedian, M.Y. AlloDerm performance in the setting of prosthetic breast surgery, infection, and irradiation. *Plast. Reconstr. Surg.* **2009**, *124*, 1743–1753. [CrossRef] [PubMed]

8. Ganske, I.; Hoyler, M.; Fox, S.E.; Morris, D.J.; Lin, S.J.; Slavin, S.A. Delayed hypersensitivity reaction to acellular dermal matrix in breast reconstruction. The Red Breast Syndrome? *Ann. Plast. Surg.* **2014**, *73* (Suppl. 2), S139–S143. [CrossRef] [PubMed]

9. Nguyen, T.H.C.; Brown, A.M.; Kulber, D.A.; Moliver, C.L.; Kuehnert, M.J. The role of endotoxin in sterile inflammation after implanted acellular dermal matrix: Red breast syndrome explained? *Aesthet. Surg.* **2020**, *40*, 392–399. [CrossRef] [PubMed]

10. Nahabedian, M.Y. Reply: AlloDerm performance in the setting of prosthetic breast surgery, infection, and irradiation. *Plast. Reconstr. Surg.* **2010**, *126*, 1120–1121. [CrossRef]

11. Wu, P.S.; Winocour, S.; Jacobson, S.R. Red Breast Syndrome: A review of available literature. *Aesth Plast. Surg.* **2015**, *39*, 227–230. [CrossRef] [PubMed]

12. Lee, J.H.; Kim, H.G.; Lee, W.J. Characterization and tissue incorporation of cross-linked human acellular dermal matrix. *Biomaterials* **2015**, *44*, 195–205. [CrossRef] [PubMed]

13. Hill, J.L.; Wong, L.; Kemper, P.; Buseman, J.; Davenport, D.L.; Vasconez, H.C. Infectious complications associated with the use of acellular dermal matrix in implant-based bilateral breast reconstruction. *Ann. Plast. Surg.* **2012**, *68*, 432–434. [CrossRef] [PubMed]

14. Hillberg, N.S.; Meesters-Caberg, M.A.; Beugels, J.; Winkens, B.; Vissers, Y.L.; van Mulken, T.J. Delay of adjuvant radiotherapy due to postoperative complications after oncoplastic breast conserving surgery. *Breast* **2018**, *39*, 110–116. [CrossRef]

15. Friberg, D.D.; Bryant, J.L.; Whiteside, T.L. Measurements of natural killer (NK) activity and NK-cell quantification. *Methods* **1996**, *9*, 316–326. [CrossRef]

16. Lewis, P.; Jewell, J.; Mattison, G.; Gupta, S.; Kim, H. Reducing postoperative infections and red breast syndrome in patients with acellular dermal matrix-based breast reconstruction. The relative roles of product sterility and lower body mass index. *Ann. Plast. Surg.* **2015**, *74* (Suppl. 1), S30–S32. [CrossRef]

17. Kim, B.R.; Chun, S.; Cho, D.; Kim, K.H. Association of neutrophil-to-lymphocyte ratio and natural killer cell activity revealed by measurement of interferon-gamma levels in a healthy population. *J. Clin. Lab. Anal.* **2019**, *33*, e22640. [CrossRef]

18. Wu, S.Y.; Fu, T.; Jiang, Y.Z.; Shao, Z.M. Natural killer cells in cancer biology and therapy. *Mol. Cancer* **2020**, *19*, 120–145. [CrossRef]

19. Rafei-Shamsabadi, D.A.; Klose, C.S.; Halim, T.Y.; Tanriver, Y.; Jakob, T. Context dependent role of type 2 innate lymphoid cells in allergic skin inflammation. *Front. Immunol.* **2019**, *10*, 2591. [CrossRef] [PubMed]

20. Cho, U.; Park, H.S.; Im, S.Y.; Yoo, C.Y.; Jung, J.H.; Suh, Y.J.; Choi, H.J. Prognostic value of systemic inflammatory markers and development of a nomogram in breast cancer. *PLoS ONE* **2018**, *13*, e0200936. [CrossRef]

21. Tiainen, S.; Rilla, K.; Hämäläinen, K.; Oikari, S.; Auvinen, P. The prognostic and predictive role of the neutrophil-to-lymphocyte ratio in early breast cancer, especially in the HER2+ subtype. *Br. Cancer Res. Treat.* **2021**, *185*, 63–72. [CrossRef]

22. Palacios-Acedo, A.L.; Mège, D.; Crescence, L.; Dignat-George, F.; Dubois, C.; Panicot-Dubois, L. Platelets, thrombo-inflammation, and cancer: Collaborating with the enemy. *Front. Immunol.* **2019**, *10*, 1805. [CrossRef] [PubMed]

23. Moldoveanu, D.; Pravongviengkham, V.; Best, G.; Martínez, C.; Hijal, T.; Meguerditchian, A.N.; Lajoie, M.; Dumitra, S.; Watson, I.; Meterissian, S. Dynamic neutrophil-to-lymphocyte ratio: A novel prognosis measure for triple-negative breast cancer. *Ann. Surg. Oncol.* **2020**, *27*, 4028–4034. [CrossRef] [PubMed]

24. Marzano, A.V.; Genovese, G. Eosinophilic dermatoses: Recognition and management. *Am. J. Clin. Dermatol.* **2020**, *21*, 525–539. [CrossRef]

25. Long, H.; Zhang, G.; Wang, L.; Lu, Q. Eosinophilic skin diseases: A comprehensive review. *Clinic Rev. Allerg. Immunol.* **2016**, *50*, 189–213. [CrossRef]

26. Sorkin, M.; Qi, J.; Kim, H.M.; Hamill, J.B.; Kozlow, J.H.; Pusic, A.L.; Wilkins, E.G. Acellular dermal matrix in immediate expander/implant breast reconstruction: A multicenter assessment of risks and benefits. *Plast. Reconstr. Surg.* **2017**, *140*, 1091–1100. [CrossRef] [PubMed]

27. Craig, E.S.; Clemens, M.W.; Koshy, J.C.; Wren, J.; Hong, Z.; Butler, C.E.; Garvey, P.B.; Selber, J.C.; Kronowitz, S.J. Outcomes of acellular dermal matrix for immediate tissue expander reconstruction with radiotherapy: A retrospective cohort study. *Aesthet. Surg. J.* **2019**, *39*, 279–288. [CrossRef]

28. Myckatyn, T.M.; Cavallo, J.A.; Sharma, K.; Gangopadhyay, N.; Dudas, J.R.; Roma, A.A.; Baalman, S.; Tenenbaum, M.M.; Matthews, B.D.; Deeken, C.R. The impact of chemotherapy and radiation on remodeling of acellular dermal matrices in staged, prosthetic breast reconstruction. *Plast. Reconstr. Surg.* **2015**, *135*, 43e–57e. [CrossRef]

29. GPSDP-03.25.19-03.26.19-FDA and ADM. Available online: https://fda.report/media/123014/GPSDP-03.25.19-03.26.19-Sientra-Briefing-Package.pdf (accessed on 29 April 2024).

30. GPSDP-03.25.19-03.26.19-FDA-Exec-Summary. Available online: https://fda.report/media/122956/GPSDP-03.25.19-03.26.19-FDA-Exec-Summary.pdf (accessed on 29 April 2024).

Article

Improved Breast Cancer Classification through Combining Transfer Learning and Attention Mechanism

Asadulla Ashurov [1], Samia Allaoua Chelloug [2,*], Alexey Tselykh [3], Mohammed Saleh Ali Muthanna [3], Ammar Muthanna [4] and Mehdhar S. A. M. Al-Gaashani [5]

[1] School of Communication and Information Engineering, Chongqing University of Posts and Telecommunications, Chongqing 400065, China; asadullahashur@gmail.com
[2] Department of Information Technology, College of Computer and Information Sciences, Princess Nourah bint Abdulrahman University, P.O. Box 84428, Riyadh 11671, Saudi Arabia
[3] Institute of Computer Technologies and Information Security, Southern Federal University, Taganrog 347922, Russia; tselykh@sfedu.ru (A.T.); muthanna@sfedu.ru (M.S.A.M.)
[4] RUDN University, 6 Miklukho-Maklaya Street, Moscow 117198, Russia; muthanna.asa@spbgut.ru
[5] College of Computer Science and Technology, Chongqing University of Posts and Telecommunications, Chongqing 400065, China; mr.mehdhar@gmail.com
* Correspondence: sachelloug@pnu.edu.sa

Abstract: Breast cancer, a leading cause of female mortality worldwide, poses a significant health challenge. Recent advancements in deep learning techniques have revolutionized breast cancer pathology by enabling accurate image classification. Various imaging methods, such as mammography, CT, MRI, ultrasound, and biopsies, aid in breast cancer detection. Computer-assisted pathological image classification is of paramount importance for breast cancer diagnosis. This study introduces a novel approach to breast cancer histopathological image classification. It leverages modified pre-trained CNN models and attention mechanisms to enhance model interpretability and robustness, emphasizing localized features and enabling accurate discrimination of complex cases. Our method involves transfer learning with deep CNN models—Xception, VGG16, ResNet50, MobileNet, and DenseNet121—augmented with the convolutional block attention module (CBAM). The pre-trained models are finetuned, and the two CBAM models are incorporated at the end of the pre-trained models. The models are compared to state-of-the-art breast cancer diagnosis approaches and tested for accuracy, precision, recall, and F1 score. The confusion matrices are used to evaluate and visualize the results of the compared models. They help in assessing the models' performance. The test accuracy rates for the attention mechanism (AM) using the Xception model on the "BreakHis" breast cancer dataset are encouraging at 99.2% and 99.5%. The test accuracy for DenseNet121 with AMs is 99.6%. The proposed approaches also performed better than previous approaches examined in the related studies.

Keywords: breast cancer; CNN; attention mechanism; transfer learning; classification; malignant; benign; magnification level; histopathology image

Citation: Ashurov, A.; Chelloug, S.A.; Tselykh, A.; Muthanna, M.S.A.; Muthanna, A.; Al-Gaashani, M.S.A.M. Improved Breast Cancer Classification through Combining Transfer Learning and Attention Mechanism. *Life* **2023**, *13*, 1945. https://doi.org/10.3390/life13091945

Academic Editors: Riccardo Autelli, Taobo Hu, Mengping Long and Lei Wang

Received: 11 August 2023
Revised: 16 September 2023
Accepted: 17 September 2023
Published: 21 September 2023

1. Introduction

Cancer, the primary cause of mortality in several nations, has become a significant global health concern. Cellulitis, or abnormal cell proliferation, is essential to developing malignancy. Tumors and cancers are characterized by the uncontrolled proliferation of cells, which results in the formation of aggregates or tumors in different organs. Lung, liver, colorectal, stomach, and breast cancers are the most prevalent forms of the disease [1]. Histopathology, the microscopic examination of biopsies by pathologists, is essential for diagnosing and comprehending the progression of organ cancer. Before microscopic examination, histological slides of tissue are prepared to facilitate examination. These transparencies display a variety of cells and tissue structures. Worldwide, breast cancer is the primary cause of mortality among female cancer patients. Its prevalence and fatality make it a significant global concern that affects a large number of women. Breast cancer is the most prevalent malignancy, accounting for approximately 14% of all cancers, with high mortality and morbidity rates among women worldwide. It increases the mortality rate among women, affecting approximately 2.1 million each year. It was estimated that 6,855,000 women would perish from breast cancer in 2020 [2]. Globally, death rates are on the rise, and they are significantly higher in wealthy nations. Tumors are masses that result from cancer's aberrant cell division. A growth can be malignant (cancerous) or benign. Cancer of the breast is a malignant neoplasm that develops from breast cells.

According to clinical evidence, the survival rate of breast cancer patients can be significantly improved by early and accurate diagnosis. Although histopathological examination has long been regarded as the gold standard for breast cancer diagnosis, its accuracy can be affected by subjective factors and image texture quality [3]. This can result in incorrect diagnoses and unnecessary patient injury. In order to resolve these issues, it is anticipated that computer-assisted categorization of histological images will reduce pathologists' burden and improve the accuracy and efficiency of pathological investigations. By leveraging technology, these advancements have the potential to yield substantial therapeutic and societal advantages in the field of breast cancer diagnosis. Deep learning techniques, specifically convolutional neural networks (CNNs), have demonstrated promise in medical image analysis tasks in recent years [4,5]. This article presents a novel method for analyzing breast cancer pathological images to improve feature extraction and diagnostic precision. The proposed method employs modified pre-trained CNN models and attention mechanisms to capture salient local features and derive an exhaustive representation of breast cancer images. Incorporating a channel attention module, which facilitates non-dimensionality reduction and local cross-channel interaction, is one of the main contributions of the proposed method. This module increases the CNN's ability to extract pertinent information from pathological images by selectively concentrating on significant local features. In addition, the attention mechanism incorporates a spatial attention block that selectively highlights essential regions within the histopathological images. This attention mechanism assures the extracted breast cancer classification features are informative and highly pertinent. By focusing the network's attention on critical regions, the proposed method accomplishes a robust global representation of features, enhancing overall diagnostic performance. Experiments are conducted on the BreakHis [6] breast cancer pathology dataset to evaluate the efficacy of the proposed method. The approach obtains comparable results to the most advanced models in the field, demonstrating its competitive performance. This paper proposes a novel breast cancer pathological image analysis approach incorporating attention mechanisms and modified pre-trained CNN models. The proposed method improves feature extraction by capturing salient local features and accentuating essential regions selectively, resulting in a comprehensive global representation of features. The experimental evaluation demonstrates the efficiency and competitiveness of the proposed breast cancer diagnosis method.

2. Related Works

A significant amount of studies have been conducted on classifying breast cancer images using deep learning methods. Two major approaches can be used to classify these investigations broadly [7]. The initial strategy utilizes representative CNN models as feature extractors and classifiers [8], followed by traditional machine learning models for classification [9].

Fatima et al. [10] review several works of literature that summarize and showcase prior studies focusing on machine learning algorithms employed in breast cancer prediction. Their work is a valuable resource for researchers seeking to analyze these algorithms and develop a fundamental understanding of deep learning research in this domain. Asri et al. [11] compare the efficacy of four machine learning algorithms using the Wisconsin Breast Cancer dataset: support vector machine (SVM), decision tree (DT), naïve Bayes (NB), and k-nearest neighbor (k-NN). The primary objective is to evaluate each algorithm's accuracy, precision, sensitivity, and specificity in efficiently classifying the data. The experimental results demonstrate that SVM obtains the utmost accuracy of 97.13% with the lowest error rate. All investigations were conducted in a simulated environment using the WEKA data mining instrument. Liu et al. [12] proposed the AlexNet-BC model, a novel framework for breast pathology classification. They introduce an improved cross-entropy loss function that effectively handles overconfident low-entropy output distributions, ensuring more suitable predictions for uniform distributions. They conducted a comprehensive series of experiments using the BreaKHis, IDC, and UCSB datasets to validate their approach. The results consistently demonstrate that the proposed method achieved slightly better results across varying magnification levels. Ramesh et al. [13] incorporated the GoogLeNet architecture for segmentation purposes. The resulting segmentations serve as inputs to enhance the performance of classifiers such as SVM, DT, RF, and NB. This approach yields notable improvements in accuracy, Jaccard and dice coefficients, sensitivity, and specificity compared to traditional architectures. The authors' proposed model achieves an accuracy score of 99.12%. Notably, it outperforms the AlexNet classifier by 3.78% in accuracy, and, on average, the improvement over existing techniques is 4.61%. Sharma and Kumar [14] explored the feasibility of leveraging a pre-trained Xception model for magnification-dependent breast cancer histopathological image classification, positioning it against conventional handcrafted methodologies. Notably, the Xception model using the 'radial basis function' kernel, in tandem with an SVM classifier, showcased a superior and more consistent performance. Across magnification levels ($40\times$, $100\times$, $200\times$, and $400\times$), accuracy figures of 96.25%, 96.25%, 95.74%, and 94.11% were achieved. Liew et al. [15] researched DLXGB (deep learning and eXtreme gradient boosting) and applied it to histopathology breast cancer images from the BreaKHis dataset to classify breast cancer. The method comprises data augmentation, stain normalization, and the use of a pre-trained DenseNet201 model for feature learning. Then, a gradient-boosting classifier is combined with these features. The classification task entails differentiating benign from malignant cases and classifying the images into eight non-overlapping/overlapping categories. The results demonstrate that DLXGB obtained a remarkable 97% accuracy for both binary and multi-class classification, superseding previous studies utilizing the same dataset. Atban at el. [16] concentrated on classifying breast cancer using optimized deep features. They used the ResNet18 architecture for feature extraction results in the production of deep features. Meta-heuristic algorithms, such as particle swarm optimization (PSO), atom search optimization (ASO), and the equilibrium optimizer (EO), are employed to improve the representativeness of these features further. Traditional ML algorithms are then used to evaluate the optimized deep features' classification effect. The experimental analysis is conducted on the widely used benchmark dataset BreakHis. The results demonstrate that the proposed method, in particular, the features obtained from ResNet18-EO, obtains an impressive F-score of 97.75% when coupled with an SVM employing Gaussian and radial-based functions (RBF). Ayana et al. [17] proposed in breast cancer classification the multistage transfer learning (MSTL) algorithm, which employs three pre-trained models—EfficientNetB2, InceptionV3, and ResNet50—combined with three optimizers: Adam, Adagrad, and stochastic gradient descent (SGD). The dataset encompasses 20,400 cancer cell images, supplemented by 200 ultrasound images from Mendeley and 400 from the MT-Small-Dataset. Highlighting the versatility of MSTL, the ResNet50-Adagrad configuration yielded significantly higher test accuracy scores of 99 ± 0.612% on the Mendeley dataset and 98.7 ± 1.1% on the MT-Small-Dataset, demonstrating consistency across five-fold cross-validation. Wang et al. [18] proposed a new method for the automated diagnosis and staging of cancer based on image analysis and machine learning algorithms. They utilized the BreaKHis dataset and applied preprocessing procedures, such as color-to-grayscale conversion, thresholding, and filtering. Nuclei are segmented utilizing the distance transform and watershed algorithms. Two distinct techniques for feature extraction are investigated. An ensemble-tagged tree classifier obtains the most remarkable accuracy (89.7%) in binary classification (benign vs. malignant). An ensemble subspace discriminant classifier obtains an accuracy of 88.1% for multiclass classification. Dubey et al. [19] proposed a hybrid convolutional neural network architecture for classifying benign and malignant breast lesions in histopathological micrographs. The architecture incorporates a ResNet50 model that has been pre-trained with additional layers that include global average pooling, dropout, and batch normalization. The proposed method obtains state-of-the-art performance on the BreakHis data set, exceeding previous benchmarks. The model obtains an AUC of 0.94 and a precision of 98.7%, demonstrating its efficacy in managing the binary classification problem and overcoming target imbalance. The innovation of the proposed method resides in the strategic incorporation of ResNet50's knowledge and training with the global average pooling layer. Singh et al. [20] proposed a hybrid deep neural network based on histopathological images for computer-assisted breast cancer diagnosis. The network integrates inception and residual blocks to capitalize on their respective benefits and achieve superior performance compared to existing algorithms. Two publicly available datasets, Breast Histopathology Images (BHI) and BreakHis, are used to train and validate the proposed method. With training precisions of 0.9642 for BreakHis and 0.8017 for BHI, the experimental results demonstrate the superiority of the proposed method. The model outperforms conventional deep neural networks and cutting-edge breast cancer detection techniques with accuracies of 0.8521 for BHI and 0.8080, 0.8276, 0.8655, and 0.8580 for distinct magnification levels ($40\times$, $100\times$, $200\times$, and $400\times$) in the BreakHis dataset. Venugopal et al. [21] proposed a hybrid deep-learning model that combines Inspection-ResNetv2 and EfficientNetV2-S with ImageNet-trained weights to classify breast cancer histopathology images. Using the BreakHis and BACH datasets, the model is evaluated. The top layer is removed from both networks, and global average pooling, dense layers, dropout, and a final classification layer are added. The individual results from the Inspection-ResNetv2 and EfficientNetV2 models are contrasted with the model's output. The final classification layer comprises four neurons for the BACH dataset and eight neurons for the BreakHis dataset. The experimental results demonstrate the proposed model's efficacy, achieving an overall precision of 98.15 percent for the BACH dataset and 99.03 percent for the BreakHis dataset. Kumari et al. [22] introduced a transfer learning-based AI system for classifying breast cancer using histopathological images. The VGG-16, Xception, and Densenet-201 deep convolutional neural network architectures are used as base models in the transfer learning approach. Using the pre-trained base models, features are extracted from each test image, and the images are categorized as benign or malignant. The proposed system achieves high classification accuracies of 99.42% (IDC dataset) and 99.12% (BreaKHis dataset). Compared to extant breast cancer classification methodologies, the outcomes demonstrate superior performance. Additionally, the proposed system is independent of image magnification levels, which increases its utility and adaptability. Joshi et al. [23] proposed a deep CNN-based breast cancer detection framework. The BreakHis and IDC datasets are used to extract breast cancer-related information from histopathology images. Three pre-trained CNN models, EfficientNetB0, ResNet50, and Xception, are assessed, with the customized Xception model producing the best results. The BreakHis dataset's 40 magnification images obtain an accuracy of 93.33 percent. The models are trained on 70 percent of the BreakHis dataset and validated on 30 percent of the remaining data. For regularization, data augmentation, dropout, and batch normalization are utilized. The enhanced Xception model is fine-tuned and tested on a subset of the IDC dataset, attaining an invasive ductal carcinoma detection accuracy of 88.08%. Adapting the pre-trained model to diverse classification tasks on the BreakHis and IDC datasets, this study demonstrates the efficacy of transfer learning. R Karthik et al. [24] introduced a novel classification method for breast cancer based on an ensemble of two deep convolutional neural network architectures, CSAResnet and DAMCNN, enhanced with Channel and Spatial attention. The framework extracts features from histopathology images in parallel using both architectures and employs ensemble learning to enhance performance. On the BreakHis dataset, the proposed method obtains a high classification accuracy of 99.55%. Zou et al. [25] proposed AHoNet with a channel attention module to capture local salient features within breast cancer pathological images, enhancing their discriminative power. Their method estimates second-order covariance statistics through matrix power normalization, providing a robust global feature representation. To assess the effectiveness of AHoNet, it was evaluated on two public breast cancer pathology datasets: BreakHis and BACH. The results demonstrate AHoNet's competitive performance, achieving optimal patient-level classification accuracies of 99.29% on BreakHis and 85% on BACH. These findings underscore the potential of attention mechanisms in enhancing breast cancer classification, aligning with the objectives of the proposed research, which combines transfer learning and attention mechanisms for breast cancer classification. Jadah et al. [26] introduced a breast cancer classification model based on the AlexNet convolutional neural network. The model is applied to histopathological images from the BreakHis dataset to diagnose breast cancer. Various modifications to the parameters and data are implemented to enhance the model's ability to recognize and classify input images as benign or malignant tumors. By optimizing the training frequency and balancing the training data, a classification accuracy of up to 96% is achieved. This research paper reviewed various approaches and advancements in breast cancer detection and classification, which are highlighted by the related works. Various techniques, such as machine learning algorithms, deep learning models, and ensemble methods, have been utilized to accurately analyze histopathological images and diagnose breast cancer. Transfer learning, data augmentation, and feature extraction techniques have been extensively used to improve classification model performance. The studies highlight the significance of image normalization, segmentation, and feature selection for increasing the accuracy of breast cancer classification systems. In addition, the use of publicly accessible datasets, such as BreakHis and IDC, has allowed researchers to compare and evaluate the efficacy of various methods. The related works demonstrate the potential of computational approaches to aid medical professionals in the early detection and accurate diagnosis of breast cancer, paving the way for enhanced patient outcomes and individualized treatment strategies. Several research voids exist in the classification of breast cancer that must be addressed. In machine learning-based approaches, there is a reliance on manual feature extraction, which can be time-consuming and subjective. To combat this, it is necessary to investigate methods of deep learning that can autonomously learn pertinent features from histopathology images. An additional area for improvement is the imbalance in training datasets, which can result in biased predictions. Effective data augmentation strategies are required to represent all classes adequately. In addition, existing methods frequently employ single-path deep learning architectures, which may lead to a more significant number of false positives and false negatives. Consider utilizing sophisticated techniques such as feature fusion and ensembling to increase classification accuracy. Motivated by these gaps, this paper proposes a novel classification method for breast cancer that tackles these limitations and enhances diagnostic accuracy.

3. Materials and Methods

This section presents the proposed image classification architecture for breast cancer using histopathology images. In addition, we investigate the specifics and fundamentals of the essential methods used in this study to assess their effect on medical image categorization.

The BreakHis dataset is selected as the sole dataset for evaluating our proposed models. This extensive collection of non-full-field breast cancer histopathology images is chosen due to its widespread recognition and accessibility. We establish a firm foundation for our medical application by utilizing the BreakHis dataset. This decision ensures the validity and generalizability of our proposed model, as the BreakHis dataset is representative of a typical dataset for image multiclassification tasks in breast cancer pathology. The 2016-created BreakHis dataset is available online via the breast cancer database. It consists of 7909 histopathological images of breast tumors from 82 patients. The dataset comprises 2480 benign tumor images (fibroadenoma, adenoma, tubular adenoma, and trichome tumors) and 5429 malignant tumor images (lobular carcinoma, ductal carcinoma, papillary carcinoma, and mucinous carcinoma) obtained at four magnifications: $40\times$, $100\times$, $200\times$, and $400\times$. In the breast tumor pathology section, each image from the BreakHis database is displayed with a 700-460-pixel resolution and RGB color format. Figure 1 depicts breast tissue segments, including malignant tumors, from the BreakHis dataset at four magnifications.

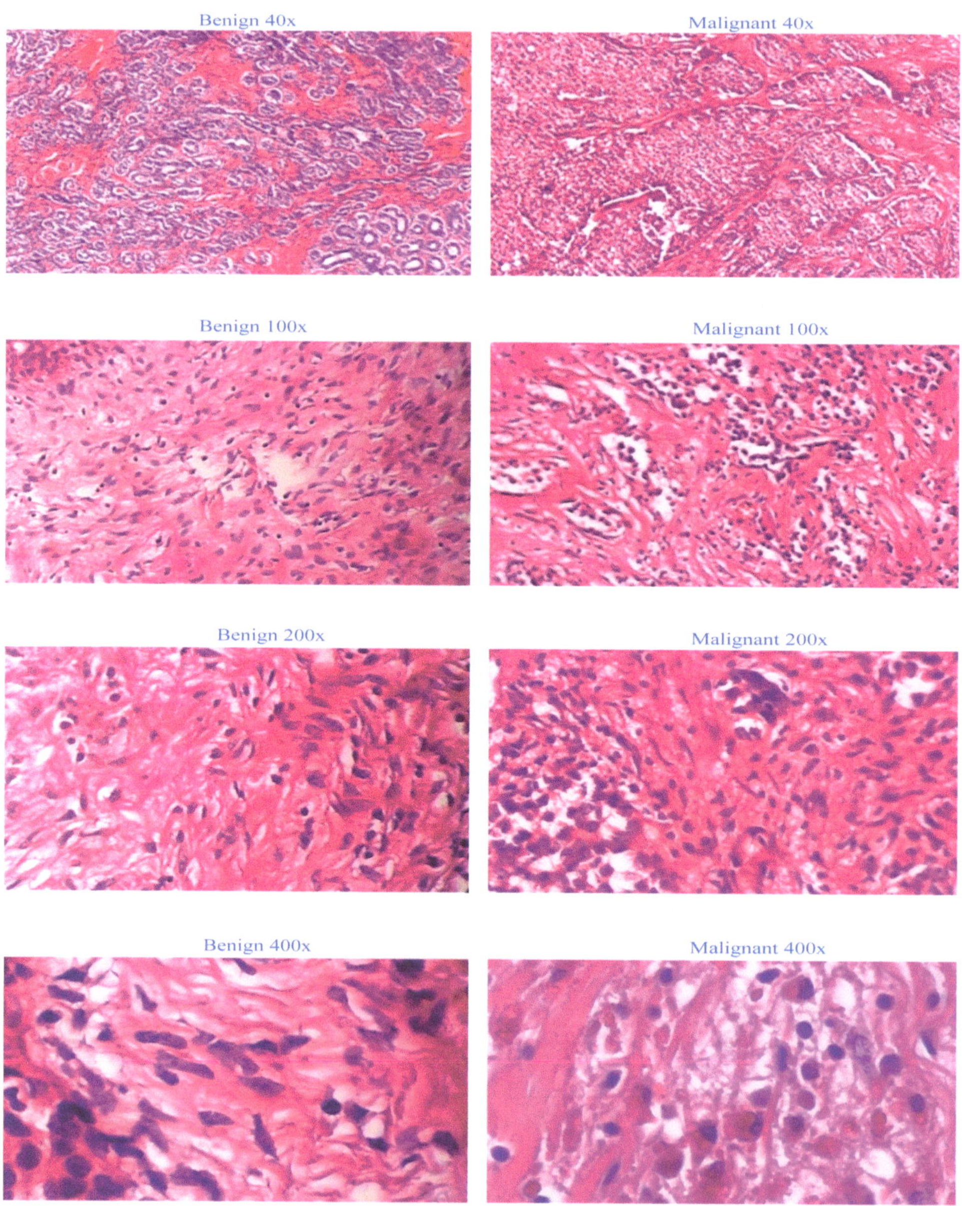

Figure 1. Representation of BreakHis dataset at four magnifications.

The dataset is divided into training and test sets based on the magnifications for the purpose of breast cancer prediction. Table 1 depicts the distribution of samples across various magnification levels ($40\times$, $100\times$, $200\times$, and $400\times$) and the corresponding division into training and testing subsets. The dataset is separated into training and test sets for each magnification level. The training set is used to train models for predicting breast cancer, while the testing set is used to evaluate the efficacy of the trained models. This division of the dataset, systematically based on the magnification level, guarantees that the trained models are evaluated using independent and representative samples from each magnification level. This enables a comprehensive evaluation of the models' performances at various magnification levels, contributing to the robustness and generalizability of the research findings.

Table 1. Splitting the dataset among the magnification levels into training and testing.

Magnification Level	$40\times$		$100\times$		$200\times$		$400\times$		Total
Splitting to Train and Test Set	**Training**	**Testing**	**Training**	**Testing**	**Training**	**Testing**	**Training**	**Testing**	**All**
Benign	370	255	383	261	368	255	351	237	2480
Malignant	880	490	938	499	901	489	814	418	5429
Total	1250	745	1321	760	1269	744	1165	655	7909

Before further modifications are made to several pre-trained models [27], namely Xception, VGG16, ResNet50, MobileNet, and DenseNet121, the BreakHis dataset is preprocessed. After each individual model, the convolutional layer is connected to construct a modified architecture for each model. Figure 2 represents the overall pipeline of the proposed system for breast cancer classification in this study.

The preprocessing stages for the BreakHis dataset likely included resizing the images to a consistent size, normalizing pixel values, and possibly applying additional transformations to improve the dataset's quality and consistency.

After preprocessing, each pre-trained model is modified by adding a convolutional layer. This additional layer permits further extraction and refinement of BreakHis-specific features. By connecting the conv layer after each model individually, the modified architectures can effectively capture and learn from the dataset's pertinent patterns and characteristics. The architecture uses pre-trained models as its backbone, leveraging prior knowledge to extract meaningful and representative features from the input images. This incorporation enables the model to leverage the ImageNet dataset's extensive feature representations. Following that, convolutional layers, max-pooling layers, channel attention layers, and spatial attention layers are included in the architecture. These components collaborate to improve the extraction of salient image features and highlight pertinent image regions. The extracted multidimensional features are then transformed into a compact 1D vector by a flattened layer, which is followed by batch normalization and a dense layer with softmax activation for efficient classification. The model is compiled utilizing the binary cross-entropy loss function and the well-known Adam optimizer [28] with a variable learning rate. We prioritize optimal performance and resource allocation by incorporating memory management techniques. It helps mitigate potential memory leaks, reduce memory usage, and enhance the overall efficiency of the deep learning model.

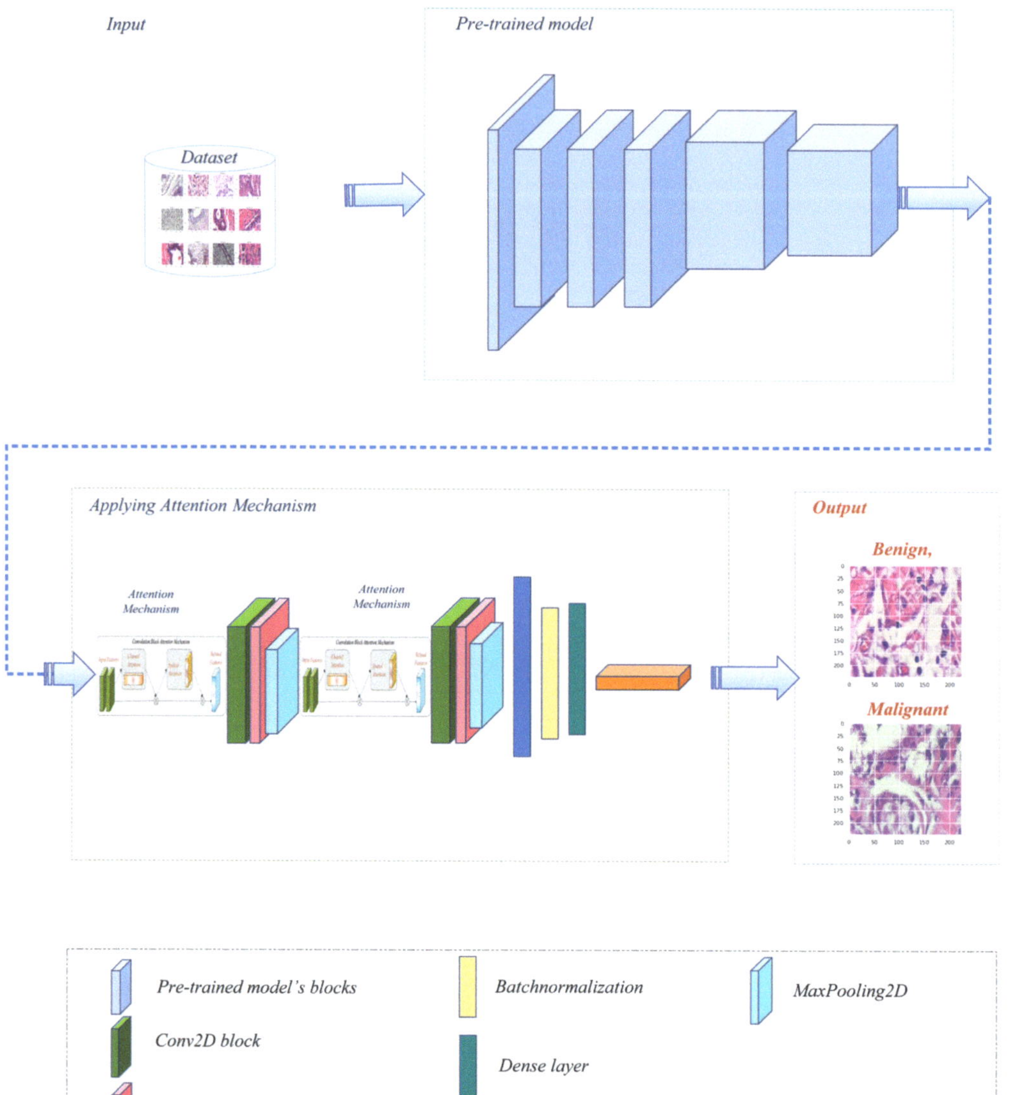

Figure 2. The proposed breast cancer detection workflow utilized a pre-trained model with an attention mechanism.

3.1. Attention Mechanism

The Convolutional Block Attention Module (CBAM) [29]: In our research, we applied the CBAM to address the challenges posed by limited data and computational constraints in breast cancer histopathology image classification. The CBAM module selectively highlights relevant features and suppresses extraneous ones, thereby improving the performance of the model [30].

The Channel Attention Module: The channel attention module focuses on capturing channel interactions. Given the input feature map F, we perform global average pooling and global max-pooling to obtain average-pooled (FC_{Avg}) and max-pooled (FC_{Max}) features, respectively. These features are then individually passed through a multilayer perceptron (MLP), and the resulting elements are added element-wise. This process generates the channel attention map ($Mc(F)$), representing inter-channel relationships within the input feature map.

To incorporate the channel attention map, we perform element-wise multiplication between the input feature map (F) and the channel attention map ($Mc(F)$), which can be expressed as:

$$FC = \sigma(M(FC_{\text{Max}}) + M(FC_{\text{Avg}})) \otimes F, \tag{1}$$

where σ denotes the sigmoid function.

The Spatial Attention Module: The spatial attention module aims to leverage inter-spatial connections between features. We apply average pooling and max-pooling along the channel axis to generate two feature maps, FS_{Avg} and FS_{Max}. These feature maps are then concatenated along the channel axis and processed through a convolution layer to generate a spatial attention map ($Ms(F)$) with dimensions $H \times W \times 1$. To integrate the inter-spatial connections, the spatial attention map ($Ms(F)$) is element-wise multiplied with the input feature map (FC), resulting in the revised feature map FCS with dimensions $H \times W \times C$. This process can be expressed as:

$$FCS = \sigma(\text{Conv}(\text{Cat}[FS_{\text{Max}}; FS_{\text{Avg}}])) \otimes FC, \tag{2}$$

where Conv() represents a convolution process with a 7×7 filter size.

By incorporating the CBAM module into our model, we enhance the ability to capture essential features and improve the performance of breast cancer histopathology image classification. Figure 3 represents the attention mechanism used in this study.

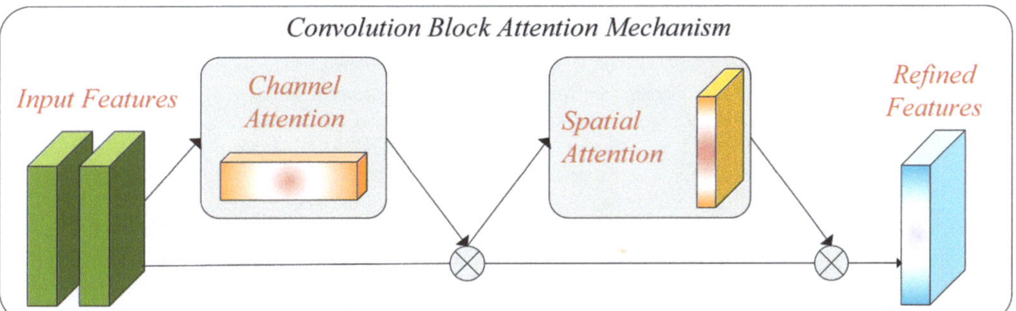

Figure 3. The attention mechanism, which includes channel and spatial attention blocks.

3.2. Feature Extraction

In our proposed method for histopathology image classification on the BreakHis dataset, we integrate attention mechanisms into the feature extraction stage. These mechanisms are intended to improve the extraction of salient image features and highlight relevant regions, thereby enhancing the model's discriminative ability [31]. This section details the feature extraction process of incorporating attention mechanisms. The feature extraction phase begins by employing pre-trained models as the foundation of our architecture. We can extract meaningful and representative features from input images by leveraging prior knowledge from models trained on the ImageNet dataset. This technique takes advantage of the extensive feature representations acquired from a large and diverse dataset. Our proposed architecture includes convolutional layers, max-pooling layers, channel attention layers, and spatial attention layers that collaborate to refine the derived characteristics. The convolutional layers employ filters to capture various image patterns, whereas the max-pooling layers downsample the feature maps to preserve the most pertinent data. The attention mechanisms are crucial to our feature extraction procedure. The channel attention layers permit the model to dynamically adjust the significance of various channels within the feature maps. The model can extract more discriminative features and suppress irrelevant data by concentrating on the most informative channels. Similarly, the spatial attention layers enable the model to highlight pertinent image regions by adaptively weighting the spatial locations within the feature maps. This mechanism focuses the model's attention on the most informative regions, which improves the localization and representation of pertinent image features. The incorporation of attention mechanisms offers several benefits. First, it enhances the model's ability to capture fine-grained details and intricate image patterns. By selectively focusing on the most informative channels and spatial locations, the model becomes more sensitive to pertinent characteristics, contributing to accurate classification. The attention mechanisms improve the interpretability of the decision-making process of the model. We gain insight into which image regions and features influence the model's predictions by highlighting the significant regions and channels. This interpretability can be advantageous in applications that require transparency and explicability. The attention mechanisms contribute to the overall efficacy of the feature extraction process. The model can reduce computational redundancy and derive more compact and discriminative representations by focusing on pertinent image regions and features. This efficacy results in quicker inference times and enhanced utilization of computational resources. The attention mechanisms improve the extraction of prominent image features, enhance interpretability, and optimize computational efficiency. By incorporating attention mechanisms, our model is able to effectively capture pertinent information and accomplish enhanced performance in image classification tasks using the BreakHis dataset.

3.3. Classification

Following the feature extraction phase, the extracted features are forwarded to the classification phase [32], where attention mechanisms are employed to refine the feature representations further and enhance the model's discriminative power. During the classification phase, we employ attention mechanisms to dynamically weight the importance of distinct features within the extracted representations. This enables the model to concentrate on the most pertinent information for accurate prediction. We apply attention layers to the extracted feature representations to integrate attention mechanisms. These layers determine the attention weights for each feature based on their importance to the classification assignment. The attention weights are then applied to the features, accentuating those most informative while downplaying those less pertinent. Our model acquires several benefits by integrating attention mechanisms into the classification stage. First, it enables the model to concentrate on most informative discriminative characteristics for the classification assignment. This selective focus improves the model's robustness and precision by minimizing the influence of irrelevant or chaotic features. Attention mechanisms improve the interpretability of the classification procedure. By displaying the attention weights, we can determine which features influence the model's decision-making. This interpretability permits us to gain insight into the model's logic and provides a method for validating its predictions. It can mitigate the effects of class disparity by concentrating on underrepresented classes. The model can enhance its performance on these classes by allocating greater attention weights to minority-class samples, resulting in a more accurate and balanced classification. Incorporating attention mechanisms into the classification phase improves the model's discriminative ability, interpretability, and robustness [33]. By dynamically balancing the significance of features, the model can prioritize pertinent information and make more accurate predictions. Incorporating attention mechanisms into the classification stage enhances the overall performance of our proposed method for image classification on the BreakHis dataset.

4. Results

This section evaluates and analyzes the proposed CNN models with attention mechanisms for histopathology image classification on the BreakHis dataset. The model's performance is evaluated based on classification accuracy, precision, recall, and F1-score. We conducted the same experimental configuration with other prominent CNN architectures, including Xception, VGG16, ResNet50, MobileNet, and DenseNet121. Regarding accuracy and other evaluation metrics, our proposed model with attention mechanisms consistently outperformed these architectures, demonstrating its efficacy in breast cancer classification.

Table 2 displays the evaluation of the performance of the proposed models with attention mechanisms on the BreakHis dataset at different magnification levels ($40\times$, $100\times$, $200\times$, and $400\times$). Included in the evaluation metrics are validation accuracy and validation loss. The Xception model obtained high accuracy rates at all magnification levels, ranging from 98.5% to 99.5%, as shown in Table 1. The loss values are consistently modest, ranging between 0.02 and 0.04. This indicates that the model can accurately classify breast histopathology images with attention mechanisms. Compared to Xception, the VGG16 model displayed accuracy rates from 92.8% to 97.2%. The model obtained acceptable performance despite the dataset's complexity. ResNet50 exhibited consistent performance across magnification levels, with 98.0% and 98.8% accuracy rates and minimal loss values between 0.05 and 0.09. This demonstrates the robustness and efficacy of the model in breast cancer classification. The MobileNet model obtained accuracy rates ranging from 92.4% to 99.2%, which are deemed satisfactory. However, its loss values are significantly greater than those of other models, ranging from 0.05 to 0.26. It may require additional optimization and finetuning to achieve optimal performance. DenseNet121 demonstrated a competitive level of performance, with accuracy rates ranging from 95.5% to 99.5%. The model consistently obtained low loss values varying from 0.02 to 0.12, demonstrating its classification accuracy. The experimental results demonstrate that incorporating attention mechanisms into proposed breast cancer classification models is effective. Xception and DenseNet121 demonstrated especially promising results. These findings validate the benefits of attention mechanisms in enhancing the ability of models to focus on pertinent features and regions, which contributes to enhanced classification performance.

Table 2. Accuracy rates and loss of the proposed models with attention mechanism.

Method\Magnification Level	$40\times$		$100\times$		$200\times$		$400\times$	
	Accuracy	Loss	Accuracy	Loss	Accuracy	Loss	Accuracy	Loss
Xception	99.2	0.02	98.5	0.05	99.2	0.04	99.5	0.04
VGG16	92.8	0.21	93.6	0.29	97.2	0.10	94.8	0.15
ResNet50	98.8	0.06	98.1	0.09	98.0	0.06	98.2	0.05
MobileNet	92.4	0.17	96.2	0.06	99.2	0.05	91.4	0.26
DenseNet121	97.2	0.07	95.5	0.12	98.8	0.11	99.6	0.02

The accuracy and loss graphs among magnification levels are to visually illustrate in Figures 4–7 the performances of the proposed models at different magnification levels. These graphs provide an intuitive representation of how the models perform in terms of accuracy and loss as the magnification level of the images varies. They provide a visual understanding of how the models' accuracy and loss values change as the level of magnification increases or decreases. Accuracy and loss graphs across magnification levels offer an informative visual representation of the model's performance, enhancing the clarity and comprehensibility of the research findings.

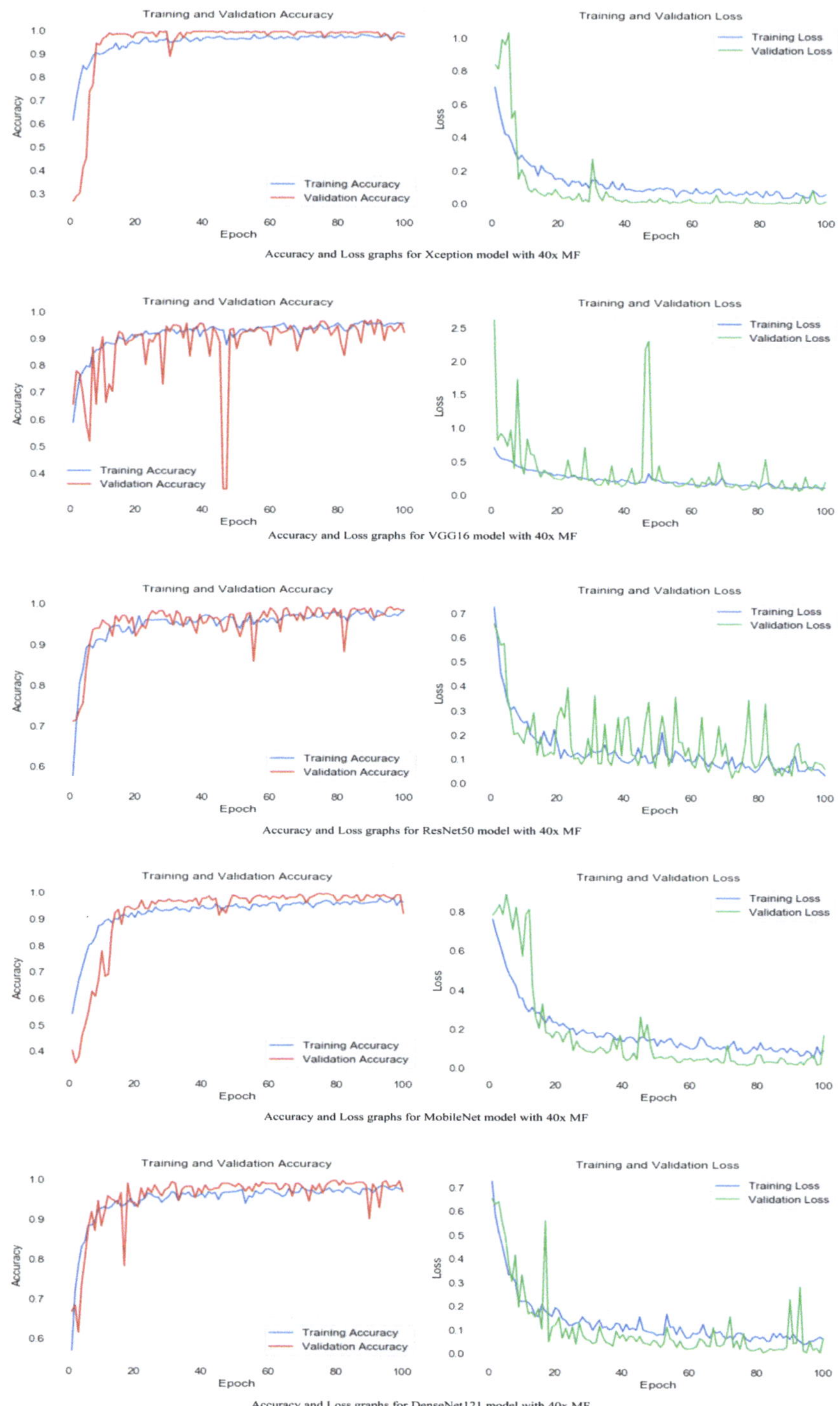

Figure 4. Accuracy and loss graphs for proposed models with 40× magnification level.

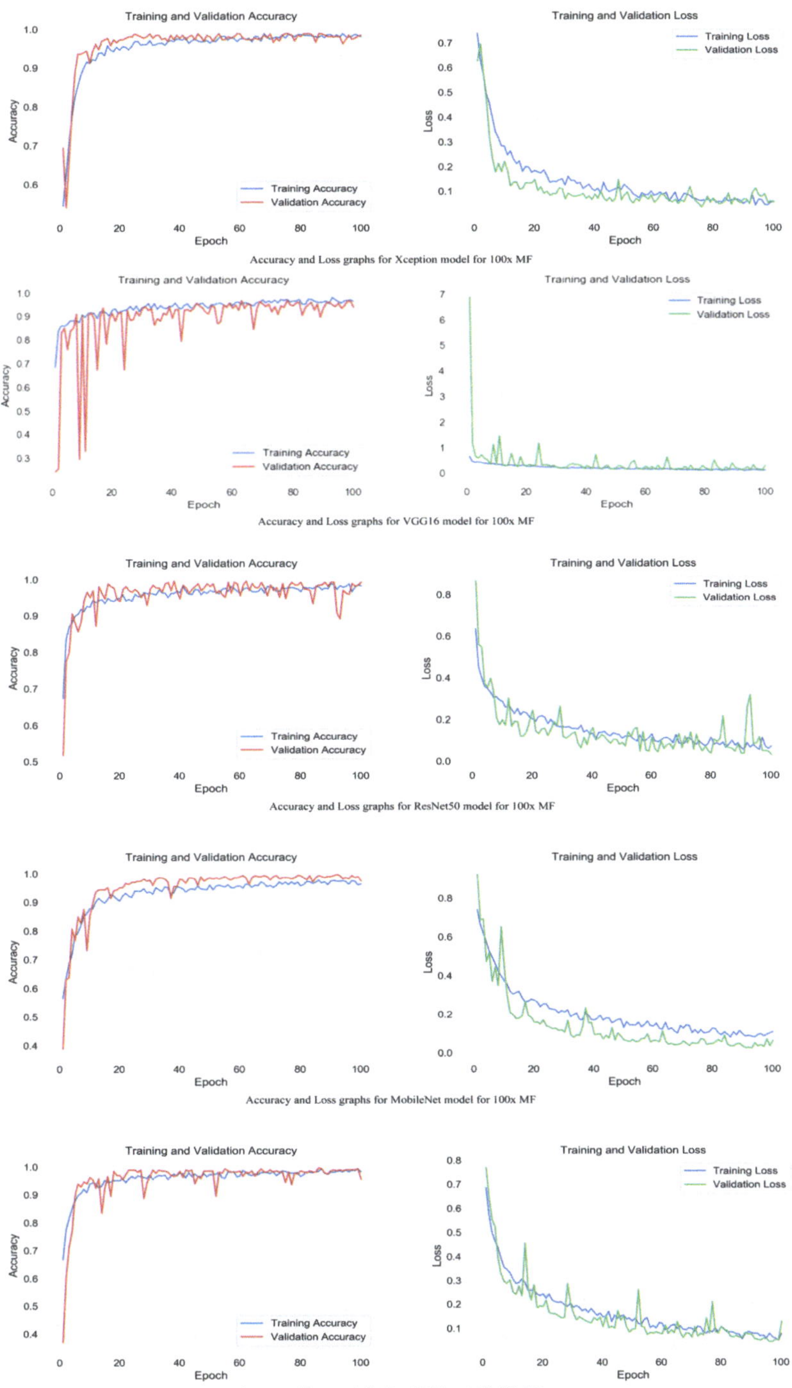

Figure 5. Accuracy and loss graphs for proposed models with $100\times$ magnification level.

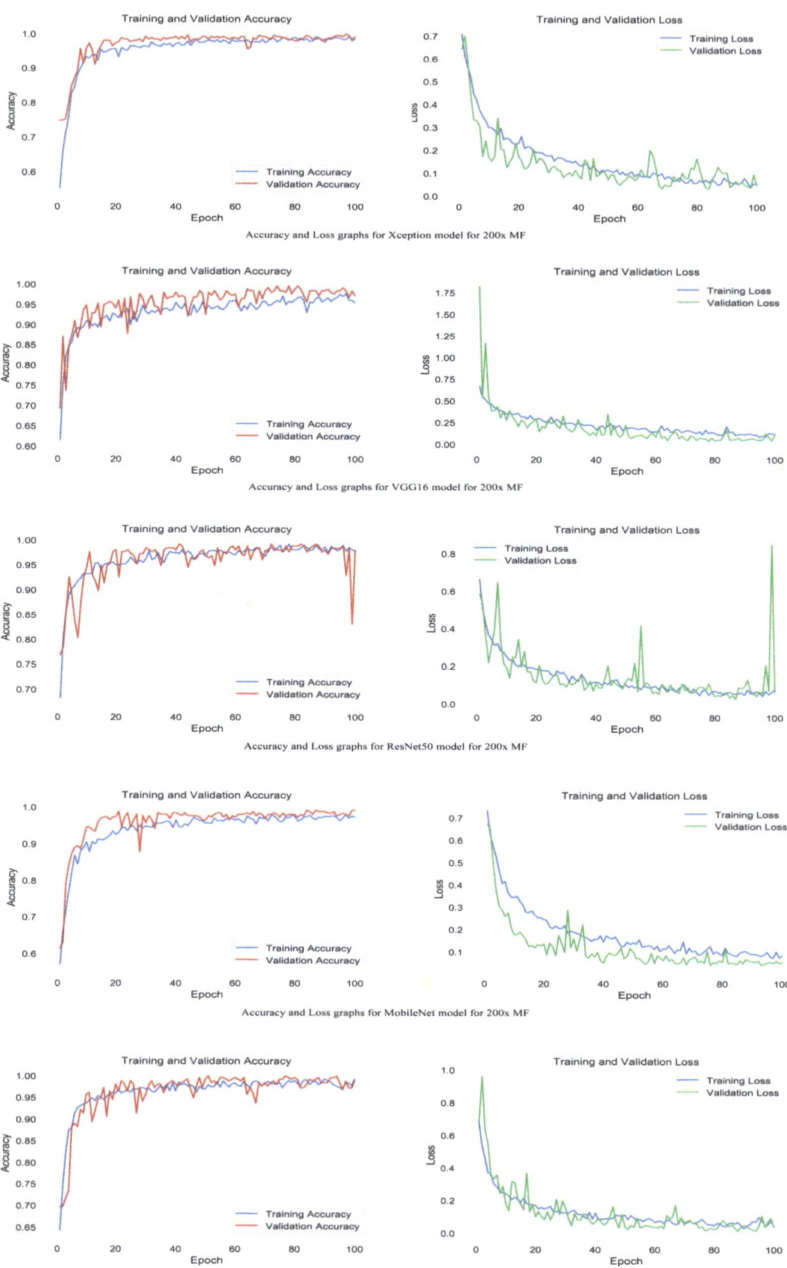

Figure 6. Accuracy and loss graphs for proposed models with $200\times$ magnification level.

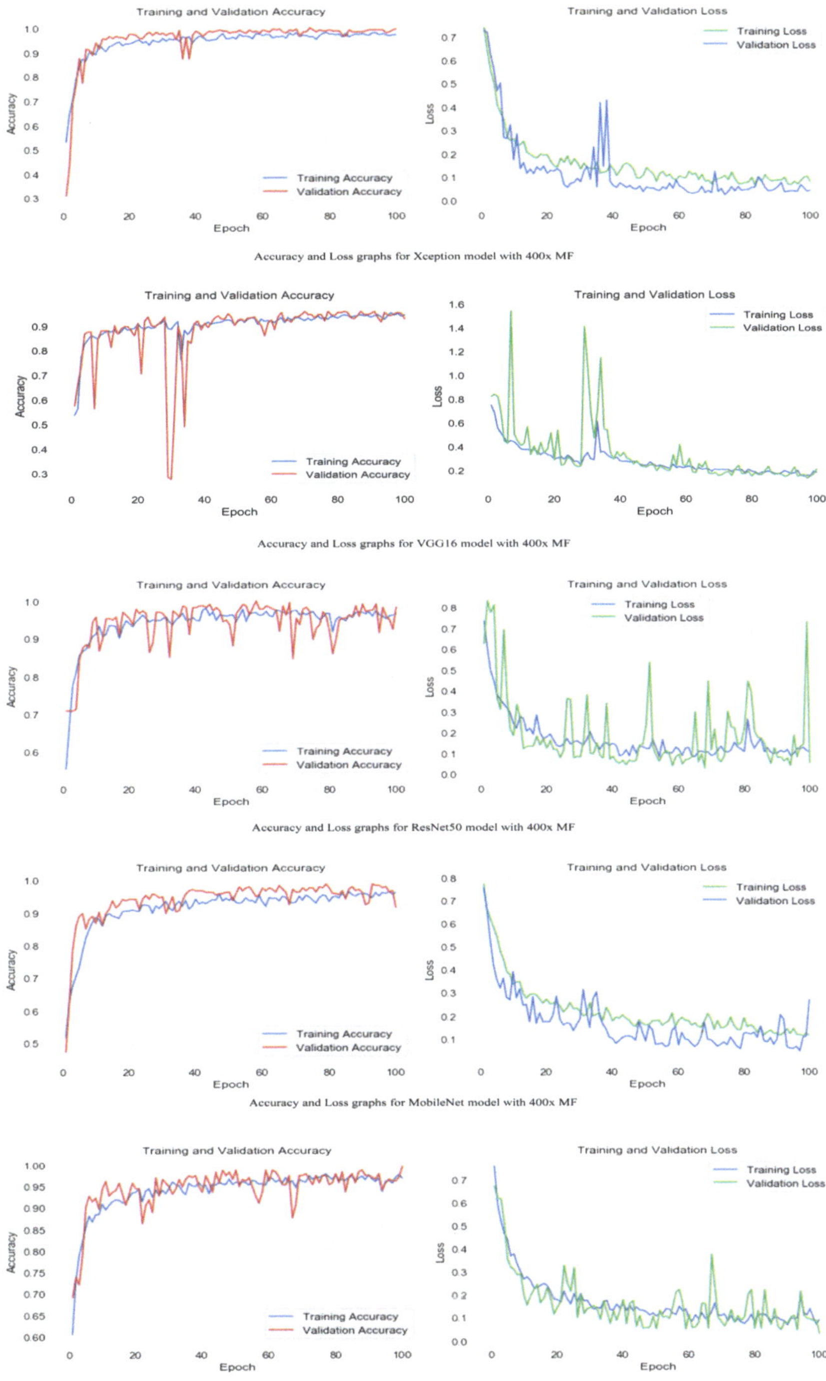

Accuracy and Loss graphs for Xception model with 400x MF

Accuracy and Loss graphs for VGG16 model with 400x MF

Accuracy and Loss graphs for ResNet50 model with 400x MF

Accuracy and Loss graphs for MobileNet model with 400x MF

Accuracy and Loss graphs for DenseNet121 model with 400x MF

Figure 7. Accuracy and loss graphs for proposed models with $400\times$ magnification level.

The precision performance results presented in this paragraph are based on our experiments with the proposed models and the BreakHis dataset. The macro average and weighted average metrics provide valuable insight into the precision performance of the models across all magnification levels, considering both equal and variable class distributions. In terms of precision, the efficacy of the proposed models at the $40\times$, $100\times$, $200\times$, and $400\times$ magnification levels are evaluated. Precision measures a model's ability to classify positive instances correctly. Due to providing a thorough evaluation, we utilize macro average metrics to assess various performance metrics. The macro average calculates the average value of these metrics across all magnification levels, assigning equal weight to each level. This approach provides an overall performance measure that treats all magnification levels equally, regardless of class distribution. It is useful for evaluating the models' general performance without considering class imbalances at different magnification levels. In contrast, the weighted average metrics consider the class distribution for each magnification level. They calculate the average value of the metrics while giving more weight to magnification levels with larger class sizes. This approach offers a more comprehensive evaluation of the models' performance, considering the varying class distributions across magnification levels. By giving appropriate weight to each magnification level based on its class distribution, the weighted average metrics provide a more representative assessment that avoids bias toward the most frequent class and reflects the true performance across different magnification levels.

Table 3 shows that Xception consistently obtained the highest precision performance across all magnification levels, with scores ranging from 86.1% to 90.4% based on the macro average. The VGG16 model's precision performance is marginally inferior, with scores ranging from 81.3% to 87.2%. Xception and ResNet50 maintained their superiority, obtaining precision scores from 86.3% to 88.2% and from 89.0% to 88.2%, respectively, when considering the weighted average. The precision ratings for VGG16 and MobileNet ranged from 85.4% to 87.2% and 86.1% to 88.2%, respectively. These results demonstrate that the precision performance of the proposed models varies with the magnification level. Xception and ResNet50 have consistently demonstrated greater precision than VGG16 and DenseNet121. The evaluation of precision using both macro average and weighted average metrics provides a comprehensive comprehension of the precision capabilities of the models for breast cancer classification tasks at varying magnification levels.

Table 3. Precision performances of the proposed models with attention mechanism across magnification levels.

Magnification Level\Model	40×		100×		200×		400×	
	Macro Average (%)	Weighted Average (%)	Macro Average (%)	Weighted Average (%)	Macro Average (%)	Weighted Average (%)	Macro Average (%)	Weighted Average (%)
Xception	89.2	89.1	90.4	90.3	86.3	87.5	86.3	88.2
VGG16	87.1	84.0	85.4	86.3	81.3	84.2	85.6	87.2
ResNet50	91.7	88.2	91.6	90.1	89.0	89.4	86.3	88.1
MobileNet	87.7	89.5	90.2	88.2	86.6	88.3	86.1	88.4
DenseNet121	86.5	87.4	85.4	86.6	88.2	88.5	87.3	88.1

The recall performance results are present based on experiments conducted on the BreakHis dataset using the proposed attention mechanism models. The macro average and weighted average metrics provide valuable insight into the overall recall performance of the models across all magnification levels. Table 4 displays the recall performance of the proposed models with the attention mechanism at various magnification levels. Recall, also known as sensitivity or true positive rate, quantifies the model's ability to accurately identify positive instances (for example, correctly classifying malignant cells in the BreakHis dataset). The overall recall performance of the models across all magnification levels is evaluated using macro average and weighted average metrics. The macro average computes the average recall value across all magnification levels, considering each level equally. The weighted average considers the class distribution at each magnification level, providing a more accurate evaluation by giving greater weight to magnification levels with larger class sizes. According to Table 4, Xception achieved the highest recall performance across all magnification levels, with macro average scores between 84.1% and 88.2%. The macro average recall performance of VGG16 ranged between 74.6% and 83.4%. ResNet50 exhibited excellent recall performance, with macro average scores ranging from 85.5 to 87.7 percent. The MobileNet model exhibits variable recall performance, with macro average scores ranging between 80.1% and 88.5%. The recall performance of DenseNet121 is relatively consistent, with macro average scores ranging from 84.4% to 86.9%. Based on the weighted average, Xception and ResNet50 maintain their superiority, with recall scores spanning from 87.3% to 90.4% and from 86.3% to 90.1%, respectively. The weighted average recall scores for the VGG16 and MobileNet models range between 81.2% and 86.1% and between 86.3% and 87.6%, respectively. The weighted average recall scores for DenseNet121 range between 86.1% and 88.6%. These results demonstrate that the recall performance of the proposed models with an attention mechanism varies across various magnification levels. Xception and ResNet50 demonstrated consistently superior recall performance, whereas VGG16 and DenseNet121 demonstrated relatively inferior performance. The evaluation of recall using both macro average and weighted average metrics provides a comprehensive insight into the ability of models.

Table 4. Recall performances of the proposed models with attention mechanism across magnification levels.

Model\Magnification Level	40×		100×		200×		400×	
	Macro Average (%)	Weighted Average (%)	Macro Average (%)	Weighted Average (%)	Macro Average (%)	Weighted Average (%)	Macro Average (%)	Weighted Average (%)
Xception	84.1	88.3	88.2	90.4	85.6	87.5	87.3	87.6
VGG16	74.6	82.5	83.4	86.1	83.0	83.3	81.2	84.5
ResNet50	85.5	89.4	87.7	90.1	86.3	89.1	87.2	87.1
MobileNet	88.4	88.2	80.1	86.3	86.2	88.2	88.5	87.6
DenseNet121	86.8	87.3	85.2	86.1	84.4	88.3	86.9	88.6

The F1-score performance results presented in this section are based on experiments performed on the BreakHis dataset using the proposed pre-trained models with an attention mechanism. The macro average and weighted average metrics provide valuable insight into the models' overall F1-score performance across magnification levels, considering both equal and variable class distributions. Table 5 illustrates the F1-score performances of the proposed pre-trained models with attention mechanisms at varying magnification levels. The F1 score is a metric that integrates precision and recall to evaluate a model's performance in binary classification tasks. Similar to the preceding sections, macro average and weighted average metrics are used to evaluate the overall F1-score performance of the models across all levels of magnification. The macro average computes the average F1-score across all magnification levels, considering each level equally. The weighted average considers the class distribution at each magnification level, providing a more accurate evaluation by giving greater weight to magnification levels with larger class sizes. According to the data presented in Table 5, Xception demonstrated strong F1-score performance at all magnification levels, with macro average scores ranging from 86.5% to 89.1%. ResNet50 also demonstrates impressive F1-score performance, with macro average scores spanning between 87.3% and 88.3%. MobileNet exhibited relatively consistent F1-score performance, with macro average scores ranging between 83.5% and 88.5%. The F1-score performance of DenseNet121 is stable, with macro average scores ranging from 85.3% to 87.6%. However, VGG16 demonstrated inferior F1-score performance to the other models, with average macro scores ranging between 77.9% and 84.6%.

Table 5. F1-score performances of the proposed models with attention mechanism across magnification levels.

Model\Magnification Level	40×		100×		200×		400×	
	Macro Average (%)	Weighted Average (%)	Macro Average (%)	Weighted Average (%)	Macro Average (%)	Weighted Average (%)	Macro Average (%)	Weighted Average (%)
Xception	86.5	88.3	89.1	90.3	85.7	87.6	87.2	88.1
VGG16	77.9	88.3	84.6	86.6	82.4	83.2	82.1	84.8
ResNet50	87.3	88.5	88.3	90.2	87.8	89.4	87.3	87.7
MobileNet	87.8	88.4	83.5	85.6	86.2	88.5	86.1	87.4
DenseNet121	86.7	87.3	85.5	86.1	85.3	87.4	87.6	88.5

Xception and ResNet50 maintain their superiority, with F1-score scores spanning from 87.6% to 90.3% and from 87.3% to 90.2%, respectively, when the weighted average is considered. MobileNet demonstrated weighted average F1-score scores ranging from 85 to 88 percent, whereas DenseNet121 demonstrated scores ranging from 86.1% to 88.5%. VGG16 demonstrated a weighted average F1 score, with scores ranging from 83.2% to 88.3%. Evaluating the F1-score using macro average and weighted average metrics comprehensively comprehends the models' overall performance in binary classification tasks at varying magnification levels. Xception and ResNet50 consistently exhibited superior F1-score performances, whereas VGG16 demonstrated relatively inferior performance. The F1-score evaluation allows us to assess the models' ability to reconcile precision and recall when identifying positive instances at different magnification levels.

We have presented the outcomes of our proposed models for breast cancer classification using histopathological images at four distinct levels of magnification: 40×, 100×, 200×, and 400×. To assess the performance of these models, we employed confusion matrices, which provide a comparison of the model's predictions to the actual disease labels. Analyzing the confusion matrices for the proposed pre-trained models with attention mechanisms in breast cancer classification provides valuable insight into their performance. The true positive (TP), true negative (TN), false positive (FP), and false negative (FN) rates provide a comprehensive comprehension of the models' ability to classify malignant and benign samples accurately. Figure 8 represents confusion matrices across all magnification levels for the proposed models. In this figure, Xception consistently demonstrates higher TP and TN rates than other models evaluated, indicating its superior performance in identifying malignant and benign samples reliably. ResNet50 demonstrates competitive TP and TN rates, indicating its efficacy in classification tasks. VGG16 and MobileNet, on the other hand, have relatively lesser TPR and TNR rates, indicating that their classification performance has room for improvement.

The disorientation matrices are organized into four columns, one for each level of magnification. Column "*(a)*" indicates a magnification level of 40×, column "*(b)*" indicates a magnification level of 100×, column "*(c)*" indicates a magnification level of 200×, and column "*(d)*" indicates a magnification level of 400×.

These outcomes are consistent with the previous results of our proposed system, in which Xception and ResNet50 outperformed VGG16 and MobileNet in terms of accuracy, precision, recall, and F1-score. Consequently, the analysis of the confusion matrices further validates the superiority of the Xception and ResNet50 models in classifying breast cancer samples accurately.

Figure 8. Confusion matrices for proposed models with four magnification levels. Column (**a**) 40×, Column (**b**) 100×, Column (**c**) 200×, Column (**d**) 400×.

5. Conclusions

This study investigates the efficacy of pre-trained models with attention mechanisms for breast cancer classification at various magnification levels. Xception, VGG16, ResNet50, MobileNet, and DenseNet121 are evaluated using accuracy, loss rates, precision, recall, and F1-score metrics among macro and weight averages. The obtained results indicate that the choice of magnification level substantially affects classification performance. Different magnification levels resulted in models with varying degrees of accuracy, loss, precision, recall, and F1-score. This is due to differences in image quality, resolution, and the presence of distinct histopathological characteristics at each magnification level. Therefore, it is essential to consider the appropriate magnification level when designing and implementing classification systems for breast cancer. Compared to other models, Xception consistently demonstrated superior performance across most evaluation metrics, including accuracy, precision, recall, and F1-score. ResNet50 and DenseNet121 also performs competitively, whereas VGG16 and MobileNet, produce relatively inferior results. These results highlight the significance of selecting a suitable pre-trained model for breast cancer classification tasks, considering the model's ability to extract pertinent features and recognize intricate patterns within the histopathological images. The incorporation of attention mechanisms into the models proved advantageous, as this improved the extraction of salient features and the overall classification performance. The attention mechanisms enabled the models to focus on significant regions and emphasize pertinent information, thereby contributing to more precise and robust predictions. This demonstrates the significance of attention mechanisms in deep learning-based medical image analysis and their potential to improve breast cancer detection and diagnosis. This study emphasizes the significance of incorporating magnification levels and attention mechanisms into breast cancer classification models. The findings contribute to the existing corpus of knowledge in the field of medical image analysis and offer valuable insights to breast cancer diagnosis researchers and practitioners. To further improve the performance of breast cancer classification models, additional research is encouraged to investigate other advanced techniques, evaluate larger data sets, and incorporate evaluation metrics.

Author Contributions: Conceptualization, A.A. and S.A.C.; Methodology, A.A., S.A.C., A.M. and M.S.A.M.A.-G.; Software, A.A.; Formal analysis, M.S.A.M.; Investigation, A.T.; Resources, A.M. and M.S.A.M.A.-G.; Data curation, A.A., A.T. and M.S.A.M.; Writing—original draft, A.A., S.A.C., A.T., M.S.A.M. and M.S.A.M.A.-G.; Writing—review & editing, A.A. and M.S.A.M.A.-G.; Visualization, A.M.; Supervision, A.M.; Funding acquisition, S.A.C. and M.S.A.M. All authors have read and agreed to the published version of the manuscript.

Funding: This work was supported in part by Princess Nourah bint Abdulrahman University, Riyadh, Saudi Arabia, through the Princess Nourah bint Abdulrahman University Researchers Supporting Project under Grant PNURSP2023R239; and in part by the Postdoctoral Fellowship granted by the Institute of Computer Technologies and Information Security, Southern Federal University, under Project P.D./22-01-KT.

Institutional Review Board Statement: Not applicable.

Informed Consent Statement: Not applicable.

Data Availability Statement: The data are contained within the article and/or available from the corresponding author upon reasonable request.

Acknowledgments: The authors express their gratitude to Princess Nourah bint Abdulrahman University Researchers Supporting Project Number (PNURSP2023R239), Princess Nourah bint Abdulrahman University, Riyadh, Saudi Arabia, and, in part, to the postdoc fellowship granted by the Institute of Computer Technologies and Information Security, Southern Federal University, Project No. P.D./22-01-KT.

Conflicts of Interest: The authors declare no conflict of interest.

References

1. Chou, L.B.; Johnson, B.; Shapiro, L.M.; Pun, S.; Cannada, L.K.; Chen, A.F.; Valone, L.C.; Van Nortwick, S.S.; Ladd, A.L.; Finlay, A.K. Increased prevalence of breast and all-cause cancer in female orthopaedic surgeons. *JAAOS Glob. Res. Rev.* **2022**, *6*, e22.00031. [CrossRef] [PubMed]
2. Siegel, R.L.; Miller, K.D.; Jemal, A. R Cancer statistics. *CA Cancer J. Clin.* **2020**, *70*, 7–30. [CrossRef] [PubMed]
3. Liu, L.; Feng, W.; Chen, C.; Liu, M.; Qu, Y.; Yang, J. Classification of breast cancer histology images using MSMV-PFENet. *Sci. Rep.* **2022**, *12*, 17447. [CrossRef] [PubMed]
4. Huang, J.; Mei, L.; Long, M.; Liu, Y.; Sun, W.; Li, X.; Shen, H.; Zhou, F.; Ruan, X.; Wang, D.; et al. Bm-net: Cnn-based mobilenet-v3 and bilinear structure for breast cancer detection in whole slide images. *Bioengineering* **2022**, *9*, 261. [CrossRef] [PubMed]
5. Samee, N.A.; Alhussan, A.A.; Ghoneim, V.F.; Atteia, G.; Alkanhel, R.; Al-Antari, M.A.; Kadah, Y.M. A Hybrid Deep Transfer Learning of CNN-Based LR-PCA for Breast Lesion Diagnosis via Medical Breast Mammograms. *Sensors* **2022**, *22*, 4938. [CrossRef]
6. Spanhol, F.A.; Oliveira, L.S.; Petitjean, C.; Heutte, L. A dataset for breast cancer histopathological image classification. *IEEE Trans. Biomed. Eng.* **2015**, *63*, 1455–1462. [CrossRef]

7. Houssein, E.H.; Emam, M.M.; Ali, A.A.; Suganthan, P.N. Deep and machine learning techniques for medical imaging-based breast cancer: A comprehensive review. *Expert Syst. Appl.* **2021**, *167*, 114161. [CrossRef]

8. Debelee, T.G.; Schwenker, F.; Ibenthal, A.; Yohannes, D. Survey of deep learning in breast cancer image analysis. *Evol. Syst.* **2020**, *11*, 143–163. [CrossRef]

9. Yassin, N.I.; Omran, S.; El Houby, E.M.; Allam, H. Machine learning techniques for breast cancer computer aided diagnosis using different image modalities: A systematic review. *Comput. Methods Programs Biomed.* **2018**, *156*, 25–45. [CrossRef]

10. Fatima, N.; Liu, L.; Hong, S.; Ahmed, H. Prediction of breast cancer, comparative review of machine learning techniques, and their analysis. *IEEE Access* **2020**, *8*, 150360–150376. [CrossRef]

11. Asri, H.; Mousannif, H.; Al Moatassime, H.; Noel, T. Using machine learning algorithms for breast cancer risk prediction and diagnosis. *Procedia Comput. Sci.* **2016**, *83*, 1064–1069. [CrossRef]

12. Liu, M.; Hu, L.; Tang, Y.; Wang, C.; He, Y.; Zeng, C.; Lin, K.; He, Z.; Huo, W. A deep learning method for breast cancer classification in the pathology images. *IEEE J. Biomed. Health Inform.* **2022**, *26*, 5025–5032. [CrossRef] [PubMed]

13. Ramesh, S.; Sasikala, S.; Gomathi, S.; Geetha, V.; Anbumani, V. Segmentation and classification of breast cancer using novel deep learning architecture. *Neural Comput. Appl.* **2022**, *34*, 16533–16545. [CrossRef]

14. Sharma, S.; Kumar, S. The Xception model: A potential feature extractor in breast cancer histology images classification. *ICT Express* **2022**, *8*, 101–108. [CrossRef]

15. Liew, X.Y.; Hameed, N.; Clos, J. An investigation of XGBoost-based algorithm for breast cancer classification. *Mach. Learn. Appl.* **2021**, *6*, 100154. [CrossRef]

16. Atban, F.; Ekinci, E.; Garip, Z. Traditional machine learning algorithms for breast cancer image classification with optimized deep features. *Biomed. Signal Process. Control* **2023**, *81*, 104534. [CrossRef]

17. Ayana, G.; Park, J.; Jeong, J.W.; Choe, S.W. A novel multistage transfer learning for ultrasound breast cancer image classification. *Diagnostics* **2022**, *12*, 135. [CrossRef]

18. Wang, J.; Zhu, T.; Liang, S.; Karthiga, R.; Narasimhan, K.; Elamaran, V. Binary and multiclass classification of histopathological images using machine learning techniques. *J. Med. Imaging Health Inform.* **2020**, *10*, 2252–2258. [CrossRef]

19. Dubey, A.; Singh, S.K.; Jiang, X. Leveraging CNN and Transfer Learning for Classification of Histopathology Images. In Proceedings of the International Conference on Machine Learning, Image Processing, Network Security and Data Sciences, Bhopal, India, 21–22 December 2022; Springer: Cham, Switzerland, 2022; pp. 3–13.

20. Singh, S.; Kumar, R. Breast cancer detection from histopathology images with deep inception and residual blocks. *Multimed. Tools Appl.* **2022**, *81*, 5849–5865. [CrossRef]

21. Venugopal, A.; Sreelekshmi, V.; Nair, J.J. Ensemble Deep Learning Model for Breast Histopathology Image Classification. In *ICT Infrastructure and Computing: Proceedings of ICT4SD 2022*; Springer: Singapore, 2022; pp. 499–509.

22. Kumari, V.; Ghosh, R. A magnification-independent method for breast cancer classification using transfer learning. *Healthc. Anal.* **2023**, *3*, 100207. [CrossRef]

23. Joshi, S.A.; Bongale, A.M.; Olsson, P.O.; Urolagin, S.; Dharrao, D.; Bongale, A. Enhanced Pre-Trained Xception Model Transfer Learned for Breast Cancer Detection. *Computation* **2023**, *11*, 59. [CrossRef]

24. Karthik, R.; Menaka, R.; Siddharth, M. Classification of breast cancer from histopathology images using an ensemble of deep multiscale networks. *Biocybern. Biomed. Eng.* **2022**, *42*, 963–976. [CrossRef]

25. Zou, Y.; Zhang, J.; Huang, S.; Liu, B. Breast cancer histopathological image classification using attention high-order deep network. *Int. J. Imaging Syst. Technol.* **2022**, *32*, 266–279. [CrossRef]

26. Jadah, Z.; Alfitouri, A.; Chantar, H.; Amarif, M.; Aeshah, A.A. Breast Cancer Image Classification Using Deep Convolutional Neural Networks. In Proceedings of the 2022 International Conference on Engineering & MIS (ICEMIS), Istanbul, Turkey, 4–6 July 2022; IEEE: Piscataway, NJ, USA, 2022; pp. 1–6.

27. Wang, X.; Chen, G.; Qian, G.; Gao, P.; Wei, X.Y.; Wang, Y.; Tian, Y.; Gao, W. Large-scale multi-modal pre-trained models: A comprehensive survey. In *Machine Intelligence Research*; Springer: Berlin/Heidelberg, Germany, 2023; pp. 1–36.

28. Zhang, Z. Improved adam optimizer for deep neural networks. In Proceedings of the 2018 IEEE/ACM 26th International Symposium on Quality of Service (IWQoS), Banff, AB, Canada, 4–6 June 2018; IEEE: Piscataway, NJ, USA, 2018; pp. 1–2.

29. Woo, S.; Park, J.; Lee, J.Y.; Kweon, I.S. Cbam: Convolutional block attention module. In Proceedings of the European Conference on Computer Vision (ECCV), Munich, Germany, 8–14 September 2018; pp. 3–19.

30. Niu, Z.; Zhong, G.; Yu, H. A review on the attention mechanism of deep learning. *Neurocomputing* **2021**, *452*, 48–62. [CrossRef]

31. Gao, F.; Wu, T.; Li, J.; Zheng, B.; Ruan, L.; Shang, D.; Patel, B. SD-CNN: A shallow-deep CNN for improved breast cancer diagnosis. *Comput. Med. Imaging Graph.* **2018**, *70*, 53–62. [CrossRef] [PubMed]

32. Zhu, Z.; Wang, S.H.; Zhang, Y.D. A survey of convolutional neural network in breast cancer. *Comput. Model. Eng. Sci. CMES* **2023**, *136*, 2127. [CrossRef] [PubMed]

33. Zhao, Y.; Zhang, J.; Hu, D.; Qu, H.; Tian, Y.; Cui, X. Application of Deep Learning in Histopathology Images of Breast Cancer: A Review. *Micromachines* **2022**, *13*, 2197. [CrossRef]

Systematic Review

Adverse Events of PD-1 or PD-L1 Inhibitors in Triple-Negative Breast Cancer: A Systematic Review and Meta-Analysis

Yixi Zhang [1], Jingyuan Wang [1], Taobo Hu [2], Huina Wang [1], Mengping Long [3] and Baosheng Liang [1,*]

1 Department of Biostatistics, School of Public Health, Peking University, Beijing 100191, China
2 Department of Breast Surgery, Peking University People's Hospital, Beijing 100044, China
3 Department of Pathology, Peking University Cancer Hospital, Beijing 100083, China
* Correspondence: liangbs@hsc.pku.edu.cn

Abstract: (1) Background: This study aimed to develop a comprehensive understanding of the treatment-related adverse events when using PD-1 or PD-L1 inhibitors in triple-negative breast cancer (TNBC). (2) Methods: We conducted a meta-analysis of Phase II/III randomized clinical trials. Studies were searched for using PubMed, Embase, and Cochrane Library from 1 March 1980 till 30 June 2022. Data on adverse events were mainly extracted from ClinicalTrials.gov and published articles. A generalized linear mixed model with the logit transformation was employed to obtain the overall incidence of adverse events across all studies. For serious adverse events with low incidences, the Peto method was used to calculate the odds ratio (OR) and 95% confidence interval (95%CI) in the PD-1 or PD-L1 inhibitors groups compared to the control groups. (3) Results: Nine studies were included in the meta-analysis, including a total of 2941 TNBC patients treated with PD-1 or PD-L1 inhibitors (including atezolizumab, pembrolizumab and durvalumab) and 2339 patients in the control groups. Chemotherapy alone was the control group in all studies. The average incidences of all serious immune-related adverse events of interest (hypothyroidism, hyperthyroidism, pneumonitis, pruritus, rash) were less than 1%, except for adrenal insufficiency (1.70%, 95%CI: 0.50–5.61%) in the PD-1 or PD-L1 groups. PD-1 or PD-L1 inhibitors significantly increased the risk of serious pneumonitis (OR = 2.52, 95%CI: 1.02–6.26), hypothyroidism (OR = 5.92, 95%CI: 1.22–28.86), alanine aminotransferase (ALT) elevation (OR = 1.66, 95%CI: 1.12–2.45), and adrenal insufficiency (OR = 18.81, 95%CI: 3.42–103.40). For non-serious adverse events, the patients treated with PD-1 or PD-L1 inhibitors had higher risk of aspartate aminotransferase (AST) elevation (OR =1.26, 95%CI: 1.02–1.57), hypothyroidism (OR = 3.63, 95%CI: 2.92–4.51), pruritus (OR = 1.84, 95%CI: 1.30–2.59), rash (OR = 1.29, 95%CI: 1.08–1.55), and fever (OR = 1.77, 95%CI: 1.13–2.77), compared with chemotherapy alone. (4) Conclusions: The incidence of serious immune-related adverse events in PD-1 or PD-L1 inhibitors groups is low but significantly higher than in chemotherapy groups. When using PD-1 or PD-L1 inhibitors for the treatment of TNBC, serious pneumonitis, hypothyroidism, ALT elevation, and adrenal insufficiency should be considered. Non-serious adverse events, such as AST elevation, rash, and fever, should also be taken into consideration.

Keywords: triple-negative breast cancer; PD-1 inhibitors; PD-L1 inhibitors; adverse events; safety

check for updates

Citation: Zhang, Y.; Wang, J.; Hu, T.; Wang, H.; Long, M.; Liang, B. Adverse Events of PD-1 or PD-L1 Inhibitors in Triple-Negative Breast Cancer: A Systematic Review and Meta-Analysis. *Life* **2022**, *12*, 1990. https://doi.org/10.3390/life12121990

Academic Editor: Shinichiro Kashiwagi

Received: 24 October 2022
Accepted: 24 November 2022
Published: 28 November 2022

Publisher's Note: MDPI stays neutral with regard to jurisdictional claims in published maps and institutional affiliations.

1. Introduction

Breast cancer is a noteworthy public health problem, with a rising global burden in many countries [1,2]. Breast cancer is the most common cancer in women, and it ranks first as cause of death [2]. Triple-negative breast cancer (TNBC) is a special type of breast cancer, which refers to the immunohistochemical examination of breast cancer cells showing that estrogen receptor (ER), progesterone receptor (PR), and human epidermal growth factor receptor 2 (HER2) all lack expression [3]. TNBC accounts for 10–20% of all breast cancer patients [4]. However, the prognosis of TNBC is worse than other types of breast cancer.

The mortality of TNBC is over 40% within the first five years and most patients develop distant metastasis [5].

Neoadjuvant chemotherapy is a primary pharmacotherapy for TNBC. Because of its negative expression of ER, PR, and HER2, TNBC is not sensitive to endocrine therapy or targeted therapy [6]. The national comprehensive cancer network guidelines recommend using combination regimens based on taxane, anthracycline, cyclophosphamide, cisplatin, and fluorouracil [7]. However, various combinations of chemotherapy drugs may lead to different outcomes and prognoses for TNBC patients. According to available clinical trial results, basal-like 1 subtype TNBC has more sensitivity to chemotherapy than other subtypes, with the highest pathologic complete response (pCR) rate of 52% [8,9]. TNBC has high heterogeneity and lacks useful targets, making it difficult to discover new targets. Despite massive chemotherapy combinations being optional, drug resistance happens inevitably in some patients [5]. Exploring and developing an effective and safe treatment for TNBC is vital. Programmed cell death protein 1 (PD-1) or programmed cell death ligand 1(PD-L1) inhibitors can block the binding of PD-1 receptor protein on the surface of tumor cells with PD-1 receptor in T cells, thereby causing T cells to kill tumor cells [10,11]. The analysis of immunohistochemistry in TNBC patients has discovered that half of the TNBC patients have high expression of PD-1 or PD-L1, which implies that PD-1 or PD-L1 could be a potential target [10,12], while other studies have considered de-glycosylated PD-L1 in TNBC cells as a biomarker [13].

In March 2019, the U.S. Food and Drug Administration (FDA) approved anti-PD-L1 therapy atezolizumab combined with chemotherapy for first-line treatment of patients with PD-L1 positive advanced or metastatic TNBC patients based on the result of IMpassion 130 [14]. This approval made the atezolizumab and abraxane combination the first cancer immunotherapy scheme for the treatment of PD-L1-positive metastatic TNBC. However, in July 2021, Roche withdrew its application to extend the use of atezolizumab to the treatment of TNBC patients in Europe because of the post-market study Impassion 131 [15]. KEYNOTE-012 was the first published result investigating the safety and efficiency of PD-1 inhibitor pembrolizumab in TNBC patients. Pembrolizumab given every 2 weeks to TNBC patients achieved an overall response rate of 18.5% and had an acceptable safety profile [16]. Further studies discovered that platinum-based chemotherapy could make the tumor cell more sensitive to PD-1 or PD-L1 inhibitors by exerting immunomodulation properties [17].

There are several meta-analyses that have evaluated the efficacy and safety of neoadjuvant immune checkpoint inhibitors [18–20]. Previously, studies have shown that PD-1 or PD-L1 inhibitors are related to high incidences of various treatment-related adverse events, such as fatigue, pruritus and hypothyroidism [21,22]. In a recent meta-analysis that evaluated the effectiveness of PD-1 and PD-L1 inhibitors combined with chemotherapy for TNBC, researchers found that the combination strategy improved the pCR rate and progression-free survival (PFS) [20], but the combination treatment increased the risk of several adverse events. There are limitations in the previous meta-analysis. Their safety analysis was only confined to three clinical trials (Impassion 130 [23], Impassion 131 [24], and Keynote-355 [25]) and failed to stratify immune checkpoint inhibitor regimens. Thus, it is essential to have comprehensive understanding of treatment-related adverse events using PD-1 or PD-L1 inhibitors in TNBC patients.

In this study, we conducted a systematic review and meta-analysis to thoroughly evaluate the adverse events and the safety of PD-1 or PD-L1 inhibitors in TNBC patients based on extensive randomized clinical trials.

2. Materials and Methods

2.1. Search Strategy

We systematically searched three databases (PubMed, Embase and Cochrane Library) regarding PD-1 or PD-L1 inhibitors in triple-negative breast cancer from 1 March 1980 to 30 June 2022, independently by two authors (Y.Z. and J.W.). The keywords used in the search strategy were "PD-1", "PD-L1", "nivolumab", "pembrolizumab", "durvalumab", "atezolizumab", "avelumab", "triple-negative breast cancer" and "TNBC". We reviewed all the abstracts of the resulting studies and full texts were retrieved.

2.2. Study Selection

Three authors (Y.Z., J.W., and H.W.) independently conducted the literature selection. Inconsistencies were resolved by consensus. We used the following criteria for study selection: (1) Patients: triple-negative breast cancer patients, (2) Intervention: using PD-1 or PD-L1 as treatment, including but not limited to monotherapy, (3) Control: we did not make any limitations to the control group, (4) Outcome: the data on adverse events should be reported in the article or website (ClinicalTrials.gov), and (5) published in English. We had the following exclusion criteria: (1) other study designs such as case reports, case series, case–control studies, cohort studies, and so on, (2) protocols and secondary research, such as systematic reviews, pooled analysis, and study protocols, (3) studies that did not focus on PD-1 or PD-L1 inhibitors, (4) animal studies, and (5) duplicates.

2.3. Outcome and Data Extraction

We paid close attention to different reported adverse events among TNBC patients treated with PD-1 or PD-L1 inhibitors. The adverse events included anemia, neutropenia, arthralgia, back pain, alanine aminotransferase (ALT) elevation, aspartate aminotransferase (AST) elevation, hypothyroid, pneumonitis, colitis, fever, headache, pruritus, rash, and so on. We first searched ClinicalTrials.gov for the submitted results. For those not available on the website, we extracted data from the published articles. From the data from ClinicalTrials.gov, we extracted both serious and non-serious adverse events. For data from the article, we classified Grade 3 or higher as serious adverse events and Grade 1–2 as non-serious adverse events. Besides the adverse events, we also extracted the author, published year, drug information, and trial name.

2.4. Quality Assessment and Data Analysis

We used Cochrane Bias Risk Evaluation Tool to assess six dimensions of bias: randomization process, deviations from intended interventions, missing outcome data, measurement of the outcome, and selection of the reported result. The bias was determined as high risk, low risk and uncertain. After bias assessment, we first conducted a proportion meta-analysis to calculate the overall incidence of serious and non-serious adverse events using the generalized linear mixed model with the logit transformation [26]. For serious adverse events with low incidence, we employed the Peto method to calculate the overall odds ratio (OR) and 95% confidence interval (95%CI) between the treatment group and the control group. We examined the heterogeneity between studies through the Q test and I^2 statistics. The random effect model was used when there was high heterogeneity ($I^2 > 50\%$). All data analysis was conducted by R (version 4.1.3).

3. Results

3.1. Features of Studies

We searched 648 studies in total, and 9 studies meeting the selection criteria were incorporated into the research. Figure 1 shows the study selection diagram. Among all the included studies, only one used PD-1 inhibitors [27], three studies used PD-1 inhibitors combined with chemotherapy [25,28,29], and five studies used PD-L1 inhibitors combined with chemotherapy [23,24,30–32]. Although we did not limit the control group in the literature search, we found that all the studies used chemotherapy as their control group. As a result, a total of 2941 patients were included in the treatment group, consisting of 1055 patients in the PD-L1 inhibitor atezolizumab group, 1721 in the PD-1 inhibitor pembrolizumab group, and 165 in the PD-1 inhibitor durvalumab group; 2339 patients were in the control group for this meta-analysis. Eight studies were registered on ClinicalTrials.gov and had published their data on adverse events. The data of the remaining study were retrieved from the literature [31]. Table 1 presents information on the included studies.

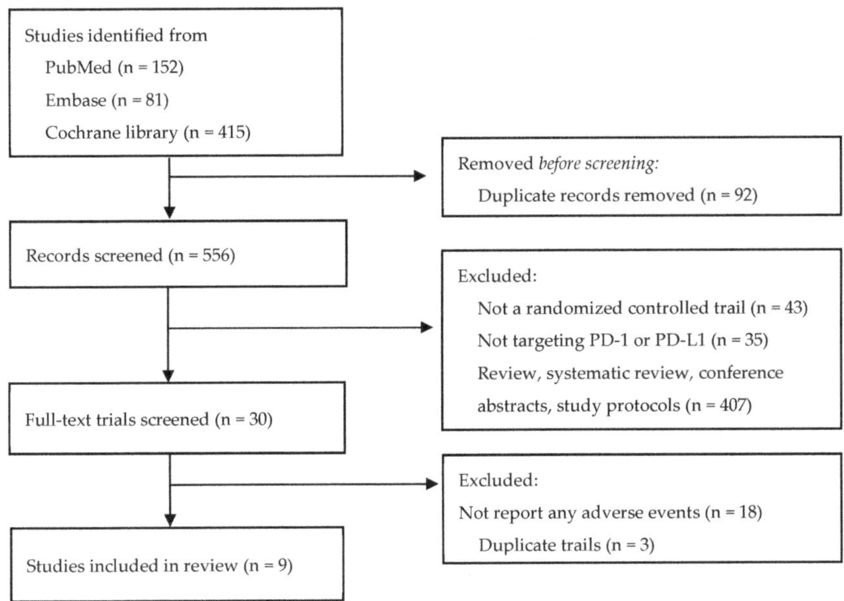

Figure 1. Flow diagram for study selection in the meta-analysis.

Table 1. Characteristics of all the included studies.

Year	Title	Authors	NCT Number	Study	Treatment
2021	First-line atezolizumab plus nab-paclitaxel for unresectable, locally advanced, or metastatic triple-negative breast cancer: IMpassion130 final overall survival analysis	Emens, L.A., et al.	NCT02425891	IMpassion130	Atezolizumab+ chemotherapy
2021	Pembrolizumab versus investigator-choice chemotherapy for metastatic triple-negative breast cancer (KEYNOTE-119): a randomized, open-label, phase 3 trial	Winer, E.P., et al.	NCT02555657	KEYNOTE-119	Pembrolizumab
2019	A randomized phase II study investigating durvalumab in addition to an anthracycline taxane-based neoadjuvant therapy in early triple-negative breast cancer: clinical results and biomarker analysis of GeparNuevo study	Loibl, S., et al.	NCT02685059	GeparNuevo	Durvalumab+ chemotherapy
2020	Pembrolizumab plus chemotherapy versus placebo plus chemotherapy for previously untreated locally recurrent inoperable or metastatic triple-negative breast cancer (KEYNOTE-355): a randomized, placebo-controlled, double-blind, phase 3 clinical trial	Cescon, D., et al.	NCT02819518	KEYNOTE-355	Pembrolizumab+ chemotherapy

Table 1. *Cont.*

Year	Title	Authors	NCT Number	Study	Treatment
2020	Pembrolizumab for early triple-negative breast cancer	Schmid, P., et al.	NCT03036488	KEYNOTE522	Pembrolizumab+ chemotherapy
2021	Primary results from IMpassion131, a double-blind, placebo-controlled, randomized phase III trial of first-line paclitaxel with or without atezolizumab for unresectable locally advanced/metastatic triple-negative breast cancer	Miles, D., et al.	NCT03125902	IMpassion131	Atezolizumab+ chemotherapy
2020	Neoadjuvant atezolizumab in combination with sequential nab-paclitaxel and anthracycline-based chemotherapy versus placebo and chemotherapy in patients with early-stage triple-negative breast cancer (IMpassion031): a randomized, double-blind, phase 3 trial	Mittendorf, E, et al.	NCT03197935	IMpassion031	Atezolizumab+ chemotherapy
2021	Durvalumab with olaparib and paclitaxel for high-risk HER2-negative stage II/III breast cancer: Results from the adaptively randomized I-SPY2 trial	Pusztai, L., et al.	-	-	Durvalumab+ chemotherapy
2020	Effect of pembrolizumab plus neoadjuvant chemotherapy on pathologic complete response in women with early-stage breast cancer an analysis of the ongoing phase 2 adaptively randomized I-SPY2 trial	Nanda, R., et al.	NCT01042379	-	Pembrolizumab+ chemotherapy

3.2. Risk of Bias Assessment

The risk of bias assessment is summarized in Table 2. We found there was a low bias of selection and outcome. For randomization and deviations, two studies, Pusztai 2021 and Nanda 2020, were considered as high risk. Because the primary aims of the included studies were not related to adverse events, collection of information on adverse events was mainly from online. Thus, we deemed Pusztai et al.'s study [31] at high risk of bias with regard to measurement of the outcome, since its data were from the article.

3.3. Meta-Analysis Results

3.3.1. Meta-Analysis Results with Serious Adverse Events

The overall incidences of serious adverse events are shown in Figure 2a. In particular, the most common serious adverse events were neutropenia (3.15%, 95% CI: 0.66–13.72%), fatigue (2.50%, 95% CI: 1.22–5.05%), and anemia (2.16%, 95% CI: 0.70–6.45%), followed by adrenal insufficiency (1.70%, 95% CI: 0.50–5.61%) and alanine aminotransferase (ALT) elevation (1.47%, 95% CI: 0.60–3.60%). The incidences of serious immune-related adverse events were lower than 1%, including pneumonitis (0.76%, 95% CI: 0.42–1.38%), hypothyroidism (0.38%, 95% CI: 0.18–0.79%), and hyperthyroidism (0.21%, 95% CI: 0.08–0.57%).

Table 2. Risk of bias assessment of included studies.

Study	Randomization Process	Deviations from Intended Interventions	Missing Outcome Data	Measurement of the Outcome	Selection of the Reported Result
Emens, L.A., 2021 [23]	low	low	low	low	low
Winer, E.P., 2021 [27]	low	low	low	low	low
Loibl, S., 2019 [32]	unclear	low	low	low	low
Cescon, D., 2020 [25]	low	low	low	low	low
Schmid, P., 2020 [29]	low	low	low	low	low
Miles, D., 2021 [24]	low	low	low	low	low
Mittendorf, E., 2020 [30]	low	low	low	low	low
Pusztai, L., 2021 [31]	unclear	high	unclear	high	low
Nanda, R., 2020 [28]	unclear	high	unclear	low	low

(a)

Adverse Event	Incidence(%) (95%CI)
Neutropenia	3.15 (0.66-13.72)
Fatigue	2.50 (1.22-5.05)
Anemia	2.16 (0.70-6.45)
Adrenal insufficiency	1.70 (0.50-5.61)
ALT elevation	1.47 (0.60-3.60)
Diarrhoea	1.31 (0.57-2.98)
Nausea	1.16 (0.53-2.50)
Fever	1.11 (0.68-1.81)
Dyspnea	0.84 (0.47-1.46)
Pneumonitis	0.76 (0.42-1.38)
AST elevation	0.65 (0.24-1.80)
Rash	0.64 (0.32-1.28)
Arthralgia	0.62 (0.18-2.20)
Voimting	0.57 (0.21-1.51)
Colitis	0.40 (0.20-0.80)
Hypothyroid	0.38 (0.18-0.79)
Headache	0.29 (0.11-0.78)
Constipation	0.24 (0.08-0.74)
Hyperthyroid	0.21 (0.08-0.57)
Abdominal Pain	0.20 (0.05-0.79)
Back pain	0.15 (0.04-0.58)

(b)

Adverse Event	Incidence(%) (95%CI)
Nausea	49.70 (35.90-63.55)
Fatigue	49.10 (30.24-68.21)
Anemia	34.74 (19.91-53.28)
Diarrhoea	32.77 (21.25-46.81)
Headache	25.14 (16.72-35.97)
Arthralgia	24.88 (13.87-40.52)
Constipation	23.20 (19.27-27.65)
Voimting	21.43 (14.39-30.67)
Fever	18.24 (13.49-24.18)
ALT elevation	18.05 (11.73-26.74)
AST elevation	17.98 (11.26-27.45)
Rash	17.64 (11.83-25.48)
Dyspnea	16.59 (11.75-22.92)
Pruritus	15.70 (11.26-21.47)
Neutropenia	14.15 (7.02-26.47)
Back pain	12.17 (8.12-17.85)
Hypothyroid	11.77 (9.83-14.04)
Abdominal Pain	10.65 (6.43-17.11)
Hyperthyroid	4.67 (3.79-5.73)
Pneumonitis	2.65 (1.18-5.88)
Colitis	1.27 (0.64-2.51)
Adrenal insufficiency	1.08 (0.58-2.00)

Figure 2. Overall incidences of adverse events in TNBC patients using PD-1/PD-L1 inhibitors: (**a**) incidences of serious adverse events; (**b**) incidences of non-serious adverse events.

Seven studies reported serious pneumonitis and integrated data showed that PD-1 or PD-L1 inhibitors combined with chemotherapy increased the risk of pneumonitis (OR = 2.52, 95% CI: 1.02–6.26), as is shown in Figure 3. There was no heterogeneity ($I^2 = 2\%$) in the overall meta-analysis. When we separated the PD-1 inhibitors and PD-L1 inhibitors, the results in subgroups were not significant. Four studies reported serious hypothyroidism, from which we found that the PD-1 or PD-L1 inhibitors had higher risk than chemotherapy (OR = 5.92, 95% CI: 1.22–28.86), especially in the subgroup with PD-1 combined with chemotherapy (see Figure S1). Figure 4 illustrates that the ALT elevation had a higher incidence rate in the group using PD-1 or PD-L1 inhibitors than in the chemotherapy group (OR = 1.66, 95% CI: 1.12–2.45). Subgroup analysis indicated that the PD-1 inhibitors group had a higher risk of ALT elevation (OR = 1.63, 95% CI: 1.06–2.52). Figure S2 shows that adrenal insufficiency also showed a significantly higher risk in PD-1 or PD-L1 inhibitors group compared to the chemotherapy group (OR = 18.81, 95% CI: 3.42–103.40).

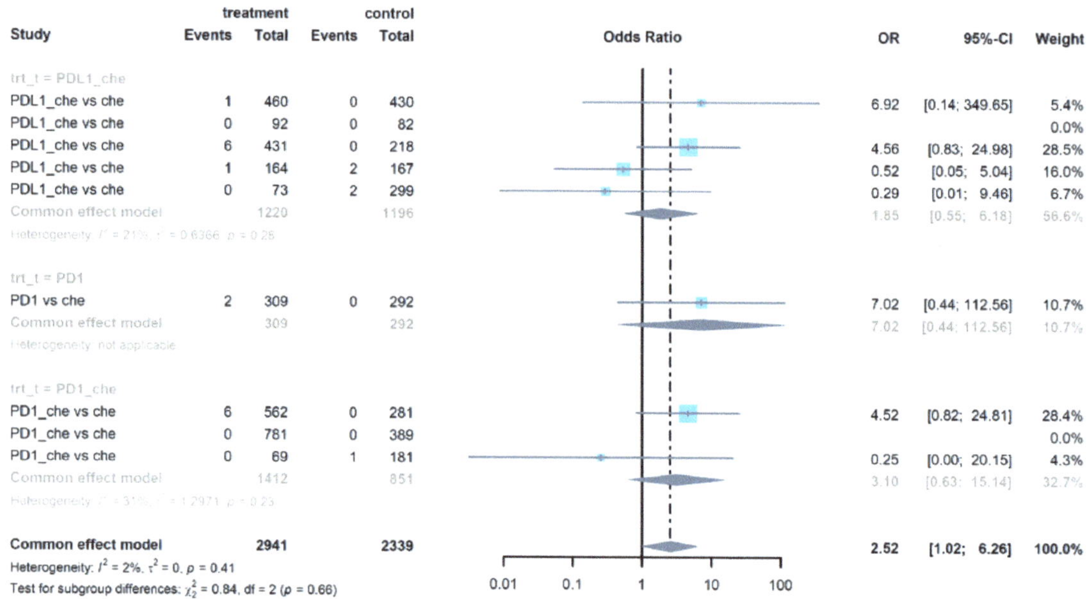

Figure 3. Forest plot of serious pneumonitis in patients treated with PD-1 or PD-L1 inhibitors versus chemotherapy (NOTE: 'PDL1_che' means patients treated with PD_L1 and chemotherapy combined therapy, 'PD1_che' means patients treated with PD1 and chemotherapy).

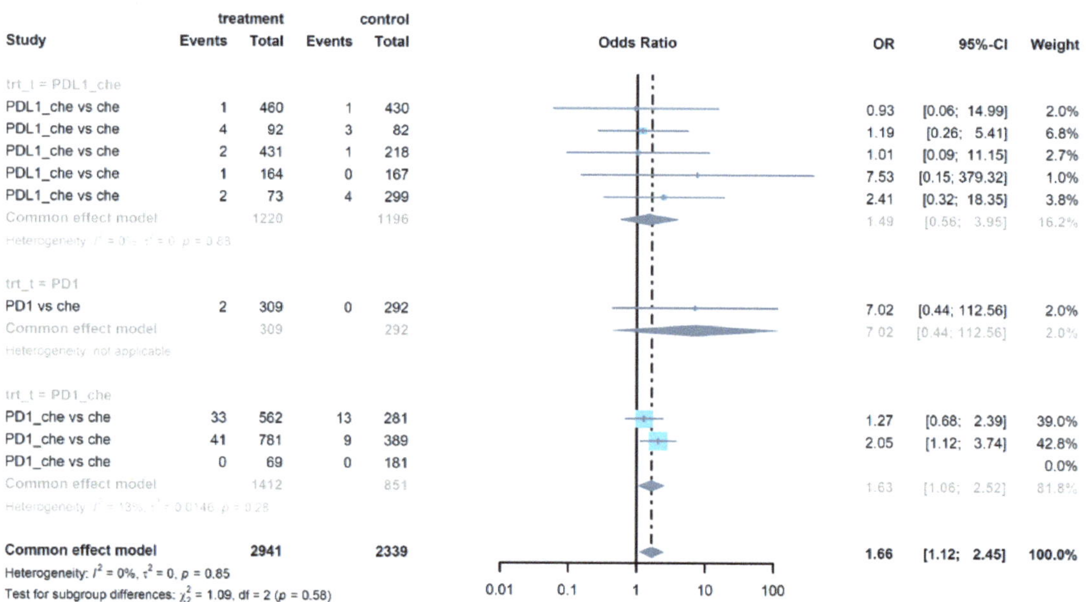

Figure 4. Forest plot of serious ALT elevation in patients treated with PD-1 or PD-L1 inhibitors versus chemotherapy.

3.3.2. Meta-Analysis Results with Non-Serious Adverse Events

Figure 2b summarizes the results of non-serious adverse events. The most frequent general adverse events were nausea (49.70%, 95% CI: 35.90–63.55%), fatigue (49.10%, 95% CI: 30.24–68.21%), anemia (34.74%, 95% CI: 19.91–53.28%), diarrhea (32.77%, 95% CI: 21.25–46.81%), headache (25.14%, 95% CI: 16.72–35.97%), and arthralgia (24.88%, 95% CI: 13.87–40.52%).

As is shown in Figure 5, patients treated with PD-1 or PD-L1 inhibitors were more likely to experience hypothyroidism (OR = 3.63, 95% CI: 2.92–4.51). This trend was clear in both PD-1 groups (OR = 5.74, 95% CI: 1.48–22.20) and PD-L1 groups (OR = 3.85, 95% CI: 2.72–5.44). Compared with patients treated in the control arms, those treated with PD-1 or PD-L1 inhibitors were at higher risk of AST elevation (OR =1.26, 95% CI: 1.02–1.57, see Figure S3). Further analysis showed that there was no report of AST elevation in PD-1 groups. Patients in PD-L1 groups were prone to AST elevation (OR = 1.30, 95% CI: 1.03–1.65). Figures 6 and S4 show that patients were more likely to report pruritus (OR = 1.84, 95% CI: 1.30–2.59) and rash (OR = 1.29, 95% CI: 1.08–1.55) in treatment arms compared with patients in the control arms. Figure S5 demonstrates that PD-L1 combined with chemotherapy groups were at increased risk of fever (OR = 2.04, 95% CI: 1.33–3.14) than the control group.

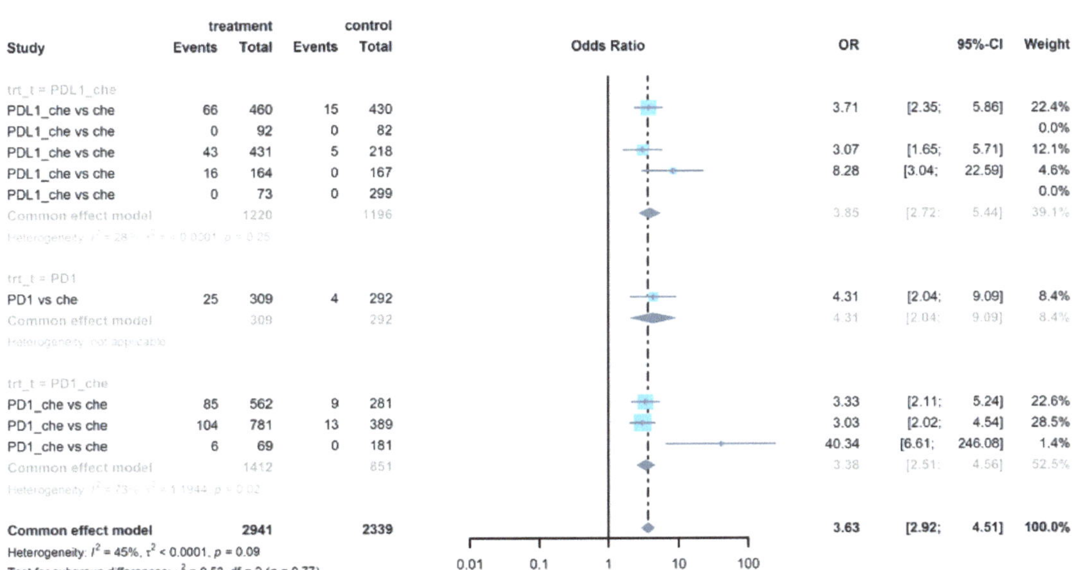

Figure 5. Forest plot of non-serious hypothyroidism in patients treated with PD-1 or PD-L1 inhibitors versus chemotherapy.

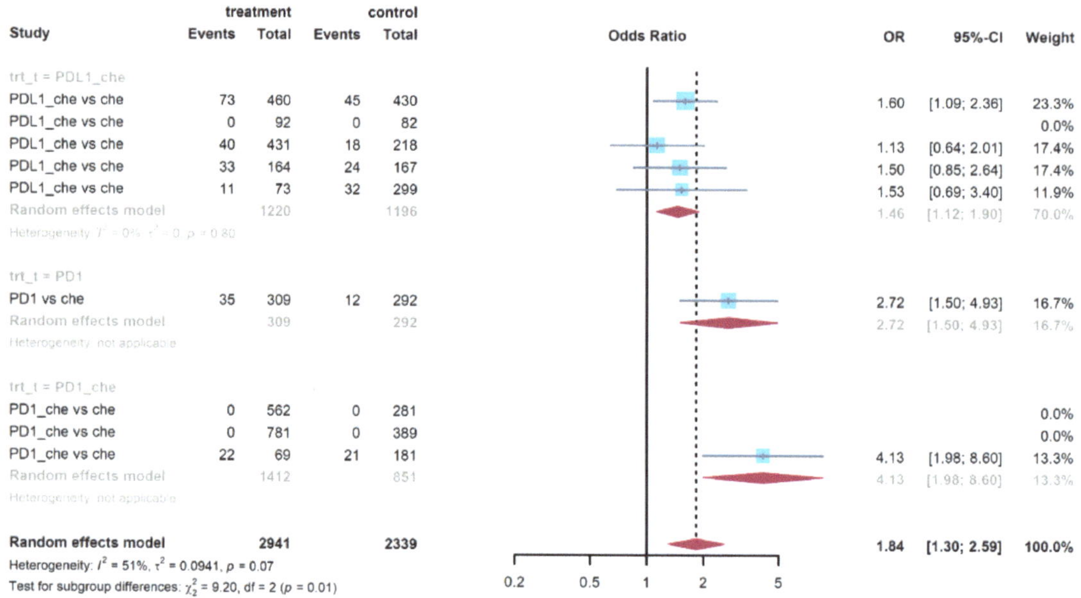

Figure 6. Forest plot of non-serious pruritus in patients treated with PD-1 or PD-L1 inhibitors versus chemotherapy.

4. Discussion

Although PD-1 or PD-L1 drugs are widely used in lung cancer and melanoma, research on their impacts on triple-negative breast cancer, a refractory breast tumor, is still very limited, as is research on their safety. Several meta-analysis results showed that PD-1 or PD-L1 inhibitors plus chemotherapy prolonged the progression-free survival in the neoadjuvant and adjuvant settings when treating TNBC patients [20,33,34]. In this meta-analysis, we first explored the incidence of adverse events in TNBC patients through the proportion meta-analysis method, and then conducted a traditional meta-analysis to compare the risk of different adverse events between the PD-1 or PD-L1 inhibitors group and the chemotherapy group. We found that, for serious adverse events, neutropenia had the highest incidence, followed by fatigue and anemia. This was consistent with the results from Zhou et al. showing that the common treatment-related adverse events in PD-1 or PD-L1 inhibitors and chemotherapy combination was anemia (45.4%) of all-grade adverse events and neutropenia (19.6%) of grade 3 or higher [21]. The difference is that our meta-analysis did not show any significant results in blood-related adverse events such as neutropenia and anemia. From the results of Keynote-119, the pembrolizumab group had less frequent anemia and neutropenia [27]. In our analysis, eight out of the nine studies used PD-1 or PD-L1 inhibitors combined with chemotherapy for TNBC patients, indicating that the blood toxicity may have been mainly related to chemotherapy.

The most common adverse events were related to digestive reactions, such as nausea (49.7%), diarrhea (32.8%), constipation (23.2%), and vomiting (21.4%). Previous studies have suggested that gastrointestinal-tract-related reactions are the most common all-grade adverse events with anti-PD-1/PD-L1 agents [35,36]. However, our meta-analysis showed that there were no significant results of gastrointestinal reactions (nausea: OR = 1, 95% CI: 0.78–1.29; diarrhea: OR = 1.04, 95% CI: 0.63–1.72; constipation: OR = 1.03, 95% CI: 0.88–1.21;) upon comparing the PD-1 or PD-L1 groups with chemotherapy groups. Doctors should also pay attention to these less severe adverse events and take precautions, since they could affect patient quality of life.

Among the immune-related adverse events, the incidences of the most serious adverse events were very low (less than 1%). For non-serious adverse events, the most common one was arthralgia (24.9%), followed by ALT elevation (18.1%), AST elevation (17.9%), rash (17.6%), and pruritus (15.7%). Previous clinical trials have reported high incidence of immune-related adverse events, up to 58.7% in Impassion130 [23–25,27]. In our analysis, pneumonitis, hypothyroidism and hyperthyroidism were less common, but they were more likely to be severe. Compared with chemotherapy, PD-1 or PD-L1 inhibitors had more risk of serious pneumonitis and hypothyroidism, which are similar results to the meta-analyses by Zhang et al. and Wang et al. [33,34]. In the tumor microenvironment, tumor immune escape is related to the role of PD-1/PD-L1 and T lymphocytes [37]. Therefore, immune checkpoint inhibition therapy could affect the balance between autoimmunity and immunity, and thus enhance the activity of the immune system and attack tumor cells [38]. These immune-related adverse events may be due to the immunosuppressive effect of the drugs and should be treated seriously [39].

It is worth noting that, among the serious immune-related adverse events, adrenal insufficiency (1.7%) was very common and the incidence of non-serious adrenal insufficiency was 1.1%. In a study by Wang et al., the incidence of all-grade adverse events was 0.69%, but they did not distinguish between cancer types [22]. In addition, the risk of adrenal insufficiency was significantly higher in the PD-1 or PD-L1 groups than in the chemotherapy groups. Adrenal insufficiency may have a great influence on patient metabolism [36]. Due to diagnostic techniques, doctors may not notice adrenal insufficiency, leading to low incidence rates [40]. Once patients are diagnosed with adrenal insufficiency, it can be so serious that hospitalization is required [41]. Thus, for future TNBC patients treated with PD-1 or PD-L1 inhibitors, adrenal insufficiency should be taken into more consideration.

We performed subgroup analysis of PD-1 and PD-L1 inhibitors. Apart from one study which used PD-1 inhibitors alone, other studies all combined PD-1 or PD-L1 inhibitors with chemotherapy. The overall trends of adverse events were consistent between the PD-1 group and the PD-L1 group, though the data were limited and some adverse events may not have been reported in some studies. Sonpavde et al.'s study showed that the incidence of grade 3 or higher adverse events was higher in the PD-1 inhibitors group compared with PD-L1 inhibitors [42]. Campelo et al. found that PD-L1 inhibitors were associated with a lower risk of adverse events that led to treatment discontinuation than PD-1 inhibitors [43]. The use of PD-1 or PD-L1 inhibitors significantly increased the risk of serious immune-related adverse events in this study [43], but the incidences were very low. Meanwhile, the risk of non-serious adverse events did not increase. Considering its efficacy, we consider the safety of PD-1 or PD-L1 inhibitors as acceptable. However, adverse events were not the primary outcome of the included studies. We cannot avoid incomplete data reports even if we extract the data from the clinical trial registration website [44]. More research on PD-1 or PD-L1 inhibitors in TNBC patients is needed.

The limitations of our study are as follows. First, we used adverse event data from clinicaltrial.gov. For those studies that had no data on this website, we extracted data from their published articles. Heterogeneity may arise from these different data sources [45]. Second, we only included II/III randomized clinical trials in our meta-analysis. The number of studies was still limited. In the future, to investigate the safety of PD-1 or PD-L1 Inhibitors more comprehensively, some observational studies such as cohort studies may also be considered. Third, we did not distinguish between chemotherapy regimens, doses of PD-1 or PD-L1 inhibitors, patient age, etc. Lastly, because the incidence of serious adverse events was usually very low, regular meta-analysis approaches were not applicable, which leads to the challenge of rare event modeling [46]. In this study, we adopted the Peto method to tackle this problem. Other meta-analysis methods for rare events may also be applicable to this study.

5. Conclusions

Compared to the treatment of chemotherapy alone, PD-1 or PD-L1 inhibitors combined with chemotherapy significantly increased the risk of immune-related adverse events in TNBC patients, including serious pneumonitis, hypothyroidism, and adrenal insufficiency, but the incidences were relatively low. For practical treatment using PD-1 or PD-L1 inhibitors in TNBC, serious adverse events, such as serious pneumonitis, hypothyroidism, ALT elevation, and adrenal insufficiency, should be considered and monitored. Non-serious adverse events, such as AST elevation, rash, and fever, should also be taken into consideration.

Supplementary Materials: The following supporting information can be downloaded at: https://www.mdpi.com/article/10.3390/life12121990/s1, Figure S1: Forest plot of serious hypothyroidism in patients treated with PD-1 or PD-L1 inhibitors versus chemotherapy; Figure S2: Forest plot of serious adrenal insufficiency in patients treated with PD-1 or PD-L1 inhibitors versus chemotherapy; Figure S3: Forest plot of other AST elevation in patients treated with PD-1 or PD-L1 inhibitors versus chemotherapy; Figure S4: Forest plot of other rashes in patients treated with PD-1 or PD-L1 inhibitors versus chemotherapy; Figure S5: Forest plot of other fevers in patients treated with PD-1 or PD-L1 inhibitors versus chemotherapy.

Author Contributions: Conceptualization, B.L. and T.H.; methodology, Y.Z. and B.L.; software, Y.Z.; validation, Y.Z., J.W., H.W. and M.L.; formal analysis, Y.Z., J.W. and B.L.; investigation, Y.Z., and B.L.; resources, B.L.; data curation, Y.Z., J.W., H.W. and M.L.; writing—original draft preparation, Y.Z. and B.L.; writing—review and editing, Y.Z., B.L., J.W., H.W. and M.L.; visualization, T.H.; supervision, B.L., M.L. and T.H.; project administration, B.L.; funding acquisition, B.L. and T.H. All authors have read and agreed to the published version of the manuscript.

Funding: This research was funded by the National Natural Science Foundation of China (grant number 11901013), the Research and Development Funds of Peking University People's Hospital (RDX2021-05), Beijing Natural Science Foundation (1204031), the Fundamental Research Funds for the Central Universities (BMU2021RCZX023), PKU-Baidu Fund (2020BD029).

Institutional Review Board Statement: Not applicable.

Informed Consent Statement: Not applicable.

Conflicts of Interest: The authors declare no conflict of interest.

References

1. Fahad Ullah, M. Breast Cancer: Current Perspectives on the Disease Status. In *Advances in Experimental Medicine and Biology*; Ahmad, A., Ed.; Springer International Publishing: Cham, Switzerland, 2019; Volume 1152, pp. 51–64.
2. Howard, F.M.; Olopade, O.I. Epidemiology of Triple-Negative Breast Cancer: A Review. *Cancer J.* **2021**, *27*, 8–16. [CrossRef] [PubMed]
3. Gluz, O.; Liedtke, C.; Gottschalk, N.; Pusztai, L.; Nitz, U.; Harbeck, N. Triple-negative breast cancer—Current status and future directions. *Ann. Oncol.* **2009**, *20*, 1913–1927. [CrossRef] [PubMed]
4. Bianchini, G.; Balko, J.M.; Mayer, I.A.; Sanders, M.E.; Gianni, L. Triple-negative breast cancer: Challenges and opportunities of a heterogeneous disease. *Nat. Rev. Clin. Oncol.* **2016**, *13*, 674–690. [CrossRef]
5. Bergin, A.R.T.; Loi, S. Triple-negative breast cancer: Recent treatment advances. *F1000Research* **2019**, *8*, 1342. [CrossRef]
6. Lv, W.; Tan, Y.; Zhou, X.; Zhang, Q.; Zhang, J.; Wu, Y. Landscape of prognosis and immunotherapy responsiveness under tumor glycosylation-related lncRNA patterns in breast cancer. *Front. Immunol.* **2022**, *13*, 9928. [CrossRef]
7. Yin, L.; Duan, J.J.; Bian, X.W.; Yu, S.C. Triple-negative breast cancer molecular subtyping and treatment progress. *Breast Cancer Res.* **2020**, *22*, 61. [CrossRef]
8. Killelea, B.K.; Yang, V.Q.; Mougalian, S.; Horowitz, N.R.; Pusztai, L.; Chagpar, A.B.; Lannin, D.R. Neoadjuvant Chemotherapy for Breast Cancer Increases the Rate of Breast Conservation: Results from the National Cancer Database. *J. Am. Coll. Surg.* **2015**, *220*, 1063–1069. [CrossRef]
9. Sikov, W.M.; Berry, D.A.; Perou, C.M.; Singh, B.; Cirrincione, C.T.; Tolaney, S.M.; Kuzma, C.S.; Pluard, T.J.; Somlo, G.; Port, E.R.; et al. Impact of the addition of carboplatin and/or bevacizumab to neoadjuvant once-per-week paclitaxel followed by dose-dense doxorubicin and cyclophosphamide on pathologic complete response rates in stage II to III triple-negative breast cancer: CALGB 40603 (Alliance). *J. Clin. Oncol.* **2015**, *33*, 13–21. [CrossRef]
10. Han, Y.; Liu, D.; Li, L. PD-1/PD-L1 pathway: Current researches in cancer. *Am. J. Cancer Res.* **2020**, *10*, 727–742.

11. Keir, M.E.; Butte, M.J.; Freeman, G.J.; Sharpe, A.H. PD-1 and its ligands in tolerance and immunity. *Annu. Rev. Immunol.* **2008**, *26*, 677–704. [CrossRef]
12. Patel, S.P.; Kurzrock, R. PD-L1 Expression as a Predictive Biomarker in Cancer Immunotherapy. *Mol. Cancer Ther.* **2015**, *14*, 847–856. [CrossRef] [PubMed]
13. Fu, O.Y.; Li, C.L.; Chen, C.C.; Shen, Y.C.; Moi, S.H.; Luo, C.W.; Xia, W.Y.; Wang, Y.N.; Lee, H.H.; Wang, L.H.; et al. De-glycosylated membrane PD-L1 in tumor tissues as a biomarker for responsiveness to atezolizumab (Tecentriq) in advanced breast cancer patients. *Am. J. Cancer Res.* **2022**, *12*, 123–137.
14. FDA Approves Atezolizumab for PD-L1 Positive Unresectable Locally Advanced or Metastatic Triple-Negative Breast Cancer. Available online: https://www.fda.gov/drugs/drug-approvals-and-databases/fda-approves-atezolizumab-pd-l1-positive-unresectable-locally-advanced-or-metastatic-triple-negative (accessed on 20 October 2022).
15. Atezolizumab TNBC Indication Withdrawn by Manufacturer after Talks with FDA. Available online: https://www.cancernetwork.com/view/atezolizumab-tnbc-indication-withdrawn-by-manufacturer-after-talks-with-fda (accessed on 15 October 2022).
16. Nanda, R.; Chow, L.Q.; Dees, E.C.; Berger, R.; Gupta, S.; Geva, R.; Pusztai, L.; Pathiraja, K.; Aktan, G.; Cheng, J.D.; et al. Pembrolizumab in Patients with Advanced Triple-Negative Breast Cancer: Phase Ib KEYNOTE-012 Study. *J. Clin. Oncol.* **2016**, *34*, 2460–2467. [CrossRef]
17. Xue, Y.; Gao, S.; Gou, J.; Yin, T.; He, H.; Wang, Y.; Zhang, Y.; Tang, X.; Wu, R. Platinum-based chemotherapy in combination with PD-1/PD-L1 inhibitors: Preclinical and clinical studies and mechanism of action. *Expert Opin. Drug Deliv.* **2021**, *18*, 187–203. [CrossRef] [PubMed]
18. Sternschuss, M.; Yerushalmi, R.; Saleh, R.R.; Amir, E.; Goldvaser, H. Efficacy and safety of neoadjuvant immune checkpoint inhibitors in early-stage triple-negative breast cancer: A systematic review and meta-analysis. *J. Cancer Res. Clin. Oncol.* **2021**, *147*, 3369–3379. [CrossRef] [PubMed]
19. Xin, Y.; Shen, G.; Zheng, Y.; Guan, Y.; Huo, X.; Li, J.; Ren, D.; Zhao, F.; Liu, Z.; Li, Z.; et al. Immune checkpoint inhibitors plus neoadjuvant chemotherapy in early triple-negative breast cancer: A systematic review and meta-analysis. *BMC Cancer* **2021**, *21*, 1261. [CrossRef] [PubMed]
20. Qi, Y.; Qi, Y.; Zhang, W.; Jiang, R.; Xu, O.; Kong, X.; Zhang, L.; Zhang, L.; Zhang, L.; Fang, Y.; et al. Efficacy and safety of PD-1 and PD-L1 inhibitors combined with chemotherapy in randomized clinical trials among triple-negative breast cancer. *Front. Pharmacol.* **2022**, *13*, 323. [CrossRef] [PubMed]
21. Zhou, X.; Yao, Z.; Bai, H.; Duan, J.; Wang, Z.; Wang, X.; Zhang, X.; Xu, J.; Fei, K.; Zhang, Z.; et al. Treatment-related adverse events of PD-1 and PD-L1 inhibitor-based combination therapies in clinical trials: A systematic review and meta-analysis. *Lancet Oncol.* **2021**, *22*, 1265–1274. [CrossRef]
22. Wang, Y.; Zhou, S.; Yang, F.; Qi, X.; Wang, X.; Guan, X.; Shen, C.; Duma, N.; Vera Aguilera, J.; Chintakuntlawar, A.; et al. Treatment-Related Adverse Events of PD-1 and PD-L1 Inhibitors in Clinical Trials: A Systematic Review and Meta-analysis. *JAMA Oncol.* **2019**, *5*, 1008–1019. [CrossRef]
23. Emens, L.A.; Adams, S.; Barrios, C.H.; Dieras, V.; Iwata, H.; Loi, S.; Rugo, H.S.; Schneeweiss, A.; Winer, E.P.; Patel, S.; et al. First-line atezolizumab plus nab-paclitaxel for unresectable, locally advanced, or metastatic triple-negative breast cancer: IMpassion130 final overall survival analysis. *Ann. Oncol.* **2021**, *32*, 1650. [CrossRef]
24. Miles, D.; Gligorov, J.; Cameron, D.; Xu, B.; Semiglazov, V.; Reinisch, M.; Patre, M.; Morales, L.; Patel, S.L.; Barata, T.; et al. Primary results from IMpassion131, a double-blind, placebo-controlled, randomised phase III trial of first-line paclitaxel with or without atezolizumab for unresectable locally advanced/metastatic triple-negative breast cancer. *Ann. Oncol.* **2021**, *32*, 994–1004. [CrossRef] [PubMed]
25. Cescon, D.W.; Nowecki, Z.; Gallardo, C.; Holgado, E.; Iwata, H.; Masuda, N.; Otero, M.T.; Loi, S.; Guo, Z.; Gonzalo, G.A.; et al. Pembrolizumab plus chemotherapy versus placebo plus chemotherapy for previously untreated locally recurrent inoperable or metastatic triple-negative breast cancer (KEYNOTE-355): A randomised, placebo-controlled, double-blind, phase 3 clinical trial. *Lancet* **2020**, *396*, 1817–1828. [CrossRef]
26. Warton, D.I.; Hui, F.K.C. The arcsine is asinine: The analysis of proportions in ecology. *Ecology* **2011**, *92*, 3–10. [CrossRef] [PubMed]
27. Winer, E.P.; Lipatov, O.; Im, S.-A.; Goncalves, A.; Muñoz-Couselo, E.; Lee, K.S.; Schmid, P.; Tamura, K.; Testa, L.; Witzel, I.; et al. Pembrolizumab versus investigator-choice chemotherapy for metastatic triple-negative breast cancer (KEYNOTE-119): A randomised, open-label, phase 3 trial. *Lancet Oncol.* **2021**, *22*, 499–511. [CrossRef] [PubMed]
28. Nanda, R.; Liu, M.C.; Yau, C.; Shatsky, R.; Pusztai, L.; Wallace, A.; Chien, A.J.; Forero-Torres, A.; Ellis, E.; Han, H.; et al. Effect of Pembrolizumab Plus Neoadjuvant Chemotherapy on Pathologic Complete Response in Women with Early-Stage Breast Cancer: An Analysis of the Ongoing Phase 2 Adaptively Randomized I-SPY2 Trial. *JAMA Oncol.* **2020**, *6*, 676–684. [CrossRef]
29. Schmid, P.; Cortes, J.; Pusztai, L.; McArthur, H.; Kümmel, S.; Bergh, J.; Denkert, C.; Park, Y.H.; Hui, R.; Harbeck, N.; et al. Pembrolizumab for Early Triple-Negative Breast Cancer. *N. Engl. J. Med.* **2020**, *382*, 810–821. [CrossRef]
30. Mittendorf, E.A.; Zhang, H.; Barrios, C.H.; Saji, S.; Jung, K.H.; Hegg, R.; Koehler, A.; Sohn, J.; Iwata, H.; Telli, M.L.; et al. Neoadjuvant atezolizumab in combination with sequential nab-paclitaxel and anthracycline-based chemotherapy versus placebo and chemotherapy in patients with early-stage triple-negative breast cancer (IMpassion031): A randomised, double-blind, phase 3 trial. *Lancet* **2020**, *396*, 1090–1100. [CrossRef] [PubMed]

31. Pusztai, L.; Yau, C.; Wolf, D.M.; Han, H.S.; Du, L.; Wallace, A.M.; String-Reasor, E.; Boughey, J.C.; Chien, A.J.; Elias, A.D.; et al. Durvalumab with olaparib and paclitaxel for high-risk HER2-negative stage II/III breast cancer: Results from the adaptively randomized I-SPY2 trial. *Cancer Cell.* **2021**, *39*, 989–998.e985. [CrossRef] [PubMed]

32. Loibl, S.; Untch, M.; Burchardi, N.; Huober, J.; Sinn, B.V.; Blohmer, J.U.; Grischke, E.M.; Furlanetto, J.; Tesch, H.; Hanusch, C.; et al. A randomised phase II study investigating durvalumab in addition to an anthracycline taxane-based neoadjuvant therapy in early triple-negative breast cancer: Clinical results and biomarker analysis of GeparNuevo study. *Ann. Oncol.* **2019**, *30*, 1279–1288. [CrossRef]

33. Wang, C. A Meta-Analysis of Efficacy and Safety of PD-1/PD-L1 Inhibitors in Triple-Negative Breast Cancer. *J. Oncol.* **2022**, *2022*, 2407211. [CrossRef]

34. Zhang, M.; Song, J.; Yang, H.; Jin, F.; Zheng, A. Efficacy and safety of PD-1/PD-L1 inhibitors in triple-negative breast cancer: A systematic review and meta-analysis. *Acta Oncol.* **2022**, *61*, 1105–1115. [CrossRef] [PubMed]

35. Michot, J.M.; Bigenwald, C.; Champiat, S.; Collins, M.; Carbonnel, F.; Postel-Vinay, S.; Berdelou, A.; Varga, A.; Bahleda, R.; Hollebecque, A.; et al. Immune-related adverse events with immune checkpoint blockade: A comprehensive review. *Eur. J. Cancer* **2016**, *54*, 139–148. [CrossRef] [PubMed]

36. Wang, P.F.; Chen, Y.; Song, S.Y.; Wang, T.J.; Ji, W.J.; Li, S.W.; Liu, N.; Yan, C.X. Immune-Related Adverse Events Associated with Anti-PD-1/PD-L1 Treatment for Malignancies: A Meta-Analysis. *Front. Pharm.* **2017**, *8*, 730. [CrossRef] [PubMed]

37. Dammeijer, F.; van Gulijk, M.; Mulder, E.E.; Lukkes, M.; Klaase, L.; van den Bosch, T.; van Nimwegen, M.; Lau, S.P.; Latupeirissa, K.; Schetters, S.; et al. The PD-1/PD-L1-Checkpoint Restrains T Cell Immunity in Tumor-Draining Lymph Nodes. *Cancer Cell.* **2020**, *38*, 685–700.e68. [CrossRef] [PubMed]

38. Bagchi, S.; Yuan, R.; Engleman, E.G. Immune Checkpoint Inhibitors for the Treatment of Cancer: Clinical Impact and Mechanisms of Response and Resistance. *Annu. Rev. Pathol.* **2021**, *16*, 223–249. [CrossRef] [PubMed]

39. Santoni, M.; Romagnoli, E.; Saladino, T.; Foghini, L.; Guarino, S.; Capponi, M.; Giannini, M.; Cognigni, P.D.; Ferrara, G.; Battelli, N. Triple negative breast cancer: Key role of Tumor-Associated Macrophages in regulating the activity of anti-PD-1/PD-L1 agents. *Biochim. Biophys. Acta Rev. Cancer* **2018**, *1869*, 78–84. [CrossRef]

40. Ferrari, S.M.; Fallahi, P.; Elia, G.; Ragusa, F.; Ruffilli, I.; Patrizio, A.; Galdiero, M.R.; Baldini, E.; Ulisse, S.; Marone, G.; et al. Autoimmune Endocrine Dysfunctions Associated with Cancer Immunotherapies. *Int. J. Mol. Sci.* **2019**, *20*, 2560. [CrossRef]

41. Bornstein, S.R.; Allolio, B.; Arlt, W.; Barthel, A.; Don-Wauchope, A.; Hammer, G.D.; Husebye, E.S.; Merke, D.P.; Murad, M.H.; Stratakis, C.A.; et al. Diagnosis and Treatment of Primary Adrenal Insufficiency: An Endocrine Society Clinical Practice Guideline. *J. Clin. Endocrinol. Metab.* **2016**, *101*, 364–389. [CrossRef]

42. Sonpavde, G.P.; Grivas, P.; Lin, Y.; Hennessy, D.; Hunt, J.D. Immune-related adverse events with PD-1 versus PD-L1 inhibitors: A meta-analysis of 8730 patients from clinical trials. *Future Oncol.* **2021**, *17*, 2545–2558. [CrossRef]

43. Campelo, M.R.G.; Arriola, E.; Balea, B.C.; Lopez-Brea, M.; Fuentes-Pradera, J.; Carpeno, J.D.; Aguado, C.; Parente, D.P.; Pulido, F.D.; Ruiz-Gracia, P.; et al. PD-L1 Inhibitors as Monotherapy for the First-Line Treatment of Non-Small-Cell Lung Cancer in PD-L1 Positive Patients: A Safety Data Network Meta-Analysis. *J. Clin. Med.* **2021**, *10*, 4583. [CrossRef]

44. Baxi, S.; Yang, A.; Gennarelli, R.L.; Khan, N.; Wang, Z.; Boyce, L.; Korenstein, D. Immune-related adverse events for anti-PD-1 and anti-PD-L1 drugs: Systematic review and meta-analysis. *BMJ* **2018**, *360*, k793. [CrossRef] [PubMed]

45. Garcia-Alamino, J.M.; Bankhead, C.; Heneghan, C.; Pidduck, N.; Perera, R. Impact of heterogeneity and effect size on the estimation of the optimal information size: Analysis of recently published meta-analyses. *BMJ Open* **2017**, *7*, e015888. [CrossRef] [PubMed]

46. Jia, P.L.; Lin, L.F.; Kwong, J.S.W.; Xu, C. Many meta-analyses of rare events in the Cochrane Database of Systematic Reviews were underpowered. *J. Clin. Epidemiol.* **2021**, *131*, 113–122. [CrossRef] [PubMed]

Review

Breast Cancer Exposomics

Anca-Narcisa Neagu [1,*], Taniya Jayaweera [2], Lilian Corrice [2], Kaya Johnson [2] and Costel C. Darie [2,*]

[1] Laboratory of Animal Histology, Faculty of Biology, "Alexandru Ioan Cuza" University of Iași, Carol I Bvd. 20A, 700505 Iasi, Romania
[2] Biochemistry & Proteomics Laboratories, Department of Chemistry and Biochemistry, Clarkson University, Potsdam, NY 13699-5810, USA; jayawetm@clarkson.edu (T.J.); corriclg@clarkson.edu (L.C.); johnsokr@clarkson.edu (K.J.)
* Correspondence: aneagu@uaic.ro (A.-N.N.); cdarie@clarkson.edu (C.C.D.)

Abstract: We are exposed to a mixture of environmental man-made and natural xenobiotics. We experience a wide spectrum of environmental exposure in our lifetime, including the effects of xenobiotics on gametogenesis and gametes that undergo fertilization as the starting point of individual development and, moreover, in utero exposure, which can itself cause the first somatic or germline mutation necessary for breast cancer (BC) initiation. Most xenobiotics are metabolized or/and bioaccumulate and biomagnify in our tissues and cells, including breast tissues, so the xenobiotic metabolism plays an important role in BC initiation and progression. Many considerations necessitate a more valuable explanation regarding the molecular mechanisms of action of xenobiotics which act as genotoxic and epigenetic carcinogens. Thus, exposomics and the exposome concept are based on the diversity and range of exposures to physical factors, synthetic chemicals, dietary components, and psychosocial stressors, as well as their associated biologic processes and molecular pathways. Existing evidence for BC risk (BCR) suggests that food-borne chemical carcinogens, air pollution, ionizing radiation, and socioeconomic status are closely related to breast carcinogenesis. The aim of this review was to depict the dynamics and kinetics of several xenobiotics involved in BC development, emphasizing the role of new omics fields related to BC exposomics, such as environmental toxicogenomics, epigenomics and interactomics, metagenomics, nutrigenomics, nutriproteomics, and nutrimiRomics. We are mainly focused on food and nutrition, as well as endocrine-disrupting chemicals (EDCs), involved in BC development. Overall, cell and tissue accumulation and xenobiotic metabolism or biotransformation can lead to modifications in breast tissue composition and breast cell morphology, DNA damage and genomic instability, epimutations, RNA-mediated and extracellular vesicle effects, aberrant blood methylation, stimulation of epithelial–mesenchymal transition (EMT), disruption of cell–cell junctions, reorganization of the actin cytoskeleton, metabolic reprogramming, and overexpression of mesenchymal genes. Moreover, the metabolism of xenobiotics into BC cells impacts almost all known carcinogenic pathways. Conversely, in our food, there are many bioactive compounds with anti-cancer potential, exerting pro-apoptotic roles, inhibiting cell cycle progression and proliferation, migration, invasion, DNA damage, and cell stress conditions. We can conclude that exposomics has a high potential to demonstrate how environmental exposure to xenobiotics acts as a double-edged sword, promoting or suppressing tumorigenesis in BC.

Keywords: breast cancer (BC); exposomics; xenobiotics; breast cancer risk (BCR); biologic pathways

Citation: Neagu, A.-N.; Jayaweera, T.; Corrice, L.; Johnson, K.; Darie, C.C. Breast Cancer Exposomics. *Life* **2024**, *14*, 402. https://doi.org/10.3390/life14030402

Academic Editors: Riccardo Autelli, Taobo Hu, Mengping Long and Lei Wang

Received: 26 February 2024
Revised: 14 March 2024
Accepted: 15 March 2024
Published: 18 March 2024

1. Introduction

The aim of this review is to deepen our understanding of the study of breast cancer as an "environmental disease", using an exposomics-based hypothesis sustaining that BC is an "ecological disorder" [1–3]. We are what we eat [4–8], we are what we breathe [9], and we are what we live in [10]. This means that food-borne chemicals, all air, soil, and water pollutants; drugs and drug-related metabolites; different types of radiation; aflatoxins; nanoparticles; noise; and many other environmental factors act, individually or synergistically, as genetic and epigenetic carcinogens, in association with inheritance, disparities, reproductive life, age at exposure, and socioeconomic status, which can also increase BCR [11]. Many studies concluded that cumulative environmental exposure and lifestyle factors account for 70% to 95% of risk factors that drive the BC incidence rate [12], whereas only 10% to 30% of chronic disease risk can be explained by individual genomic landscape [13]. The effects of different types of environmental exposure on BC development, recurrence, overall survival, or treatment resistance [14–16] have been reviewed by many authors. Some studies suggest that even climate change will affect women's cancers [17]. Cell/mobile phone or smartphone use can result in increased BCR, due to the emission of radiofrequency energy that is absorbed by human tissues situated in the proximity, including breast tissue [18]. Many occupational habits, such as heat or night-light exposure, as well as dysregulation of the circadian rhythm, can result in moderate or increased BCR [19,20]. Hair dyes [21], cigarette smoking [22,23], radiofrequency radiation [24], laptops, tablets, and other devices [25], hormone-based treatments [26,27], residential and road traffic noise [28,29], and dust [30] were significantly associated with tumorigenesis and invasive BCR. Last but not least, oncogenic viruses have an important role in BC initiation and development [31].

Exposomics is a modern exposome analysis that characterizes all exposures in an untargeted and comprehensive manner [13]. Thus, the exposome concept is based on the diversity of exposures to physical factors, synthetic chemicals, dietary components, and psychosocial stressors, as well as their associated biological responses [32]. More than two decades ago, Ziegler et al. (1997) reported that BC incidence rates were 4–7 times higher in the United States compared to China and Japan; moreover, when Japanese, Chinese, or Filipino women migrate to the United States, their BCR rates increase over several generations, becoming almost similar with the BCR among American whites [33]. Many studies emphasize that the BCR is elevated compared to countries of origin, mainly due to the exposure to a Western lifestyle [33]. It is known that exposure to a Western diet is a risk factor for the development and maintenance of chronic and systemic tissue inflammation associated with reprogramming of innate immune cells [34]. This lifestyle-associated inflammation is an important cause of multiple cancers, including BC [35]. Recently, the concept of "metaflammation" was used to describe a crosstalk between immune and metabolic pathways that connect obesity to metabolic syndrome (MetS), chronic inflammation, and insulin resistance [36]. It is well-known that MetS is more prevalent in BC patients and is an independent risk factor or predictor for BC [37–39].

Consequently, numerous exogenous risk factors influence the growth, proliferation, and differentiation of breast tissue and BC development. A total of 50% of all cancers in women are hormonally mediated, with both estrogen and androgen playing key roles in initiation and BC development [40]. Of all xenobiotic classes, we chose to detail in this review EDCs and food components that can interact with endocrine receptors (ERs) to disturb the normal hormonal equilibrium in BC cells [41]. EDCs can be also ingested with food, so increasing and convincing evidence associates food and food-based dietary patterns with BCR [42]. Moreover, other food components that act as mutagens [43] can be involved in nutritional regulation of the mammary tumor microenvironment (TME) [44], and also impact growth and proliferation of cancer cells [45]. Conversely, food can contain many bioactive compounds with anti-BC potential, exerting a pro-apoptotic role and inhibiting cell cycle progression/cancer cell proliferation, migration, invasion, DNA damage, and cell stress conditions.

It is known that EDC exposure could elevate BCR [46]. Most studies assessed environmental EDC exposure, which includes pesticides, plasticizers, pharmaceutical agents, personal care products, food products, and food packaging, via biomarker measurements [46], so that hundreds of EDCs have been assessed as entering human breast tissue from a wide range of environmental sources, enabling all the hallmarks of cancer to develop in human BC cells [47]. Furthermore, diets comprising energy-dense and nutrient-poor foods have been associated with an increased BCR [48]. Food and food-related/dietary habits, including excessive alcohol use [49], deregulate many signals and metabolic pathways that stimulate the epithelial–mesenchymal transition (EMT), oxidative stress, and reactive oxygen species [50]; dioxin contamination [51], sweetened and highly processed coffee [52] and food [53], meat [54], sweetened drinks [55], EDCs [56], polycyclic aromatic hydrocarbons (PAHs) [57] present in our food, and an inadequate water/liquid daily intake [58] were significantly correlated with carcinogenesis and invasive BC.

Study of absorption, distribution, metabolism/biotransformation, excretion/elimination, and toxicity (ADME-Tox), as well as the bioaccumulation and biomagnification of xenobiotics in cells and liquid or solid tissues emphasize complex interactions with different structures of the human body, such as cellular components (i.e., membranes and proteins), molecular pathways, biological processes, and intra-/extracellular environments [59]. Several exposomics-related omics have been developed as a consequence of advances in molecular sciences and analytical techniques based on high-throughput sequencing and mass spectrometry (MS). Thus, environmental toxicogenomics, epigenomics, and interactomics, metagenomics, nutrigenomics and nutriproteomics, micromiRomics, and nutrimiRomics are several new omics fields related with BC exposomics and are involved in molecular characterization of the complex relationship between the human body, environmental exposure, and breast cancer.

2. Advances and Trends in Omics Fields Related to BC Exposomics

Advances in molecular approaches and analytical techniques based on high-throughput sequencing and mass spectrometry (MS) have generated multi-omics data that can be successfully used to understand the underlying molecular mechanisms involved in BC exposomics [60]. BC is mainly caused by mutations in multiple oncogenes and tumor suppressor genes, accompanying epigenetic aberrations of genes and protein pathways [61]. Thus, first of all, environmental toxicogenomics aims to collect, analyze, and interpret data on the changes in genes or protein expression, resulting from exposure to xenobiotics, using high-throughput technologies [62]. Evidence suggests that various pollutants, such as particulate matter involved in air pollution, act as carcinogenic factors in humans, inducing high rates of genomic instability [63], which is known as an initiator of BC development [61]. In addition, environmental epigenomics focuses on environmental factors that induce aberrant DNA methylation of cancer-related genes, even in developing embryos, when result in epigenetic mosaicism that can increase the oncogenic risk later in life [64]. Moreover, metagenomics, the study of genetic information of microorganisms present in an environment [65], is involved in the assessment of the human microbiome as a biomarker that experiences long-term exposure to numerous organic contaminants, known as xenobiotics [66]. Zhang et al. (2028), using liquid chromatography MS-based global metabolomics coupled with targeted metabolomics, demonstrated that the human microbiome can be significantly perturbed by exposure to xenobiotic mixtures, resulting in dysbiosis and metabolite-modified profiles that play an important role in the host's health [66]. With regard to BC, it is well-known that human microbiome-related disturbance may contribute to BC development by producing toxins or promoting inflammation, while certain types of bacteria may have positive effects against BC [67]. Recently, network biology techniques were used to identify xenobiotics that target hub proteins in the human interactomes, mainly in disease-associated proteins and contaminant-sensitive biomarkers [68], suggesting a new omics field, environmental interactomics. To exemplify, Moslehi et al. (2021) confirmed the role of arsenic as an ED or xenoestrogen involved in breast carcinogenesis, highlighting the complex arsenic-responsive BC interactome [69]. Nutrigenetics studies the effects of nutrition at the gene level, while nutrigenomics is focused on the effects of nutrients on the genome and transcriptome patterns [70]. Thus, based on the complex interaction between food components and human genome/proteome, nutrigenomics and nutriproteomics provide new opportunities for development of personalized diets in patients at risk of developing BC [71].

Tissue or circulating microRNA (miRNA) can serve as a novel toxicological biomarker involved in gene activation or suppression, being associated with several key epigenetic mechanisms involved in xenobiotic toxicity [72–74]. miRNAs are also studied and validated as biomarkers for various diseases, as in the case of miR-423, which is highly expressed in BC and promotes cancer cell proliferation, migration, and invasion by activating NF-κB signaling [75,76]. Thus, miRomics is focused on the study of the role of miRNAs in a variety of human diseases, including BC [73]. Evidence suggests that organic pollutant exposure, like bisphenol A (BPA), can alter miRNA expression in response to toxicity [77]. Recently, nutrimiRomics has been defined as a new omics field focused on the influence of diet components on the dysregulation of gene expression due to epigenetic modification that involves miRNAs, resulting in a higher risk for the development of chronic diseases [78]. Thus, Venkatadri et al. (2016) demonstrated that resveratrol, a dietary compound found in a wide variety of plants, can inhibit BC progression by controlling miRNA, regulating the expression of several proteins involved in apoptosis and the cell cycle [79]. These authors emphasized the key role in BC cell death in response to resveratrol for miR-542-3p in MCF7 cell line and miR-122-5p in MDA-MB-231 BC cells [79]. All these new omics fields complement the traditional approach of genomics, proteomics, transcriptomics, and metabolomics, in order to depict the complicated molecular mechanisms studied by BC exposomics.

3. Absorption, Distribution, Metabolism/Biotransformation, Bioaccumulation, and Excretion/Bioelimination of Xenobiotics Involved in Breast Cancer

Xenobiotics are substances that are foreign to the intrinsic metabolism of a biological system that has the capacity to bioaccumulate or remove xenobiotics by xenobiotic metabolism, which consists of the deactivation and excretion of xenobiotics and their metabolites [80,81]. The human body is exposed to 1–3 million foreign chemical compounds that form a cocktail/mixture of xenobiotics during a lifetime [82]. In BC, genotoxic carcinogens include dietary or environmental xenobiotics—heterocyclic amines, aromatic amines, PAHs, and nitropolycyclic aromatic hydrocarbons (NPAHs) [83]. Also, many cytotoxic compounds used as anti-cancer drugs for chemotherapy can cause high levels of DNA damage [84], undergo metabolic activation, and are subject to drug metabolism, including uptake, efflux, and detoxification [85].

3.1. Absorption

Generally, environmental xenobiotics enter the human body through different absorption surfaces/barriers from input compartments: skin and its appendages, by topical application and absorption, gastro-intestinal mucosa, by ingestion and absorption, and the pulmonary alveolar–capillary membrane, by inhalation. To begin with, EDCs from personal care products are easily absorbed by the skin into systemic circulation after topical application, and can be detected in blood, urine, and breast milk [86,87]. However, Rylander et al. (2019) concluded that intensive use of skin care products did not increase the BCR [86]. On the other hand, 70–100% of patients receiving radiation therapy following BC experienced radiation-induced skin toxicity [88] comparable to UV exposure, which was associated with decreased postmenopausal BCR, due to higher circulating concentration of a precursor to the active form of vitamin D [89]. In addition, the gut absorbs dietary nutrients and provides a barrier to many xenobiotics and microbiome-derived metabolites, so the intestinal epithelium becomes one of the most rapidly proliferating tissues in the body, assuring a rapid and effective elimination of some xenobiotics that bioaccumulate in enterocytes [90]. Consequently, the gastro-intestinal tract is also an important route by which drugs, chemicals, pesticides, environmental pollutants, and metabolites of other species are absorbed in the human body [91]. Last but not least, air pollution is known as a human carcinogen, especially by gaseous components, as well as through particulate matter, including fine, inhalable particles that can be vectors for radioactive isotopes [92,93]. Air polluting agents on their way to the bloodstream pass through the lung barriers [93]. White et al. (2022) showed that higher exposure to ambient particle radioactivity (PR-β) was associated with an elevated risk of ER– BC [92]. Moreover, Smotherman et al. (2023) found a positive association of particulate matter with postmenopausal BCR [94].

3.2. Distribution

The distribution compartment, mainly represented by the systemic bloodstream, transports xenobiotics and their metabolites to all tissues and organs, so that blood is the most used liquid biopsy for biomonitoring of xenobiotics, such as persistent organic pollutants (POPs) [95]. From blood, xenobiotics/drugs enter cells, including breast epithelial cells or different cell populations from their ECM or TME. Distribution in cells depend on the chemical nature of xenobiotics, the binding to different receptors or exertion of effects without cellular entry, or using membrane transporters that allow for their entry into the intracellular compartment [85]. Moreover, Ish et al. (2023) showed that changes in breast tissue composition may be a potential pathway by which outdoor air pollution impacts BCR [96]. Thus, quantitative changes in the relative amount of fibro-glandular tissue can represent a biomarker of BCR that can be used to emphasize the potential biologic pathways underlying the association between environmental exposures and BC [96]. In addition, Segovia-Mendoza et al. (2020) showed that the environmental bisphenols, BPA and BBS, induce alteration of the proteomic landscape of different human BC cell lines [97]. After bisphenol exposure, vascular endothelial growth factor (VEGF) secretion, CD44, as a biomarker of stemness, and metalloproteinase MMP-14, as a biomarker for invasion, were overexpressed in ER+ BC cell lines, whereas the epidermal growth factor receptor (EGFR) and transforming growth factor beta (TGF-β) were upregulated in ER− BC lines [97]. Overall, cell and tissue accumulation of xenobiotics, such as EDCs/POPs, could lead to cellular DNA damage and genomic instability [98], epimutations induced by DNA methylation, acetylation, histone posttranslational modifications (PTMs), RNA-mediated effects, and extracellular vesicle effects [99], alteration of DNA methylation during adipocyte differentiation [100] as well as blood methylation [101], epithelial–mesenchymal transition (EMT) by formation of lamellipodia, disruption of cell–cell junctions, E-cadherin downregulation, reorganization of the actin cytoskeleton in stress fibers as well as overexpression of mesenchymal genes, such as vimentin and fibronectin [102,103], FOXA1 repression and phosphorylation of ERK1/2, p48-MAPK, PI3K/AKT signaling in ER-BC cells [104], and upregulation of Snail and Slug in MCF7 ER+ BC cell line [103].

3.3. Biotransformation/Metabolism

Many bioreactive compartments are involved in biotransformation and elimination of xenobiotics. Consequently, many chemicals undergo metabolism and detoxification to produce various metabolites that can cause, in turn, harmful effects such as toxicity [105]. Xenobiotic metabolism and detoxification involve xenobiotic-metabolizing enzymes/proteins that are mainly expressed in the liver, but some are also expressed in breast tissue, so that intratumoral xenobiotics or metabolites generated in the liver can undergo further transformation in the breast tissue [83,106]. Thus, many enzymes such as mammary-expressed enzymes metabolically activate or detoxify potential genotoxic BC carcinogens, acting in mammary lipid, nipple aspirate, breast milk, and mammary epithelial cells, where most BCs originate [83]. Bieche et al. (2004) pointed out that the intratumoral dysregulation of genes coding for major xenobiotic-metabolizing enzymes has a role in breast tumorigenesis and drug resistance; thus, N-acetyltransferase 1 (*NAT1*) was proposed as a candidate biomarker for antiestrogen responsiveness [106]. These authors maintained that one-half of the patients with ER+ BC fail to respond favorably to antiestrogen treatment with tamoxifen due to the altered tamoxifen metabolism or bioavailability following the intratumoral alteration in expression of genes coding for xenobiotic-metabolizing enzymes. Moreover, it is known that, in solid tumors, the extracellular and intracellular distribution of xenobiotics and drugs presents a high degree of variability, and is controlled by drug/xenobiotic-metabolizing enzymes (DXMEs) as well as cellular influx and efflux systems that transport xenobiotics and drugs into and out of cells [107].

3.4. Bioelimination/Excretion

The main routes of elimination of xenobiotics and their metabolites are renal excretion, bile and fecal elimination, and pulmonary exhalation, but there are also secondary routes, such as sweat, hair and nails, breast milk, and tears [105]. For example, cadmium was detected at high concentration in BC tissue [108], as well as in the urine of patients with BC, urinary Cd being correlated with the expression of hypoxia-inducible factor 1 alpha (HIF1A) in BC tissues [109]. Other heavy metals, such as arsenic, chromium, lead, and mercury are considered to be carcinogens or co-carcinogens and have been detected in the urine of BC patients, even markedly increased [108]. Moreover, the environmental exposure to these heavy metals could influence the urine level of metabolites, in association with BC development [108]. Human breast milk, a specific breast secretion that reflects the molecular landscape of the normal or pathological mammary gland, contains secretions of the mother's body, in which there are compounds bioaccumulated in her organism, such as organic contaminants (polychlorinated biphenyls, brominated flame retardants, parabens, bisphenols, and perfluoroalkyl and polyfluoroalkyl substances) as well as heavy metals, mycotoxins, and pharmaceuticals residues [110]. Thus, many POPs, such as polychlorinated dibenzo-p-dioxins (PCDDs), polychlorinated dibenzofurans (PCDFs), polychlorinated biphenyls (PCBs), and organochlorine pesticides, such as DDT, have been detected in human blood, adipose tissue, and breast milk and tend to become magnified in the food chain over time; breastfeeding infants becoming the final target of POPs [111]. Moreover, POPs have been correlated to an increased incidence of hormone-dependent BCs [112].

3.5. Bioaccumulation

Usually, many xenobiotics, such as POPs and heavy metals, bioaccumulate within adipose tissue, considered to be widely contaminated with lipophilic xenobiotics in modern society and, consequently, acting as a significant site of xenobiotic storage or sequestration [113]. Adipose tissue can play a protective role against xenobiotic effects, because xenobiotic storage in fat can reduce the burden in other critical organs [114]. However, female breast adipose tissue is abundant in and in close contact with epithelial cells, representing a major component of the BC TME, which contributes to the development, growth, and invasion of tumor cells [115]. Weight loss and insulin resistance are involved in xenobiotic release from adipose tissue into bloodstream [114]. Heavy metals, one of the most harmful classes of environmental compounds [116], also stimulate BC progression, exerting a role of DNA methylation level in cancer cells [117]. Heavy metals are very difficult to metabolize or decompose, and accumulate in all tissue and organs over the lifetime [116]. Evidence suggests that obese people accumulate more heavy metals compared to healthy people [118]. Thus, cadmium, among other heavy metals, is a widely spread compound that exerts estrogenic effects, acts as an endocrine disruptor, and accumulates in BC cells over time [109].

4. Food and Nutrition

Dietary nutritional intake is a key environmental factor with a vital role in cancer prevention and care [70]. One-third of cancers in Western high-income societies are associated with food and nutrition, in correlation with physical activity [45], so that increasing and convincing evidence associates food-based dietary patterns with BCR [42]. Thus, poor nutrition and foods with a higher energy density have been associated with an increased risk of obesity as well as BC [48,119]. Thus, Jacobs et al. (2021), analyzing dietary patterns correlated to BCR in Black urban South African women, concluded that both traditional and cereal–dairy-based meals may reduce the BCR in this population [48].

Overall, thirteen cancers, including BC, have been estimated to be associated with obesity and are known as "obesity-associated cancers" [120]. The female breast is rich in adipose tissue [121], so that, in postmenopausal women, the adipose tissue becomes a significant source of estrogen, this obesity-associated estrogen likely playing an essential role in BC growth, mainly in ER+ BC tumors [120]. Conversely, caloric restriction or intermittent fasting, a period of voluntary abstention from all food or specific food products [122], can negatively impact BC development, reduce the treatment-induced adverse effects, cytotoxicity, and DNA damage, and increase optimal glycemic regulation, improving serum glucose, insulin, and insulin-like growth factor 1 (IGF-1) levels [123]. Insulin and the IGF-1 pathway regulate lifespan and longevity [124]. IGF-1 is known as a potent mitogen of high importance in the mammary gland that binds to the cognate receptor, IGF-1R, triggering a signaling intracellular cascade, which increases the proliferative and anti-apoptotic pathways in cancer cells [125]. It is known that the Western diet, characterized by high intake of hyperglycemic carbohydrates and insulinotropic dairy, stimulates IGF-1 signaling [124]. GH, IGF-1, and insulin have BC-promoting actions, due to increased IGF-1 levels, which have been associated with increased BCR [124].

Food components may act as mutagens, such as N-nitroso-derivatives, polycyclic aromatic hydrocarbons (PAHs), and heterocyclic aromatic amines [43], which can be involved in nutritional regulation of the mammary tumor microenvironment (TME) [44], and impact the growth and proliferation of cancer cells [45]. Nutritional stimuli modulate interactions between different cell populations within the TME, such as immune cells, adipocytes, vascular cells, and mammary epithelial and BC stem cells, so that both obesity, a chronic over-nutritional condition, as well as excess caloric consumption, disrupt mammary gland homeostasis and increase BCR [44,123]. EDCs has been reported in aquatic macroinvertebrates, mussels and seawater or freshwater fish [126], pork, beef, and chicken meat [127], vegetables [128], as well as in milk and dairy products [129]. Heavy metals, such as cadmium, mercury, and lead, act as EDCs and bioaccumulate mainly in fish and seafood products [130]. Fish product consumption acts as a double-edged sword. There are studies that emphasize the protective effect of omega-3 fatty acid in fish consumption against BC [131], while the human exposure to fat from milk, eggs, fish, and meat can enhance mammary gland susceptibility to carcinogenesis [132]. Alcohol consumption has been related to higher BCR, principally for estrogen receptor-positive (ER+) BCs [133], through stimulation of migration and invasion of MCF7 human BC cells [133], EMT, angiogenesis, OS and ROS production [49,50], decreasing E-cadherin, α, β, and γ catenin expression, as well as *BRCA1* tumor suppressor gene expression [133].

Fortunately, in our food, there are many bioactive compounds that are able to exert an anti-cancer potential, re-inducing apoptosis or targeting multiple signaling pathways that allow for cancer cell survival, proliferation, growth, and metastatic progression of BC cells [134]. Many dietary compounds are also considered epigenetic modulating agents in cancer [135]. Thus, both green or black tea, as well as green or dark coffee, have been associated with a reduced BCR [136,137]. Chlorogenic acid (CGA) from coffee exerts an inhibitory role on signaling pathways, such as NF-κB/EMT [138]. Epigallocatchin-3-gallate from green tea significantly reduces BCR by decreasing ROS and oxidative DNA damage, mutagenesis, and tumor progression [137]. Resveratrol from grapes, berries, and nuts can reduce specific cancer stem cell (CSC) biomarkers in BC cells [139]. Piperine inhibits the growth of human BC cells, cell cycle progression, and BC cell migration [140]. Carotenoids have been associated with several metabolites involved in membrane signaling, immune regulation, redox balance, and epigenetic regulation [141]. One of the most active components of garlic (*Allium sativum*), allicin (diallylthiosulfinate), induces cell cycle arrest and has pro-apoptotic effects in BC cells, through p53 pathway activation [142], exerting antiproliferative, anticlonogenic, and senolytic effects, inducing the selective death of senescent cells [143]. Last but not least, the omega-3 polyunsaturated fatty acids, eicosapentaenoic acid (EPA) and docosahexaenoic acid (DHA), decreased tumor cell proliferation by downregulation of proliferation-associated protein expression (proliferating cell nuclear antigen (PCNA) and proliferation-related kinase (PRK), induced apoptosis by increasing caspase activity and DNA fragmentation, and decreased signal transduction through the Akt/NF-κB cell survival pathway [144].

5. Exposure to Endocrine-Disrupting Chemicals (EDCs)

EDCs are man-made chemicals ubiquitously found in the atmosphere as aerosols and particulate matter [145], water [146], pesticides [147], metals such as cadmium (Cd), mercury (Hg), arsenic (As), lead (Pb), manganese (Mn), and zinc (Zn) [148], additives or contaminated food such as dairy products, fish, meat, eggs, and vegetables, bottled water and canned food [149], and cosmetics and personal care products [150]. EDCs arrive in the human body through ingestion, inhalation, and/or the transdermal route, bioaccumulate, and interfere with endocrine, immune, and other systems, leading to a disruption of the endocrine signaling and metabolic pathways, and inducing life-long effects and negative consequences even for the next generation [151]. EDC exposure also interferes with placental function [152], can interfere with gamete quality, embryo implantation, and fetal development, with serious consequences for offspring viability and health [153]. EDCs affect epigenetic markers such as DNA methylation and histone posttranslational modifications (PTMs) [154]. In addition, EDCs increase incidence of BC [151].

EDCs are heterogeneous natural or synthetic compounds that include pharmaceutical agents (diethylstilbestrol (DES)), fungicides and pesticides (dichlorodiphenyltrichloroethane (DDT)), plastics (bisphenol A (BPA)), plasticizers (phthalates), and industrial solvents/lubricants (polychlorinated biphenyls (PCBs), polybrominated biphenyls (PBBs), and dioxins). Many EDCs are persistent organic pollutants (POPs), known as lipophilic toxicants that persist in the environment due to their resistance to biodegradation and, moreover, biomagnify or move up the food webs and increase in concentration [113]. POPs affect the production of estrogens and estrogenic signals, so that, measured in breast adipose tissue, POP levels were associated with higher BCR and worse prognosis [112].

Several pathogenic effects of EDC exposure are presented in Table 1. Thus, BPA stimulates the proliferation and malignancy of cancer cells through the activation of the Wnt/β-catenin pathway [155], which is widely implicated in the pathogenesis of metastatic BC [156]. Significant deregulated gene expression and transcriptional reprogramming in adult fibroblasts exposed to in utero BPA and DES, and specifically, changes in extracellular matrix (ECM) composition due to increased collagen deposition in adult mammary glands, lead to molecular alterations, which develop over time and contribute to increased BCR in adulthood [157]. Consequently, in utero exposure of the embryo to high maternal synthetic estrogens/EDCs could be associated with an increased BCR later in life [158]. Thus, BC may start in the womb, EDCs affecting the early development of mammary glands [159,160].

It is known that African Americans (AAs) are disproportionately exposed to elevated levels of BPA, so that the urinary BPA level among Black adults and children are statistically significantly higher compared to the non-Black population [161]. Recently, Zhang et al. (2023) used a metabolomics-based approach based on both ultra-performance and high-performance liquid chromatography tandem mass spectrometry (UPLC/HPLC-MS/MS) and demonstrated a high connection between tetrabromobisphenol A (TBBPA), a brominated derivative of bisphenol A (BPA) that is extensively present in the environment, with BC development [115]. In male and female rats and Rhesus monkey, low-dose exposure to BPA can affect mammary gland development, resulting in significant alterations in the gland morphology, inducing intraductal hyperplasia that could be associated with an increased BCR [162,163]. The normal-like human breast epithelial cell line, MCF-10F, after exposure to BPA, showed an increased expression of breast cancer genes *BRCA1/2*, *BRCA1* associated RING domain 1 (*BARD1*), choline transporter-like protein (*CtlP*), RAD51 recombinase (*RAD51*), and BRCA1/2-containing complex subunit 3 (*BRCC3*), which are all involved in DNA repair, as well as the silencing of programmed cell death protein 5 (*PDCD5*) and Bcl-2-like 11 (*BCL2L11* (BIM)), which are involved in apoptosis [164].

For example, the breasts are particularly susceptible to polycyclic aromatic hydrocarbons (PAHs) that can affect cell morphology, cell division, growth, and repair, cell–cell junctions, and the number of p53 mutations [165]. Moreover, Korsh et al. (2015) investigated the link between PAHs and BC based on the use of biomarkers in measuring PAH-DNA adducts to assess the exposure level [165]. Polychlorinated biphenyls (PCBs) are persistent industrial pollutants that have been linked to BC progression [166]. Thus, many authors concluded that early life exposure to PCBs is a factor of BCR [12,167,168]. The highly reactive PCB metabolite, 2,3,5-trichloro-6-phenyl-[1,4]-benzoquinone (PCB29-pQ), induces metastasis of BC and increases cancer stem cell (CSC) biomarker expression, resulting in an increase in EMT in MDA-MB-231 BC cells; the Wnt/β-catenin pathway is also activated by PCB29-pQ, due to overproduction of ROS [166]. Many authors concluded that early life exposure to PCBs is a factor of BCR [12,167,168].

Phthalates, phenols, and parabens are considered non-persistent EDCs that have been associated with BC [169]. Biomarker concentrations of non-persistent EDCs tend to be higher among women than men, and among Black Americans compared to White Americans, especially based on inconsistent access to healthy food or use of certain products with higher concentration of phthalates, such hair relaxers and skin lightening topical products, that specifically target Black consumers [169]. Some phthalates that mimic estradiol may promote BC, as in the case of dibutyl phthalate (DBP) exposure, which is associated with a two-fold increase in the rate of ER+ BC [170].

Parabens, such as methylparaben (MeP), ethylparaben (EtP), propylparaben (PrP), and butylparaben (BuP), are a group of alkyl esters of the para-hydroxybenzoic acid esters [171] that can mimic estrogen in the body [172]. These chemicals are used as broad-spectrum antimicrobial preservatives in lotions/creams, skin foundation, eye makeup products, deodorants/perfumes, hair care products, shaving products, toothpastes, shampoos/conditioners, pharmaceuticals, textiles, clothes, and processed foods [173,174]. Parabens are absorbed by the dermal route or ingested and are systematically distributed and metabolized, being detected in human normal and tumoral tissues [175], hair [176], blood [171], saliva [177], breast milk [178,179], placenta [180], and urine [181]. Parabens can be found intact in the human breast [175] and preferentially accumulate in metastatic breast tumors compared to benign breast tumors [174]. Tapia et al. (2023) reported altered ER target gene expression and cell viability that was paraben- and cell line-specific [172].

Table 1. Pathological effects of the exposure to ECDs.

EDCs	Pathological Effects of EDC Exposure	References
DES	in utero exposure dysregulates gene expression and transcriptional reprogramming in adult fibroblasts, ECM composition and collagen deposition in adult mammary gland, molecular alteration develops over time and contributes to increased BCR in adulthood, induces epigenetic alterations/epimutations with intergenerational/transgenerational effects	[157,182]
PAHs (BaP and DB(ah)A)	in mammary gland, affect cellular morphology, cell-cell junctions, division, growth, repair, and number of p53 mutations, increase EVs production, changes in exosome content and gene expression control	[99,165]

Table 1. *Cont.*

EDCs	Pathological Effects of EDC Exposure	References
BPA and other bisphenols (AF, F, S) and TBBPA	affect mammary gland development, resulting in precancerous and cancerous lesions in adulthood, exert estrogenic effects, activate the expression of genes associated with cell proliferation and BC; associated with EMT and BC progression; activate VEGF associated with angiogenesis, MAPK signaling pathway, Wnt/β-catenin pathway, STAT3 signaling, and DNA repair; induce changes in genes associated with apoptosis and DNA methylation; inactivate p53; increase expression of *BRCA1/2, BARD1, CtIP, RAD51,* and *BRCC3* involved in DNA repair; downregulate *PDCD5* and *BCL2L11* involved in apoptosis	[103,155,162–164,183,184]
Phtalates (DBP)	mimic estradiol, interact with ER and PR, promote BC, especially ER+ BC, interfere with DNA methylation and DNA damage	[170,185]
PCBs (PCB-153, PCB-180, PCB29-pQ)	BC cell proliferation by regulating ERK1/2 activation; induce cancer cell stemness and EMT via Wnt/β-catenin signaling	[166,184]
Organochlorine insecticides (DDT)	increase in utero BCR, BC progression by interfering with androgen signaling pathways, BC cells proliferation, negative effects on OS	[184,186]
Parabens (MeP, EtP, PrP, BuP) and their metabolites	promote protumorigenic effects in BC; modulate local estrogen-converting enzymes and increase local estrogen levels; cross-talk with HER2 pathway and affect ER signaling to increase pro-oncogenic c-Myc expression in ER+/HER2+ BC cells; alter ER target gene expression and cell viability	[172,173,181]

6. Conclusions

We are living in close interaction with a cocktail of man-made and natural xenobiotics. We are experiencing a wide spectrum of exposure during our lifetime, including the effects of xenobiotics on gametogenesis and gametes that undergo fertilization as the starting point of individual development and, moreover, in utero exposure that can initiate BC development. We are what we eat, we are what we breathe, and we are what we live. Most xenobiotics are metabolized or/and bioaccumulate and biomagnify in our tissues and cells, including breast tissues, so xenobiotic metabolism can play an important role in BC initiation and progression. This review pointed out the main mechanisms involved in the absorption, distribution, metabolism, bioaccumulation, biomagnification toxicity, and excretion of xenobiotics associated with BC risk, incidence, mortality, initiation, and progression. This association necessitates more valuable explanations at the biomolecular level to highlight the effects of genotoxic and epigenetic carcinogens. However, the accumulated xenobiotics, including their metabolites that arise as a consequence of biotransformation phases, such as heavy metals, endocrine-disrupting chemicals, or food contaminants, as well as a plethora of biomarkers of exposure, can be detected in breast tumoral tissues, adipose tissue, hair, blood, saliva, breast milk, placenta, and urine. In BC tissue biopsies and non-invasive liquid biopsies, xenobiotic exposure has been associated with changes in breast tissue composition and breast cell morphology, genomic instability, DNA damage, alteration of DNA repair, epimutations and epigenetic regulation, cell migration and invasion, angiogenesis, anti-apoptosis, cell adhesion, and cytoskeletal rearrangements, OS and ROS, metabolic reprogramming, immune regulation and metaflammation, membrane transport and signaling, extracellular matrix (ECM) and tumor microenvironment (TME) modifications, or extracellular vesicle (EV) production and content, with consequences in intercellular communication. At a biologic pathway level, most xenobiotics interact with endocrine signaling, adipogenesis, angiogenesis, DNA repair, inflammatory response, IGF-1 and NF-κB signaling, epithelial–mesenchymal transition (EMT), Wnt/β-catenin pathway, PI3K/Akt signaling, fatty acid metabolism (FAM) and glycolysis, MAPK, STAT3, p53 pathway, MYC targets, xenobiotic metabolism, and other cancer-related pathways. Fortunately, in our food, there are also many bioactive compounds with anti-tumor potential, which re-induce apoptosis by activation of caspases or target multiple signaling pathways, such as EMT migration-related pathway, Akt/NF-βB cell survival pathway, or p53 tumor suppressor signaling, that allow for cell survival, proliferation, growth, and metastatic progression of BC cells.

Consequently, BC can be characterized as an environmental disease or an ecological disorder. Evidence for BC risk suggests that food-borne chemical carcinogens, air pollution, ionizing radiation, and socioeconomic status are closely related to breast carcinogenesis. Thus, exposomics and the exposome concept are based on the diversity and range of exposures to physical factors, synthetic chemicals, dietary components, and psychosocial stressors, as well as their associated biological responses. Advances in molecular sciences and analytical techniques based on high-throughput sequencing and mass spectrometry (MS) have generated multi-omics data that can be successfully used to understand the complexity of molecular mechanisms involved in BC exposomics. Thus, environmental toxicogenomics, epigenomics, and interactomics, as well as nutrigenomics and nutriproteomics, metagenomics, micromiRomics, and nutrimiRomics are several new omics fields related to BC exposomics, which can contribute to molecular characterization of the complex relationship between the human body, environmental exposure, and breast cancer.

Author Contributions: Conceptualization, A.-N.N. and C.C.D.; literature search, A.-N.N., T.J., L.C., K.J. and C.C.D.; writing—original draft preparation, A.-N.N., T.J., L.C., K.J. and C.C.D.; writing—review and editing, A.-N.N., T.J., L.C., K.J. and C.C.D.; project administration, C.C.D.; funding acquisition, C.C.D. All authors have read and agreed to the published version of the manuscript.

Funding: This publication was supported in part by the National Cancer Institute of the National Institutes of Health under Award Number R15CA260126. The content is solely the responsibility of the authors and does not necessarily represent the official views of the National Institutes of Health.

Institutional Review Board Statement: Not applicable.

Informed Consent Statement: Not applicable.

Data Availability Statement: Data sharing is not applicable.

Acknowledgments: The authors thank the members of the Biochemistry and Proteomics Laboratories for the pleasant working environment.

Conflicts of Interest: The authors declare no conflicts of interest.

Abbreviations

BaP—benz(a)pyrene; BCR—breast cancer risk; BPA—bisphenol A; BuP—butylparaben; DB(ah)A—dibenz(ah)anthracene; DBP—dibutyl phthalate; DES—diethylstilbestrol; DDT—dichlorodiphenyl trichloroethane; ECM—extracellular matrix; EDC—endocrine-disrupting chemicals; EMT—epithelial–mesenchymal transition; ER—estrogen receptor; EtP—ethylparaben; EVs—extracellular vesicles;

MeP—methylparaben; OS—overall survival; PAHs—polycyclic aromatic hydrocarbons; PCBs—polychlorinated biphenyls; PCB-29-pQ—polychlorinated biphenyl quinone; PR—progesterone receptor; PrP—propylparaben; TBBPA—tetrabromobisphenol A; VEGF—vascular endothelial growth factor.

References

1. Lynn, H.; Ward, D.; Burton, D.; Day, J.; Craig, A.; Parnell, M.; Dimmer, C. Breast Cancer: An Environmental Disease. The Case for Primary Prevention. 2005. Available online: https://www.researchgate.net/publication/275209371_Breast_Cancer_an_environmental_disease_The_case_for_primary_prevention (accessed on 14 March 2024).
2. Hiatt, R.A.; Brody, J.G. Environmental Determinants of Breast Cancer. *Annu. Rev. Public Health* **2018**, *39*, 113–133. [CrossRef]
3. Neagu, A.-N.; Whitham, D.; Bruno, P.; Arshad, A.; Seymour, L.; Morrissiey, H.; Hukovic, A.I.; Darie, C.C. Onco-Breastomics: An Eco-Evo-Devo Holistic Approach. *Int. J. Mol. Sci.* **2024**, *25*, 1628. [CrossRef]
4. Mathipa, E.R.; Semuli, Q.K. We are what we eat. In *Rethinking Teaching and Learning in the 21st Century, Proceedings of the South Africa International Conference on Education, Pretoria, South Africa, 21–23 September 2015*; Manhattan Hotel Pretoria: Pretoria, South Africa, 2015.
5. Pretty, J. We are what we eat. *New Sci.* **2004**, *184*, 44–47.
6. Rumiati, R.I.; Foroni, F. We are what we eat: How food is represented in our mind/brain. *Psychon. Bull. Rev.* **2016**, *23*, 1043–1054. [CrossRef]
7. Hull, S.C.; Charles, J.; Caplan, A.L. Are We What We Eat? The Moral Imperative of the Medical Profession to Promote Plant-Based Nutrition. *Am. J. Cardiol.* **2023**, *188*, 15–21. [CrossRef]
8. Rajkhowa, S. "ARE WE WHAT WE EAT?": Understanding Identities through Food. Master's Thesis, Ambedkar University Delhi, Delhi, India, 2021.
9. Miller, M.R.; Shah, A.S.V.; Newby, D.E. We all breathe the same air . . . and we are all mortal. *Cardiovasc. Res.* **2020**, *116*, 1797–1799. [CrossRef]
10. Sorin Mihalache, A. How do We Live and what is the World We Live in Like? Some Possible Neuroscientific Evaluations on the Anthropology of the Spiritual Life in the Context of the Contemporary Society. *Glob. J. Anthropol. Res.* **2018**, *4*, 55–65. [CrossRef]
11. Strumylaitė, L.; Mechonošina, K.; Tamašauskas, Š. Environmental factors and breast cancer. *Medicina* **2010**, *46*, 867. [CrossRef] [PubMed]
12. Leng, L.; Li, J.; Luo, X.-M.; Kim, J.-Y.; Li, Y.-M.; Guo, X.-M.; Chen, X.; Yang, Q.-Y.; Li, G.; Tang, N.-J. Polychlorinated biphenyls and breast cancer: A congener-specific meta-analysis. *Environ. Int.* **2016**, *88*, 133–141. [CrossRef] [PubMed]
13. Bucher, M.L.; Anderson, F.L.; Lai, Y.; Dicent, J.; Miller, G.W.; Zota, A.R. Exposomics as a tool to investigate differences in health and disease by sex and gender. *Exposome* **2023**, *3*, osad003. [CrossRef] [PubMed]
14. Koual, M.; Tomkiewicz, C.; Cano-Sancho, G.; Antignac, J.-P.; Bats, A.-S.; Coumoul, X. Environmental chemicals, breast cancer progression and drug resistance. *Environ. Health* **2020**, *19*, 117. [CrossRef]
15. Feng, Y.; Spezia, M.; Huang, S.; Yuan, C.; Zeng, Z.; Zhang, L.; Ji, X.; Liu, W.; Huang, B.; Luo, W.; et al. Breast cancer development and progression: Risk factors, cancer stem cells, signaling pathways, genomics, and molecular pathogenesis. *Genes Dis.* **2018**, *5*, 77–106. [CrossRef] [PubMed]
16. Terry, M.B.; on behalf of Breast Cancer and the Environment Research Program (BCERP); Michels, K.B.; Brody, J.G.; Byrne, C.; Chen, S.; Jerry, D.J.; Malecki, K.M.C.; Martin, M.B.; Miller, R.L.; et al. Environmental exposures during windows of susceptibility for breast cancer: A framework for prevention research. *Breast Cancer Res.* **2019**, *21*, 96. [CrossRef]
17. Hiatt, R.A.; Beyeler, N. Women's cancers and climate change. *Int. J. Gynecol. Obstet.* **2023**, *160*, 374–377. [CrossRef]
18. Shih, Y.-W.; Hung, C.-S.; Huang, C.-C.; Chou, K.-R.; Niu, S.-F.; Chan, S.; Tsai, H.-T. The Association Between Smartphone Use and Breast Cancer Risk Among Taiwanese Women: A Case-Control Study. *Cancer Manag. Res.* **2020**, *12*, 10799–10807. [CrossRef] [PubMed]
19. Garcia-Saenz, A.; De Miguel, A.S.; Espinosa, A. Evaluating the Association between Artificial Light-at-Night Exposure and Breast and Prostate Cancer Risk in Spain (MCC-Spain Study). *Environ. Health Perspect.* **2018**, *126*, 047011. [CrossRef]
20. Hinchliffe, A.; Kogevinas, M.; Pérez-Gómez, B.; Ardanaz, E.; Amiano, P.; Marcos-Delgado, A.; Castaño-Vinyals, G.; Llorca, J.; Moreno, V.; Alguacil, J.; et al. Occupational Heat Exposure and Breast Cancer Risk in the MCC-Spain Study. *Cancer Epidemiol. Biomark. Prev.* **2021**, *30*, 364–372. [CrossRef]
21. Gera, R.; Mokbel, R.; Igor, I.; Mokbel, K. Does the Use of Hair Dyes Increase the Risk of Developing Breast Cancer? A Meta-analysis and Review of the Literature. *Anticancer Res.* **2018**, *38*, 707–716. [CrossRef] [PubMed]
22. Jones, M.E.; Schoemaker, M.J.; Wright, L.B.; Ashworth, A.; Swerdlow, A.J. Smoking and risk of breast cancer in the Generations Study cohort. *Breast Cancer Res.* **2017**, *19*, 118. [CrossRef]
23. Huynh, D.; Huang, J.; Le, L.T.T.; Liu, D.; Liu, C.; Pham, K.; Wang, H. Electronic cigarettes promotes the lung colonization of human breast cancer in NOD-SCID-Gamma mice. *Int. J. Clin. Exp. Pathol.* **2020**, *13*, 2075–2081.
24. Shih, Y.W.; O'brien, A.P.; Hung, C.S.; Chen, K.H.; Hou, W.H.; Tsai, H.T. Exposure to radiofrequency radiation increases the risk of breast cancer: A systematic review and meta-analysis. *Exp. Ther. Med.* **2021**, *21*, 23. [CrossRef] [PubMed]

25. Mortazavi, A.R.; Mortazavi, S.M.J. Women with hereditary breast cancer predispositions should avoid using their smartphones, tablets and laptops at night. *Iran. J. Basic Med. Sci.* **2018**, *21*, 112–115. [PubMed]

26. Vinogradova, Y.; Coupland, C.; Hippisley-Cox, J. Use of hormone replacement therapy and risk of breast cancer: Nested case-control studies using the QResearch and CPRD databases. *BMJ* **2020**, *371*, m3873. [CrossRef]

27. de Blok, C.J.M.; Wiepjes, C.M.; Nota, N.M.; van Engelen, K.; Adank, M.A.; Dreijerink, K.M.A.; Barbé, E.; Konings, I.R.H.M.; den Heijer, M. Breast cancer risk in transgender people receiving hormone treatment: Nationwide cohort study in The Netherlands. *BMJ* **2019**, *365*, l1652. [CrossRef] [PubMed]

28. Sørensen, M.; Poulsen, A.H.; Kroman, N.; Hvidtfeldt, U.A.; Thacher, J.D.; Roswall, N.; Brandt, J.; Frohn, L.M.; Jensen, S.S.; Levin, G.; et al. Road and railway noise and risk for breast cancer: A nationwide study covering Denmark. *Environ. Res.* **2021**, *195*, 110739. [CrossRef]

29. Andersen, Z.J.; Jørgensen, J.T.; Elsborg, L.; Lophaven, S.N.; Backalarz, C.; Laursen, J.E.; Pedersen, T.H.; Simonsen, M.K.; Bräuner, E.V.; Lynge, E. Long-term exposure to road traffic noise and incidence of breast cancer: A cohort study. *Breast Cancer Res.* **2018**, *20*, 119. [CrossRef]

30. Xiang, P.; Wang, K.; Bi, J.; Li, M.; He, R.-W.; Cui, D.; Ma, L.Q. Organic extract of indoor dust induces estrogen-like effects in human breast cancer cells. *Sci. Total Environ.* **2020**, *726*, 138505. [CrossRef] [PubMed]

31. Afzal, S.; Fiaz, K.; Noor, A.; Sindhu, A.S.; Hanif, A.; Bibi, A.; Asad, M.; Nawaz, S.; Zafar, S.; Ayub, S.; et al. Interrelated Oncogenic Viruses and Breast Cancer. *Front. Mol. Biosci.* **2022**, *9*, 781111. [CrossRef]

32. Vermeulen, R.; Schymanski, E.L.; Barabási, A.-L.; Miller, G.W. The exposome and health: Where chemistry meets biology. *Science* **2020**, *367*, 392–396. [CrossRef]

33. Ziegler, R.G.; Hoover, R.N.; Pike, M.C.; Hildesheim, A.; Nomura, A.M.Y.; West, D.W.; Wu-Williams, A.H.; Kolonel, L.N.; Horn-Ross, P.L.; Rosenthal, J.F.; et al. Migration Patterns and Breast Cancer Risk in Asian-American Women. *J. Natl. Cancer Inst.* **1993**, *85*, 1819–1827. [CrossRef]

34. Christ, A.; Latz, E. The Western lifestyle has lasting effects on metaflammation. *Nat. Rev. Immunol.* **2019**, *19*, 267–268. [CrossRef]

35. Danforth, D.N. The Role of Chronic Inflammation in the Development of Breast Cancer. *Cancers* **2021**, *13*, 3918. [CrossRef]

36. Itoh, H.; Ueda, M.; Suzuki, M.; Kohmura-Kobayashi, Y. Developmental Origins of Metaflammation; A Bridge to the Future Between the DOHaD Theory and Evolutionary Biology. *Front. Endocrinol.* **2022**, *13*, 839436. [CrossRef] [PubMed]

37. Wani, B.; Aziz, S.A.; Ganaie, M.A.; Mir, M.H. Metabolic Syndrome and Breast Cancer Risk. *Indian J. Med. Paediatr. Oncol.* **2017**, *38*, 434–439. [CrossRef] [PubMed]

38. Dong, S.; Wang, Z.; Shen, K.; Chen, X. Metabolic Syndrome and Breast Cancer: Prevalence, Treatment Response, and Prognosis. *Front. Oncol.* **2021**, *11*, 629666. [CrossRef] [PubMed]

39. Choi, I.Y.; Chun, S.; Shin, D.W.; Han, K.; Jeon, K.H.; Yu, J.; Chae, B.J.; Suh, M.; Park, Y.-M. Changes in Metabolic Syndrome Status and Breast Cancer Risk: A Nationwide Cohort Study. *Cancers* **2021**, *13*, 1177. [CrossRef] [PubMed]

40. Starek-Świechowicz, B.; Budziszewska, B.; Starek, A. Endogenous estrogens—Breast cancer and chemoprevention. *Pharmacol. Rep.* **2021**, *73*, 1497–1512. [CrossRef]

41. Autrup, H.; Barile, F.A.; Berry, S.C.; Blaauboer, B.J.; Boobis, A.; Bolt, H.; Borgert, C.J.; Dekant, W.; Dietrich, D.; Domingo, J.L.; et al. Human exposure to synthetic endocrine disrupting chemicals (S-EDCs) is generally negligible as compared to natural compounds with higher or comparable endocrine activity: How to evaluate the risk of the S-EDCs? *Arch. Toxicol.* **2020**, *94*, 2549–2557. [CrossRef] [PubMed]

42. Kazemi, A.; Barati-Boldaji, R.; Soltani, S.; Mohammadipoor, N.; Esmaeilinezhad, Z.; Clark, C.C.T.; Babajafari, S.; Akbarzadeh, M. Intake of Various Food Groups and Risk of Breast Cancer: A Systematic Review and Dose-Response Meta-Analysis of Prospective Studies. *Adv. Nutr.* **2021**, *12*, 809–849. [CrossRef]

43. Merugu, N.K.; Manapuram, S.; Chakraborty, T.; Karanam, S.K.; Imandi, S.B. Mutagens in commercial food processing and its microbial transformation. *Food Sci. Biotechnol.* **2023**, *32*, 599–620. [CrossRef]

44. Thakkar, N.; Bin Shin, Y.; Sung, H.-K. Nutritional Regulation of Mammary Tumor Microenvironment. *Front. Cell Dev. Biol.* **2022**, *10*, 803280. [CrossRef]

45. Wiseman, M. The Second World Cancer Research Fund/American Institute for Cancer Research Expert Report. Food, Nutrition, Physical Activity, and the Prevention of Cancer: A Global Perspective: Nutrition Society and BAPEN Medical Symposium on 'Nutrition support in cancer therapy'. *Proc. Nutr. Soc.* **2008**, *67*, 253–256.

46. Wan, M.L.Y.; Co, V.A.; El-Nezami, H. Endocrine disrupting chemicals and breast cancer: A systematic review of epidemiological studies. *Crit. Rev. Food Sci. Nutr.* **2022**, *62*, 6549–6576. [CrossRef]

47. Darbre, P.D. Chapter Thirteen—Endocrine disrupting chemicals and breast cancer cells. In *Advances in Pharmacology*; Vandenberg, L.N., Turgeon, J.L., Eds.; Academic Press: Cambridge, MA, USA, 2021; pp. 485–520.

48. Jacobs, I.; Taljaard-Krugell, C.; Wicks, M.; Cubasch, H.; Joffe, M.; Laubscher, R.; Romieu, I.; Biessy, C.; Rinaldi, S.; Huybrechts, I. Dietary Patterns and Breast Cancer Risk in Black Urban South African Women: The SABC Study. *Nutrients* **2021**, *13*, 4106. [CrossRef]

49. McDonald, J.A.; Goyal, A.; Terry, M.B. Alcohol Intake and Breast Cancer Risk: Weighing the Overall Evidence. *Curr. Breast Cancer Rep.* **2013**, *5*, 208–221. [CrossRef]

50. Wang, Y.; Xu, M.; Ke, Z.-J.; Luo, J. Cellular and Molecular Mechanism Underlying Alcohol-induced Aggressiveness of Breast Cancer. *Pharmacol. Res.* **2017**, *115*, 299–308. [CrossRef] [PubMed]

51. VoPham, T.; Bertrand, K.A.; Jones, R.R.; Deziel, N.C.; DuPré, N.C.; James, P.; Liu, Y.; Vieira, V.M.; Tamimi, R.M.; Hart, J.E.; et al. Dioxin exposure and breast cancer risk in a prospective cohort study. *Environ. Res.* **2020**, *186*, 109516. [CrossRef] [PubMed]

52. Lee, P.M.Y.; Chan, W.C.; Kwok, C.C.-H.; Wu, C.; Law, S.-H.; Tsang, K.-H.; Yu, W.-C.; Yeung, Y.-C.; Chang, L.D.J.; Wong, C.K.M.; et al. Associations between Coffee Products and Breast Cancer Risk: A Case-Control study in Hong Kong Chinese Women. *Sci. Rep.* **2019**, *9*, 12684. [CrossRef]

53. Fiolet, T.; Srour, B.; Sellem, L.; Kesse-Guyot, E.; Allès, B.; Méjean, C.; Deschasaux, M.; Fassier, P.; Latino-Martel, P.; Touvier, M.; et al. Consumption of ultra-processed foods and cancer risk: Results from NutriNet-Santé prospective cohort. *BMJ* **2018**, *360*, k322. [CrossRef] [PubMed]

54. Lo, J.J.; Park, Y.-M.; Sinha, R.; Sandler, D.P. Association between meat consumption and risk of breast cancer: Findings from the Sister Study. *Int. J. Cancer* **2019**, *146*, 2156–2165. [CrossRef]

55. Chazelas, E.; Srour, B.; Desmetz, E.; Kesse-Guyot, E.; Julia, C.; Deschamps, V.; Druesne-Pecollo, N.; Galan, P.; Hercberg, S.; Latino-Martel, P.; et al. Sugary drink consumption and risk of cancer: Results from NutriNet-Santé prospective cohort. *BMJ* **2019**, *366*, l2408. [CrossRef]

56. Eve, L.; Fervers, B.; Le Romancer, M.; Etienne-Selloum, N. Exposure to Endocrine Disrupting Chemicals and Risk of Breast Cancer. *Int. J. Mol. Sci.* **2020**, *21*, 9139. [CrossRef]

57. Shen, J.; Liao, Y.; Hopper, J.L.; Goldberg, M.; Santella, R.M.; Terry, M.B. Dependence of cancer risk from environmental exposures on underlying genetic susceptibility: An illustration with polycyclic aromatic hydrocarbons and breast cancer. *Br. J. Cancer* **2017**, *116*, 1229–1233. [CrossRef]

58. Keren, Y.; Magnezi, R.; Carmon, M.; Amitai, Y. Investigation of the Association between Drinking Water Habits and the Occurrence of Women Breast Cancer. *Int. J. Environ. Res. Public Health* **2020**, *17*, 7692. [CrossRef] [PubMed]

59. Maltarollo, V.G.; Gertrudes, J.C.; Oliveira, P.R.; Honorio, K.M. Applying machine learning techniques for ADME-Tox prediction: A review. *Expert Opin. Drug Metab. Toxicol.* **2015**, *11*, 259–271. [CrossRef]

60. Neagu, A.-N.; Whitham, D.; Bruno, P.; Morrissiey, H.; Darie, C.A.; Darie, C.C. Omics-Based Investigations of Breast Cancer. *Molecules* **2023**, *28*, 4768. [CrossRef] [PubMed]

61. Rasool, R.; Ullah, I.; Mubeen, B.; Alshehri, S.; Imam, S.S.; Ghoneim, M.M.; Alzarea, S.I.; Al-Abbasi, F.A.; Murtaza, B.N.; Kazmi, I.; et al. Theranostic Interpolation of Genomic Instability in Breast Cancer. *Int. J. Mol. Sci.* **2022**, *23*, 1861. [CrossRef] [PubMed]

62. Portugal, J.; Mansilla, S.; Piña, B. Perspectives on the Use of Toxicogenomics to Assess Environmental Risk. *Front. Biosci.* **2022**, *27*, 294. [CrossRef] [PubMed]

63. Santibáñez-Andrade, M.; Quezada-Maldonado, E.M.; Osornio-Vargas, Á.; Sánchez-Pérez, Y.; García-Cuellar, C.M. Air pollution and genomic instability: The role of particulate matter in lung carcinogenesis. *Environ. Pollut.* **2017**, *229*, 412–422. [CrossRef] [PubMed]

64. Coppedè, F. Genes and the Environment in Cancer: Focus on Environmentally Induced DNA Methylation Changes. *Cancers* **2023**, *15*, 1019. [CrossRef] [PubMed]

65. Offiong, N.-A.O.; Edet, J.B.; Shaibu, S.E.; Akan, N.E.; Atakpa, E.O.; Sanganyado, E.; Okop, I.J.; Benson, N.U.; Okoh, A. Metagenomics: An emerging tool for the chemistry of environmental remediation. *Front. Environ. Chem.* **2023**, *4*, 1052697. [CrossRef]

66. Zhang, Y.; Keerthisinghe, T.P.; Han, Y.; Liu, M.; Wanjaya, E.R.; Fang, M. "Cocktail" of Xenobiotics at Human Relevant Levels Reshapes the Gut Bacterial Metabolome in a Species-Specific Manner. *Environ. Sci. Technol.* **2018**, *52*, 11402–11410. [CrossRef] [PubMed]

67. Kartti, S.; Bendani, H.; Boumajdi, N.; Bouricha, E.M.; Zarrik, O.; EL Agouri, H.; Fokar, M.; Aghlallou, Y.; EL Jaoudi, R.; Belyamani, L.; et al. Metagenomics Analysis of Breast Microbiome Highlights the Abundance of Rothia Genus in Tumor Tissues. *J. Pers. Med.* **2023**, *13*, 450. [CrossRef] [PubMed]

68. Iida, M.; Takemoto, K. A network biology-based approach to evaluating the effect of environmental contaminants on human interactome and diseases. *Ecotoxicol. Environ. Saf.* **2018**, *160*, 316–327. [CrossRef] [PubMed]

69. Moslehi, R.; Stagnar, C.; Srinivasan, S.; Radziszowski, P.; Carpenter, D.O. The possible role of arsenic and gene-arsenic interactions in susceptibility to breast cancer: A systematic review. *Rev. Environ. Health* **2021**, *36*, 523–534. [CrossRef] [PubMed]

70. Lim CJ, C.; Lim PP, C.; Pizarro RR, M.; Segocio HG, B.; Ratta, K. 8—Nutrigenomics in the management and prevention of cancer. In *Role of Nutrigenomics in Modern-Day Healthcare and Drug Discovery*; Dable-Tupas, G., Egbuna, C., Eds.; Elsevier: Amsterdam, The Netherlands, 2023; pp. 177–208.

71. Sellami, M.; Bragazzi, N.L. Nutrigenomics and Breast Cancer: State-of-Art, Future Perspectives and Insights for Prevention. *Nutrients* **2020**, *12*, 512. [CrossRef]

72. Schraml, E.; Hackl, M.; Grillari, J. MicroRNAs and toxicology: A love marriage. *Toxicol. Rep.* **2017**, *4*, 634–636. [CrossRef]

73. Singh, R.; Mo, Y.-Y. Role of microRNAs in breast cancer. *Cancer Biol. Ther.* **2013**, *14*, 201–212. [CrossRef]

74. Balasubramanian, S.; Gunasekaran, K.; Sasidharan, S.; Mathan, V.J.; Perumal, E. MicroRNAs and Xenobiotic Toxicity: An Overview. *Toxicol. Rep.* **2020**, *7*, 583–595. [CrossRef] [PubMed]

75. Dai, T.; Zhao, X.; Li, Y.; Yu, L.; Li, Y.; Zhou, X.; Gong, Q. miR-423 Promotes Breast Cancer Invasion by Activating NF-κB Signaling. *OncoTargets Ther.* **2020**, *13*, 5467–5478. [CrossRef] [PubMed]

76. Morales-Pison, S.; Jara, L.; Carrasco, V.; Gutiérrez-Vera, C.; Reyes, J.M.; Gonzalez-Hormazabal, P.; Carreño, L.J.; Tapia, J.C.; Contreras, H.R. Genetic Variation in MicroRNA-423 Promotes Proliferation, Migration, Invasion, and Chemoresistance in Breast Cancer Cells. *Int. J. Mol. Sci.* **2022**, *23*, 380. [CrossRef]

77. Farahani, M.; Rezaei-Tavirani, M.; Arjmand, B. A systematic review of microRNA expression studies with exposure to bisphenol A. *J. Appl. Toxicol.* **2021**, *41*, 4–19. [CrossRef]

78. Quintanilha, B.J.; Reis, B.Z.; Duarte, G.B.S.; Cozzolino, S.M.F.; Rogero, M.M. Nutrimiromics: Role of microRNAs and Nutrition in Modulating Inflammation and Chronic Diseases. *Nutrients* **2017**, *9*, 1168. [CrossRef]

79. Venkatadri, R.; Muni, T.; Iyer, A.K.V.; Yakisich, J.S.; Azad, N. Role of apoptosis-related miRNAs in resveratrol-induced breast cancer cell death. *Cell Death Dis.* **2016**, *7*, e2104. [CrossRef]

80. Santoro, K.L.; Yakah, W.; Singh, P.; Ramiro-Cortijo, D.; Medina-Morales, E.; Freedman, S.D.; Martin, C.R. Acetaminophen and Xenobiotic Metabolites in Human Milk and the Development of Bronchopulmonary Dysplasia and Retinopathy of Prematurity in a Cohort of Extremely Preterm Infants. *J. Pediatr.* **2022**, *244*, 224–229.e3. [CrossRef] [PubMed]

81. Rosen, M.; Bulucea, C.; Brindusa, C.; Mastorakis, N. Approaching Resonant Absorption of Environmental Xenobiotics Harmonic Oscillation by Linear Structures. *Sustainability* **2012**, *4*, 561–573.

82. Idle, J.R.; Gonzalez, F.J. Metabolomics. *Cell Metab.* **2007**, *6*, 348–351. [CrossRef]

83. Williams, J.A.; Phillips, D.H. Mammary Expression of Xenobiotic Metabolizing Enzymes and Their Potential Role in Breast Cancer. *Cancer Res.* **2000**, *60*, 4667–4677. [PubMed]

84. Swift, L.H.; Golsteyn, R.M. Genotoxic Anti-Cancer Agents and Their Relationship to DNA Damage, Mitosis, and Checkpoint Adaptation in Proliferating Cancer Cells. *Int. J. Mol. Sci.* **2014**, *15*, 3403–3431. [CrossRef]

85. Zahreddine, H.; Borden, K.L.B. Mechanisms and insights into drug resistance in cancer. *Front. Pharmacol.* **2013**, *4*, 28. [CrossRef] [PubMed]

86. Rylander, C.; Veierød, M.B.; Weiderpass, E.; Lund, E.; Sandanger, T.M. Use of skincare products and risk of cancer of the breast and endometrium: A prospective cohort study. *Environ. Health* **2019**, *18*, 105. [CrossRef]

87. Filipiuc, S.-I.; Neagu, A.-N.; Uritu, C.M.; Tamba, B.-I.; Filipiuc, L.-E.; Tudorancea, I.M.; Boca, A.N.; Hâncu, M.F.; Porumb, V.; Bild, W. The Skin and Natural Cannabinoids–Topical and Transdermal Applications. *Pharmaceuticals* **2023**, *16*, 1049. [CrossRef]

88. Andersen, E.R.; Eilertsen, G.; Myklebust, A.M.; Eriksen, S. Women's experience of acute skin toxicity following radiation therapy in breast cancer. *J. Multidisci. Healthc.* **2018**, *11*, 139–148. [CrossRef] [PubMed]

89. Gregoire, A.M.; VoPham, T.; Laden, F.; Yarosh, R.; O'Brien, K.M.; Sandler, D.P.; White, A.J. Residential ultraviolet radiation and breast cancer risk in a large prospective cohort. *Environ. Int.* **2022**, *159*, 107028. [CrossRef] [PubMed]

90. Chee, Y.C.; Pahnke, J.; Bunte, R.; Adsool, V.A.; Madan, B.; Virshup, D.M. Intrinsic Xenobiotic Resistance of the Intestinal Stem Cell Niche. *Dev. Cell* **2018**, *46*, 681–695.e5. [CrossRef] [PubMed]

91. Wen, L.; Han, Z. Identification and validation of xenobiotic metabolism-associated prognostic signature based on five genes to evaluate immune microenvironment in colon cancer. *J. Gastrointest. Oncol.* **2021**, *12*, 2788–2802. [CrossRef] [PubMed]

92. White, A.J.; Gregoire, A.M.; Fisher, J.A.; Medgyesi, D.N.; Li, L.; Koutrakis, P.; Sandler, D.P.; Jones, R.R. Exposure to Particle Radioactivity and Breast Cancer Risk in the Sister Study: A U.S.-Wide Prospective Cohort. *Environ. Health Perspect.* **2022**, *130*, 047701. [CrossRef]

93. Olesiejuk, K.; Chałubiński, M. How does particulate air pollution affect barrier functions and inflammatory activity of lung vascular endothelium? *Allergy* **2023**, *78*, 629–638. [CrossRef] [PubMed]

94. Smotherman, C.; Sprague, B.; Datta, S.; Braithwaite, D.; Qin, H.; Yaghjyan, L. Association of air pollution with postmenopausal breast cancer risk in UK Biobank. *Breast Cancer Res.* **2023**, *25*, 83. [CrossRef] [PubMed]

95. Moriceau, M.-A.; Cano-Sancho, G.; Kim, M.; Coumoul, X.; Emond, C.; Arrebola, J.-P.; Antignac, J.-P.; Audouze, K.; Rousselle, C. Partitioning of Persistent Organic Pollutants between Adipose Tissue and Serum in Human Studies. *Toxics* **2023**, *11*, 41. [CrossRef]

96. Ish, J.L.; Abubakar, M.; Fan, S.; Jones, R.R.; Niehoff, N.M.; Henry, J.E.; Gierach, G.L.; White, A.J. Outdoor air pollution and histologic composition of normal breast tissue. *Environ. Int.* **2023**, *176*, 107984. [CrossRef]

97. Segovia-Mendoza, M.; de León, C.T.G.; García-Becerra, R.; Ambrosio, J.; Nava-Castro, K.E.; Morales-Montor, J. The chemical environmental pollutants BPA and BPS induce alterations of the proteomic profile of different phenotypes of human breast cancer cells: A proposed interactome. *Environ. Res.* **2020**, *191*, 109960. [CrossRef]

98. Yuan, J.; Liu, Y.; Wang, J.; Zhao, Y.; Li, K.; Jing, Y.; Zhang, X.; Liu, Q.; Geng, X.; Li, G.; et al. Long-term Persistent Organic Pollutants Exposure Induced Telomere Dysfunction and Senescence-Associated Secretary Phenotype. *J. Gerontol. Ser. A* **2018**, *73*, 1027–1035. [CrossRef]

99. Montjean, D.; Neyroud, A.-S.; Yefimova, M.G.; Benkhalifa, M.; Cabry, R.; Ravel, C. Impact of Endocrine Disruptors upon Non-Genetic Inheritance. *Int. J. Mol. Sci.* **2022**, *23*, 3350. [CrossRef] [PubMed]

100. Dungen, M.W.v.D.; Murk, A.J.; Kok, D.E.; Steegenga, W.T. Persistent organic pollutants alter DNA methylation during human adipocyte differentiation. *Toxicol. Vitr.* **2017**, *40*, 79–87. [CrossRef] [PubMed]

101. Wielsøe, M.; Tarantini, L.; Bollati, V.; Long, M.; Bonefeld-Jørgensen, E.C. DNA methylation level in blood and relations to breast cancer, risk factors and environmental exposure in Greenlandic Inuit women. *Basic Clin. Pharmacol. Toxicol.* **2020**, *127*, 338–350. [CrossRef] [PubMed]

102. Zucchini-Pascal, N.; Peyre, L.; de Sousa, G.; Rahmani, R. Organochlorine pesticides induce epithelial to mesenchymal transition of human primary cultured hepatocytes. *Food Chem. Toxicol.* **2012**, *50*, 3963–3970. [CrossRef] [PubMed]

103. Lee, H.-M.; Hwang, K.-A.; Choi, K.-C. Diverse pathways of epithelial mesenchymal transition related with cancer progression and metastasis and potential effects of endocrine disrupting chemicals on epithelial mesenchymal transition process. *Mol. Cell. Endocrinol.* **2016**, *457*, 103–113. [CrossRef] [PubMed]
104. Zhang, X.-L.; Wang, H.-S.; Liu, N.; Ge, L.-C. Bisphenol A stimulates the epithelial mesenchymal transition of estrogen negative breast cancer cells via FOXA1 signals. *Arch. Biochem. Biophys.* **2015**, *585*, 10–16. [CrossRef] [PubMed]
105. Johnson, C.H.; Patterson, A.D.; Idle, J.R.; Gonzalez, F.J. Xenobiotic Metabolomics: Major Impact on the Metabolome. *Annu. Rev. Pharmacol. Toxicol.* **2012**, *52*, 37–56. [CrossRef]
106. Bièche, I.; Girault, I.; Urbain, E.; Tozlu, S.; Lidereau, R. Relationship between intratumoral expression of genes coding for xenobiotic-metabolizing enzymes and benefit from adjuvant tamoxifen in estrogen receptor alpha-positive postmenopausal breast carcinoma. *Breast Cancer Res.* **2004**, *6*, R252. [CrossRef] [PubMed]
107. Li, Y.; Steppi, A.; Zhou, Y.; Mao, F.; Miller, P.C.; He, M.M.; Zhao, T.; Sun, Q.; Zhang, J. Tumoral expression of drug and xenobiotic metabolizing enzymes in breast cancer patients of different ethnicities with implications to personalized medicine. *Sci. Rep.* **2017**, *7*, 4747. [CrossRef] [PubMed]
108. Men, Y.; Li, L.; Zhang, F.; Kong, X.; Zhang, W.; Hao, C.; Wang, G. Evaluation of heavy metals and metabolites in the urine of patients with breast cancer. *Oncol. Lett.* **2020**, *19*, 1331–1337. [CrossRef] [PubMed]
109. Tarhonska, K.; Janasik, B.; Roszak, J.; Kowalczyk, K.; Lesicka, M.; Reszka, E.; Wieczorek, E.; Braun, M.; Kolacinska-Wow, A.; Skokowski, J.; et al. Environmental exposure to cadmium in breast cancer—Association with the Warburg effect and sensitivity to tamoxifen. *Biomed. Pharmacother.* **2023**, *161*, 114435. [CrossRef]
110. Gadzała-Kopciuch, R.; Pajewska-Szmyt, M. Human Milk and Xenobiotics. In *Handbook of Bioanalytics*; Buszewski, B., Baranowska, I., Eds.; Springer International Publishing: Cham, Switzerland, 2022; pp. 295–308.
111. Mead, M. Contaminants in Human Milk: Weighing the Risks against the Benefits of Breastfeeding. *Environ. Health Perspect.* **2008**, *116*, A427–A434. [CrossRef] [PubMed]
112. Ennour-Idrissi, K.; Ayotte, P.; Diorio, C. Persistent Organic Pollutants and Breast Cancer: A Systematic Review and Critical Appraisal of the Literature. *Cancers* **2019**, *11*, 1063. [CrossRef] [PubMed]
113. Jackson, E.; Shoemaker, R.; Larian, N.; Cassis, L. Adipose Tissue as a Site of Toxin Accumulation. *Compr. Physiol.* **2017**, *7*, 1085–1135.
114. Lee, Y.; Kim, K.; Jacobs, D.R.; Lee, D. Persistent organic pollutants in adipose tissue should be considered in obesity research. *Obes. Rev.* **2017**, *18*, 129–139. [CrossRef]
115. Zhang, A.; Wang, R.; Liu, Q.; Yang, Z.; Lin, X.; Pang, J.; Li, X.; Wang, D.; He, J.; Li, J.; et al. Breast adipose metabolites mediates the association of tetrabromobisphenol a with breast cancer: A case-control study in Chinese population. *Environ. Pollut.* **2023**, *316*, 120701. [CrossRef]
116. Liu, T.; Liang, X.; Lei, C.; Huang, Q.; Song, W.; Fang, R.; Li, C.; Li, X.; Mo, H.; Sun, N.; et al. High-Fat Diet Affects Heavy Metal Accumulation and Toxicity to Mice Liver and Kidney Probably via Gut Microbiota. *Front. Microbiol.* **2020**, *11*, 1604. [CrossRef]
117. Romaniuk, A.; Lyndin, M.; Sikora, V.; Lyndina, Y.; Romaniuk, S.; Sikora, K. Heavy metals effect on breast cancer progression. *J. Occup. Med. Toxicol.* **2017**, *12*, 32. [CrossRef]
118. Wang, X.; Mukherjee, B.; Park, S.K. Associations of cumulative exposure to heavy metal mixtures with obesity and its comorbidities among U.S. adults in NHANES 2003–2014. *Environ. Int.* **2018**, *121*, 683–694. [CrossRef] [PubMed]
119. Teo, P.S.; van Dam, R.M.; Whitton, C.; Tan, L.W.L.; Forde, C.G. Consumption of Foods With Higher Energy Intake Rates is Associated With Greater Energy Intake, Adiposity, and Cardiovascular Risk Factors in Adults. *J. Nutr.* **2021**, *151*, 370–378. [CrossRef]
120. Bernard, J.J.; Wellberg, E.A. The Tumor Promotional Role of Adipocytes in the Breast Cancer Microenvironment and Macroenvironment. *Am. J. Pathol.* **2021**, *191*, 1342–1352. [CrossRef] [PubMed]
121. Kothari, C.; Diorio, C.; Durocher, F. The Importance of Breast Adipose Tissue in Breast Cancer. *Int. J. Mol. Sci.* **2020**, *21*, 5760. [CrossRef] [PubMed]
122. Tiwari, S.; Sapkota, N.; Han, Z. Effect of fasting on cancer: A narrative review of scientific evidence. *Cancer Sci.* **2022**, *113*, 3291–3302. [CrossRef]
123. Anemoulis, M.; Vlastos, A.; Kachtsidis, V.; Karras, S.N. Intermittent Fasting in Breast Cancer: A Systematic Review and Critical Update of Available Studies. *Nutrients* **2023**, *15*, 532. [CrossRef] [PubMed]
124. Melnik, B.C.; John, S.M.; Schmitz, G. Over-stimulation of insulin/IGF-1 signaling by Western diet may promote diseases of civilization: Lessons learnt from Laron syndrome. *Nutr. Metab.* **2011**, *8*, 41. [CrossRef]
125. Christopoulos, P.F.; Msaouel, P.; Koutsilieris, M. The role of the insulin-like growth factor-1 system in breast cancer. *Mol. Cancer* **2015**, *14*, 43. [CrossRef]
126. Cunha, S.C.; Menezes-Sousa, D.; Mello, F.V.; Miranda, J.A.; Fogaca, F.H.; Alonso, M.B.; Torres, J.P.M.; Fernandes, J.O. Survey on endocrine-disrupting chemicals in seafood: Occurrence and distribution. *Environ. Res.* **2022**, *210*, 112886. [CrossRef]
127. Law, A.Y.S.; Wei, X.; Zhang, X.; Mak, N.K.; Cheung, K.C.; Wong, M.H.; Giesy, J.P.; Wong, C.K.C. Biological analysis of endocrine-disrupting chemicals in animal meats from the Pearl River Delta, China. *J. Expo. Sci. Environ. Epidemiol.* **2012**, *22*, 93–100. [CrossRef] [PubMed]
128. Mukherjee, R.; Pandya, P.; Baxi, D.; Ramachandran, A.V. Endocrine Disruptors–'Food' for Thought. *Proc. Zool. Soc.* **2021**, *74*, 432–442. [CrossRef] [PubMed]

129. Chang, J.; Zhou, J.; Gao, M.; Zhang, H.; Wang, T. Research Advances in the Analysis of Estrogenic Endocrine Disrupting Compounds in Milk and Dairy Products. *Foods* **2022**, *11*, 3057. [CrossRef] [PubMed]

130. Djedjibegovic, J.; Marjanovic, A.; Tahirovic, D.; Caklovica, K.; Turalic, A.; Lugusic, A.; Omeragic, E.; Sober, M.; Caklovica, F. Heavy metals in commercial fish and seafood products and risk assessment in adult population in Bosnia and Herzegovina. *Sci. Rep.* **2020**, *10*, 13238. [CrossRef]

131. Nindrea, R.D.; Aryandono, T.; Lazuardi, L.; Dwiprahasto, I. Protective Effect of Omega-3 Fatty Acids in Fish Consumption Against Breast Cancer in Asian Patients: A Meta-Analysis. *Asian Pac. J. Cancer Prev.* **2019**, *20*, 327–332. [CrossRef] [PubMed]

132. Rodgers, K.M.; Udesky, J.O.; Rudel, R.A.; Brody, J.G. Environmental chemicals and breast cancer: An updated review of epidemiological literature informed by biological mechanisms. *Environ. Res.* **2018**, *160*, 152–182. [CrossRef] [PubMed]

133. Meng, Q.; Gao, B.; Goldberg, I.D.; Rosen, E.M.; Fan, S. Stimulation of Cell Invasion and Migration by Alcohol in Breast Cancer Cells. *Biochem. Biophys. Res. Commun.* **2000**, *273*, 448–453. [CrossRef] [PubMed]

134. Muniraj, N.; Siddharth, S.; Sharma, D. Bioactive Compounds: Multi-Targeting Silver Bullets for Preventing and Treating Breast Cancer. *Cancers* **2019**, *11*, 1563. [CrossRef]

135. Carlos-Reyes, Á.; López-González, J.S.; Meneses-Flores, M.; Gallardo-Rincón, D.; Ruíz-García, E.; Marchat, L.A.; la Vega, H.A.-D.; de la Cruz, O.N.H.; López-Camarillo, C. Dietary Compounds as Epigenetic Modulating Agents in Cancer. *Front. Genet.* **2019**, *10*, 79. [CrossRef]

136. Björner, S.; Rosendahl, A.H.; Tryggvadottir, H.; Simonsson, M.; Jirström, K.; Borgquist, S.; Rose, C.; Ingvar, C.; Jernström, H. Coffee Is Associated With Lower Breast Tumor Insulin-Like Growth Factor Receptor 1 Levels in Normal-Weight Patients and Improved Prognosis Following Tamoxifen or Radiotherapy Treatment. *Front. Endocrinol.* **2018**, *9*, 306. [CrossRef]

137. Hayakawa, S.; Ohishi, T.; Miyoshi, N.; Oishi, Y.; Nakamura, Y.; Isemura, M. Anti-Cancer Effects of Green Tea Epigallocatchin-3-Gallate and Coffee Chlorogenic Acid. *Molecules* **2020**, *25*, 4553. [CrossRef]

138. Zeng, A.; Liang, X.; Zhu, S.; Liu, C.; Wang, S.; Zhang, Q.; Zhao, J.; Song, L. Chlorogenic acid induces apoptosis, inhibits metastasis and improves antitumor immunity in breast cancer via the NF-κB signaling pathway. *Oncol. Rep.* **2021**, *45*, 717–727. [CrossRef]

139. Deus, C.M.; Serafim, T.L.; Magalhães-Novais, S.; Vilaça, A.; Moreira, A.C.; Sardão, V.A.; Cardoso, S.M.; Oliveira, P.J. Sirtuin 1-dependent resveratrol cytotoxicity and pro-differentiation activity on breast cancer cells. *Arch. Toxicol.* **2017**, *91*, 1261–1278. [CrossRef] [PubMed]

140. Greenshields, A.L.; Doucette, C.D.; Sutton, K.M.; Madera, L.; Annan, H.; Yaffe, P.B.; Knickle, A.F.; Dong, Z.; Hoskin, D.W. Piperine inhibits the growth and motility of triple-negative breast cancer cells. *Cancer Lett.* **2015**, *357*, 129–140. [CrossRef] [PubMed]

141. Peng, C.; Zeleznik, O.A.; Shutta, K.H.; Rosner, B.A.; Kraft, P.; Clish, C.B.; Stampfer, M.J.; Willett, W.C.; Tamimi, R.M.; Eliassen, A.H. A Metabolomics Analysis of Circulating Carotenoids and Breast Cancer Risk. *Cancer Epidemiol. Biomark. Prev.* **2022**, *31*, 85–96. [CrossRef] [PubMed]

142. Maitisha, G.; Aimaiti, M.; An, Z.; Li, X. Allicin induces cell cycle arrest and apoptosis of breast cancer cells in vitro via modulating the p53 pathway. *Mol. Biol. Rep.* **2021**, *48*, 7261–7272. [CrossRef]

143. Rosas-González, V.C.; Téllez-Bañuelos, M.C.; Hernández-Flores, G.; Bravo-Cuellar, A.; Aguilar-Lemarroy, A.; Jave-Suárez, L.F.; Haramati, J.; Solorzano-Ibarra, F. Differential effects of alliin and allicin on apoptosis and senescence in luminal A and triple-negative breast cancer: Caspase, ΔΨm, and pro-apoptotic gene involvement. *Fundam. Clin. Pharmacol.* **2020**, *34*, 671–686. [CrossRef] [PubMed]

144. Schley, P.D.; Jijon, H.B.; Robinson, L.E.; Field, C.J. Mechanisms of omega-3 fatty acid-induced growth inhibition in MDA-MB-231 human breast cancer cells. *Breast Cancer Res. Treat.* **2005**, *92*, 187–195. [CrossRef] [PubMed]

145. Annamalai, J.; Namasivayam, V. Endocrine disrupting chemicals in the atmosphere: Their effects on humans and wildlife. *Environ. Int.* **2015**, *76*, 78–97. [CrossRef]

146. Gonsioroski, A.; Mourikes, V.E.; Flaws, J.A. Endocrine Disruptors in Water and Their Effects on the Reproductive System. *Int. J. Mol. Sci.* **2020**, *21*, 1929. [CrossRef]

147. Mnif, W.; Hassine AI, H.; Bouaziz, A.; Bartegi, A.; Thomas, O.; Roig, B. Effect of Endocrine Disruptor Pesticides: A Review. *Int. J. Environ. Res. Public Health* **2011**, *8*, 2265–2303. [CrossRef] [PubMed]

148. Iavicoli, I.; Fontana, L.; Bergamaschi, A. The Effects of Metals as Endocrine Disruptors. *J. Toxicol. Environ. Health Part B* **2009**, *12*, 206–223. [CrossRef]

149. Peivasteh-Roudsari, L.; Barzegar-Bafrouei, R.; Sharifi, K.A.; Azimisalim, S.; Karami, M.; Abedinzadeh, S.; Asadinezhad, S.; Tajdar-Oranj, B.; Mahdavi, V.; Alizadeh, A.M.; et al. Origin, dietary exposure, and toxicity of endocrine-disrupting food chemical contaminants: A comprehensive review. *Heliyon* **2023**, *9*, e18140. [CrossRef]

150. Peinado, F.M.; Iribarne-Durán, L.M.; Ocón-Hernández, O.; Olea, N.; Artacho-Cordón, F. Endocrine Disrupting Chemicals in Cosmetics and Personal Care Products and Risk of Endometriosis. In *Endometriosis*; Courtney, M., Ed.; IntechOpen: Rijeka, Croatia, 2020; Chapter 2.

151. Monneret, C. What is an endocrine disruptor? *Comptes Rendus Biol.* **2017**, *340*, 403–405. [CrossRef]

152. Marinello, W.P.; Patisaul, H.B. Chapter Nine—Endocrine disrupting chemicals (EDCs) and placental function: Impact on fetal brain development. In *Advances in Pharmacology*; Vandenberg, L.N., Turgeon, J.L., Eds.; Academic Press: Cambridge, MA, USA, 2021; pp. 347–400.

153. Schjenken, J.E.; Green, E.S.; Overduin, T.S.; Mah, C.Y.; Russell, D.L.; Robertson, S.A. Endocrine Disruptor Compounds—A Cause of Impaired Immune Tolerance Driving Inflammatory Disorders of Pregnancy? *Front. Endocrinol.* **2021**, *12*, 607539. [CrossRef] [PubMed]
154. Alavian-Ghavanini, A.; Rüegg, J. Understanding Epigenetic Effects of Endocrine Disrupting Chemicals: From Mechanisms to Novel Test Methods. *Basic Clin. Pharmacol. Toxicol.* **2018**, *122*, 38–45. [CrossRef]
155. Zeng, W. Bisphenol A triggers the malignancy of nasopharyngeal carcinoma cells via activation of Wnt/β-catenin pathway. *Toxicol. Vitr.* **2020**, *66*, 104881. [CrossRef] [PubMed]
156. Cowin, P.; Wysolmerski, J. Molecular Mechanisms Guiding Embryonic Mammary Gland Development. *Cold Spring Harb. Perspect. Biol.* **2010**, *2*, a003251. [CrossRef]
157. Wormsbaecher, C.; Hindman, A.R.; Avendano, A.; Cortes-Medina, M.; Jones, C.E.; Bushman, A.; Onua, L.; Kovalchin, C.E.; Murphy, A.R.; Helber, H.L.; et al. In utero estrogenic endocrine disruption alters the stroma to increase extracellular matrix density and mammary gland stiffness. *Breast Cancer Res.* **2020**, *22*, 41. [CrossRef]
158. Speroni, L.; Voutilainen, M.; Mikkola, M.L.; Klager, S.A.; Schaeberle, C.M.; Sonnenschein, C.; Soto, A.M. New insights into fetal mammary gland morphogenesis: Differential effects of natural and environmental estrogens. *Sci. Rep.* **2017**, *7*, 40806. [CrossRef]
159. Soto, A.M.; Vandenberg, L.N.; Maffini, M.V.; Sonnenschein, C. Does Breast Cancer Start in the Womb? *Basic Clin. Pharmacol. Toxicol.* **2008**, *102*, 125–133. [CrossRef] [PubMed]
160. Soto, A.M.; Brisken, C.; Schaeberle, C.; Sonnenschein, C. Does Cancer Start in the Womb? Altered Mammary Gland Development and Predisposition to Breast Cancer due to in Utero Exposure to Endocrine Disruptors. *J. Mammary Gland Biol. Neoplasia* **2013**, *18*, 199–208. [CrossRef]
161. Tchen, R.; Tan, Y.; Barr, D.B.; Ryan, P.B.; Tran, V.; Li, Z.; Hu, Y.-J.; Smith, A.K.; Jones, D.P.; Dunlop, A.L.; et al. Use of high-resolution metabolomics to assess the biological perturbations associated with maternal exposure to Bisphenol A and Bisphenol F among pregnant African American women. *Environ. Int.* **2022**, *169*, 107530. [CrossRef] [PubMed]
162. Mandrup, K.; Boberg, J.; Isling, L.K.; Christiansen, S.; Hass, U. Low-dose effects of bisphenol A on mammary gland development in rats. *Andrology* **2016**, *4*, 673–683. [CrossRef] [PubMed]
163. Tharp, A.P.; Maffini, M.V.; Hunt, P.A.; VandeVoort, C.A.; Sonnenschein, C.; Soto, A.M. Bisphenol A alters the development of the rhesus monkey mammary gland. *Proc. Natl. Acad. Sci. USA* **2012**, *109*, 8190–8195. [CrossRef]
164. Fernandez, S.V.; Huang, Y.; Snider, K.E.; Zhou, Y.; Pogash, T.J.; Russo, J. Expression and DNA methylation changes in human breast epithelial cells after bisphenol A exposure. *Int. J. Oncol.* **2012**, *41*, 369–377. [CrossRef] [PubMed]
165. Korsh, J.; Shen, A.; Aliano, K.; Davenport, T. Polycyclic Aromatic Hydrocarbons and Breast Cancer: A Review of the Literature. *Breast Care* **2015**, *10*, 316–318. [CrossRef]
166. Qin, Q.; Yang, B.; Liu, Z.; Xu, L.; Song, E.; Song, Y. Polychlorinated biphenyl quinone induced the acquisition of cancer stem cells properties and epithelial-mesenchymal transition through Wnt/β-catenin. *Chemosphere* **2021**, *263*, 128125. [CrossRef]
167. Guo, J.-Y.; Wang, M.-Z.; Wang, M.-S.; Sun, T.; Wei, F.-H.; Yu, X.-T.; Wang, C.; Xu, Y.-Y.; Wang, L. The Undervalued Effects of Polychlorinated Biphenyl Exposure on Breast Cancer. *Clin. Breast Cancer* **2020**, *20*, 12–18. [CrossRef]
168. Parada, H.; Sun, X.; Tse, C.-K.; Engel, L.S.; Hoh, E.; Olshan, A.F.; Troester, M.A. Plasma levels of polychlorinated biphenyls (PCBs) and breast cancer mortality: The Carolina Breast Cancer Study. *Int. J. Hyg. Environ. Health* **2020**, *227*, 113522. [CrossRef]
169. Schildroth, S.; Wise, L.A.; Wesselink, A.K.; Bethea, T.N.; Fruh, V.; Taylor, K.W.; Calafat, A.M.; Baird, D.D.; Henn, B.C. Correlates of non-persistent endocrine disrupting chemical mixtures among reproductive-aged Black women in Detroit, Michigan. *Chemosphere* **2022**, *299*, 134447. [CrossRef] [PubMed]
170. Ahern, T.P.; Broe, A.; Lash, T.L.; Cronin-Fenton, D.P.; Ulrichsen, S.P.; Christiansen, P.M.; Cole, B.F.; Tamimi, R.M.; Sørensen, H.T.; Damkier, P. Phthalate Exposure and Breast Cancer Incidence: A Danish Nationwide Cohort Study. *J. Clin. Oncol. Off. J. Am. Soc. Clin. Oncol.* **2019**, *37*, 1800–1809. [CrossRef] [PubMed]
171. Mao, W.; Jin, H.; Guo, R.; Chen, P.; Zhong, S.; Wu, X. Distribution of parabens and 4-HB in human blood. *Sci. Total Environ.* **2024**, *914*, 169874. [CrossRef] [PubMed]
172. Tapia, J.L.; McDonough, J.C.; Cauble, E.L.; Gonzalez, C.G.; Teteh, D.K.; Treviño, L.S. Parabens Promote Protumorigenic Effects in Luminal Breast Cancer Cell Lines With Diverse Genetic Ancestry. *J. Endocr. Soc.* **2023**, *7*, bvad080. [CrossRef]
173. Hager, E.; Chen, J.; Zhao, L. Minireview: Parabens Exposure and Breast Cancer. *Int. J. Environ. Res. Public Health* **2022**, *19*, 1873. [CrossRef] [PubMed]
174. Downs, C.A.; Amin, M.M.; Tabatabaeian, M.; Chavoshani, A.; Amjadi, E.; Afshari, A.; Kelishadi, R. Parabens preferentially accumulate in metastatic breast tumors compared to benign breast tumors and the association of breast cancer risk factors with paraben accumulation. *Environ. Adv.* **2023**, *11*, 100325. [CrossRef]
175. Darbre, P.D.; Aljarrah, A.; Miller, W.R.; Coldham, N.G.; Sauer, M.J.; Pope, G.S. Concentrations of parabens in human breast tumours. *J. Appl. Toxicol.* **2004**, *24*, 5–13. [CrossRef]
176. Robin, J.; Binson, G.; Albouy, M.; Sauvaget, A.; Pierre-Eugène, P.; Migeot, V.; Dupuis, A.; Venisse, N. Analytical method for the biomonitoring of bisphenols and parabens by liquid chromatography coupled to tandem mass spectrometry in human hair. *Ecotoxicol. Environ. Saf.* **2022**, *243*, 113986. [CrossRef]
177. Park, Y.; Jang, J.; Park, J.; Kim, J.H.; Kim, E.; Song, Y.; Kwon, H. Analysis of parabens in dentifrices and the oral cavity. *Biomed. Chromatogr.* **2014**, *28*, 1692–1700. [CrossRef]

178. Dualde, P.; Pardo, O.; Corpas-Burgos, F.; Kuligowski, J.; Gormaz, M.; Vento, M.; Pastor, A.; Yusà, V. Biomonitoring of parabens in human milk and estimated daily intake for breastfed infants. *Chemosphere* **2020**, *240*, 124829. [CrossRef]
179. Park, N.-Y.; Cho, Y.H.; Choi, K.; Lee, E.-H.; Kim, Y.J.; Kim, J.H.; Kho, Y. Parabens in breast milk and possible sources of exposure among lactating women in Korea. *Environ. Pollut.* **2019**, *255*, 113142. [CrossRef]
180. Andersen, M.H.G.; Zuri, G.; Knudsen, L.E.; Mathiesen, L. Placental transport of parabens studied using an ex-vivo human perfusion model. *Placenta* **2021**, *115*, 121–128. [CrossRef]
181. Zhang, H.; Quan, Q.; Li, X.; Sun, W.; Zhu, K.; Wang, X.; Sun, X.; Zhan, M.; Xu, W.; Lu, L.; et al. Occurrence of parabens and their metabolites in the paired urine and blood samples from Chinese university students: Implications on human exposure. *Environ. Res.* **2020**, *183*, 109288. [CrossRef] [PubMed]
182. Zamora-León, P. Are the Effects of DES Over? A Tragic Lesson from the Past. *Int. J. Environ. Res. Public Health* **2021**, *18*, 10309. [CrossRef]
183. Stillwater, B.J.; Bull, A.C.; Romagnolo, D.F.; Neumayer, L.A.; Donovan, M.G.; Selmin, O.I. Bisphenols and Risk of Breast Cancer: A Narrative Review of the Impact of Diet and Bioactive Food Components. *Front. Nutr.* **2020**, *7*, 581388. [CrossRef] [PubMed]
184. Dumitrascu, M.C.; Mares, C.; Petca, R.-C.; Sandru, F.; Popescu, R.-I.; Mehedintu, C.; Petca, A. Carcinogenic effects of bisphenol A in breast and ovarian cancers (Review). *Oncol. Lett.* **2020**, *20*, 282. [CrossRef]
185. Yang, P.-J.; Hou, M.-F.; Ou-Yang, F.; Hsieh, T.-H.; Lee, Y.-J.; Tsai, E.-M.; Wang, T.-N. Association between recurrent breast cancer and phthalate exposure modified by hormone receptors and body mass index. *Sci. Rep.* **2022**, *12*, 2858. [CrossRef] [PubMed]
186. Cohn, B.A.; La Merrill, M.; Krigbaum, N.Y.; Yeh, G.; Park, J.S.; Zimmermann, L.; Cirillo, P.M. DDT Exposure in Utero and Breast Cancer. *J. Clin. Endocrinol. Metab.* **2015**, *100*, 2865–2872.

Article

Clinicopathological Characteristics and Prognostic Profiles of Breast Carcinoma with Neuroendocrine Features

Yue Qiu [1,2,†], Yongjing Dai [2,†], Li Zhu [2], Xiaopeng Hao [2], Liping Zhang [3], Baoshi Bao [2], Yuhui Chen [2,*] and Jiandong Wang [2,*]

1 Graduate School, Chinese PLA Medical School, Beijing 100853, China
2 Department of General Surgery, The First Medical Center, Chinese PLA General Hospital, 28 Fuxing Road, Haidian District, Beijing 100853, China
3 Graduate School, Inner Mongolia Medical University, Hohhot 010110, China
* Correspondence: cyhnju@sina.com (Y.C.); wangjiandong@188.com (J.W.)
† These authors contributed equally to this work.

Abstract: Background: Breast carcinoma with neuroendocrine features includes neuroendocrine neoplasm of the breast and invasive breast cancer with neuroendocrine differentiation. This study aimed to investigate the clinicopathological features and prognosis of this disease according to the fifth edition of the World Health Organization classification of breast tumors. Materials and Methods: A total of 87 patients with breast carcinoma with neuroendocrine features treated in the First Medical Center, Chinese PLA General Hospital from January 2001 to January 2022 were retrospectively enrolled in this study. Results: More than half of the patients were postmenopausal patients, especially those with neuroendocrine neoplasm (62.96%). There were more patients with human epidermal growth factor receptor 2 negative and hormone receptor positive tumors, and most of them were Luminal B type (71.26%). The multivariate analysis showed that diabetes and stage IV disease were related to the progression-free survival of breast carcinoma with neuroendocrine features patients ($p = 0.004$ and $p < 0.001$, respectively). Conclusion: Breast carcinoma with neuroendocrine features tended to be human epidermal growth factor receptor 2 negative and hormone receptor positive tumors, most of them were Luminal B type, and the related factors of progression-free survival were diabetes and stage IV disease.

Keywords: neuroendocrine tumors; breast neoplasms; prognosis

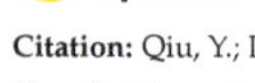

Citation: Qiu, Y.; Dai, Y.; Zhu, L.; Hao, X.; Zhang, L.; Bao, B.; Chen, Y.; Wang, J. Clinicopathological Characteristics and Prognostic Profiles of Breast Carcinoma with Neuroendocrine Features. *Life* **2023**, *13*, 532. https://doi.org/10.3390/life13020532

Academic Editors: Riccardo Autelli, Taobo Hu, Mengping Long and Lei Wang

Received: 31 January 2023
Revised: 11 February 2023
Accepted: 13 February 2023
Published: 15 February 2023

1. Introduction

Breast carcinoma with neuroendocrine features consists of a group of diseases with high heterogeneity. It was reported that the incidence rate of breast carcinoma with neuroendocrine features ranged from 0.1% to 20% [1–3]. Breast carcinoma with neuroendocrine features included neuroendocrine neoplasm (NEN) of the breast and invasive breast cancer (IBC) with neuroendocrine differentiation. According to the latest World Health Organization (WHO) classification of breast tumors, NEN was divided into neuroendocrine tumor (NET) and neuroendocrine carcinoma (NEC) based on the degree of differentiation [3]. IBC with neuroendocrine differentiation was classed into breast carcinoma of no special type and breast carcinoma of special types, such as solid papillary carcinoma and the hypercellular subtype of mucinous carcinoma.

Since the third edition of the WHO classification of breast tumors was published, the definition and classification of this disease had changed greatly in different editions. As a result, there have been controversies surrounding the definition and classification of breast carcinoma with neuroendocrine features. The diagnostic criteria for subjects included in existing studies were not identical. As well, the research results were not completely consistent or were even contradictory [4,5]. Further, due to the rarity of the disease, few studies had been conducted, and those which have were mainly case reports [6–8]. At present, the treatment strategy of IBC of no special type is used directly in breast carcinoma with neuroendocrine features. The general practice guidelines for breast carcinoma with neuroendocrine features are still not formed. The TNM stage of breast carcinoma with neuroendocrine features was defined by the eighth version of the America joint committee on cancer staging systems [9]. According to the guidelines of the Chinese Society of Clinical Oncology published in 2020 [10], the minimum positive threshold of estrogen receptor (ER), progesterone receptor (PR), and Ki-67 are 1%, 1%, and 14%, respectively, and human epidermal growth factor receptor 2 (Her-2) (3+) or ISH positivity meant Her-2 positivity. Breast carcinoma with neuroendocrine features was divided into Luminal A (ER/PR positive, Her-2 negative with low Ki-67 index) disease, Luminal B (ER/PR positive, Her-2 negative with high Ki-67 index, or ER/PR positive, Her-2 positive) disease, Her-2 positive (ER and PR negative, Her-2 positive) disease, and Triple-negative (ER, PR, and Her-2 negative) disease according to molecular subtyping. To investigate the clinicopathological features and prognosis of this disease under the fifth edition of the WHO classification of breast tumors, we designed this study.

2. Materials and Methods

2.1. Study Groups

The data of 87 patients with breast carcinoma with neuroendocrine features treated in the First Medical Center, Chinese PLA General Hospital from January 2001 to January 2022 were retrospectively collected. Patients with breast carcinoma derived from other organs were excluded. Pregnant patients and patients who were breastfeeding were excluded as well. There were no patients without definite pathological diagnosis or without complete medical records. All procedures performed in this study involving human participants were in accordance with the Declaration of Helsinki (as revised in 2013). The study was approved by the ethics committee of the Chinese PLA General Hospital (NO.: S2022-746). Individual consent for this retrospective analysis was waived.

2.2. Study Variables

General information on the patients was collected, such as age at diagnosis, gender, laterality, smoking history, drinking history, body mass index (BMI), family history, menopause status, hypertension, diabetes, hyperlipidemia, T stage, Her-2 status, Ki-67, ER, PR, molecular typing, vessel carcinoma embolus, N stage, skin or chest wall invasion, distant metastasis, and stage.

Unique clinical features of the patients were analyzed, including clinical symptoms and history of thyroid diseases. Pathological characteristics of the patients were also described, such as detailed classification, expression of neuroendocrine markers, and ductal carcinoma in situ composition. The treatment strategy of the patients, such as neoadjuvant chemotherapy, surgery, and adjuvant therapy, was discussed. All patients enrolled in this study were followed up. The 5-year overall survival (OS), 5-year progression-free survival (PFS), and 5-year disease-specific survival (DSS) of patients in the study group were described. Finally, the factors related to 5-year PFS were analyzed.

2.3. Statistical Analysis

All statistical analyses of this study were performed using Stata Statistical Software version 15.1 (StataCorp LLC, College Station, TX, USA). The measurement data were described by median (inter-quartile range, IQR). Frequency was used to show the counting data. Comparison of counting data between two groups was conducted by Pearson chi-square test. Kruskal–Wallis H test was used to examine multiple comparisons of ranked counting data between groups. Kaplan–Meier method was used in survival analysis, and Log-rank test was used to compare different survival curves. Univariate and multivariate analysis were performed using Cox model. A two-tailed $p < 0.05$ was considered statistically significant, and all confidence intervals (CI) were expressed at 95% confidence level.

3. Results

3.1. General Clinicopathological Characteristics

The median age at diagnosis of the patients in the study group was 53 (42–64) years old. More than half of the patients were postmenopausal patients, especially those with neuroendocrine neoplasm (62.96%). The proportion of patients with Her-2 negative and hormone receptor (HR) positive tumors was high, and most of them were Luminal B type (71.26% vs. 28.74%). Around 49.43% of the patients had stage II disease. There was no difference between the NEN group and IBC with neuroendocrine differentiation group except in diabetes when analyzing general characteristics (p = 0.039) (Table 1).

Table 1. General clinicopathological characteristics of breast carcinoma with neuroendocrine features patients.

Category	Breast Carcinoma with Neuroendocrine Features (n = 87) (%)	NEN Group (n = 29) (%)	IBC with Neuroendocrine Differentiation (n = 58) (%)	p Value
Age at diagnosis (years)				
≤60	55 (63.22%)	18 (62.07%)	37 (63.79%)	0.875
>60	32 (36.78%)	11 (37.93%)	21 (36.21%)	
Gender				
Female	85 (97.7%)	27 (93.10%)	58 (100.00%)	0.109
Male	2 (2.30%)	2 (2.30%)	0 (0.00%)	
Laterality				
Left	44 (50.57%)	18 (62.07%)	26 (44.83%)	0.129
Right	43 (49.43%)	11 (37.93%)	32 (55.17%)	
Smoking history				
Yes	2 (2.30%)	1 (3.45%)	1 (1.72%)	1
No	85 (97.70%)	28 (96.55%)	57 (98.28%)	
Drinking history				
Yes	1 (1.15%)	1 (3.45%)	0 (0.00%)	0.333
No	86 (98.85%)	28 (96.55%)	58 (100.00%)	
BMI (kg/m^2)				
≥24	47 (54.02%)	19 (65.52%)	28 (48.28%)	0.128
<24	40 (45.98%)	10 (34.48%)	30 (51.72%)	
Family history				
Yes	1 (1.15%)	0 (0.00%)	1 (1.72%)	1
No	86 (98.85%)	29 (100.00%)	57 (98.28%)	
Menopausal status [a]				
Yes	47 (55.29%)	17 (62.96%)	30 (51.72%)	0.332
No	38 (44.71%)	10 (37.04%)	28 (48.28%)	
Hypertension				
Yes	23 (26.44%)	9 (31.03%)	14 (24.14%)	0.492
No	64 (73.56%)	20 (68.97%)	44 (75.86%)	
Diabetes				
Yes	7 (8.05%)	5 (17.24%)	2 (3.45%)	0.039
No	80 (91.95%)	24 (82.76%)	56 (96.55%)	
Hyperlipidemia				
Yes	20 (22.99%)	5 (17.24%)	15 (25.86%)	0.368
No	67 (77.01%)	24 (82.76%)	43 (74.14%)	
T stage (AJCC 8th)				
1	45 (51.72%)	18 (62.07%)	27 (46.55%)	
2	35 (40.23%)	9 (31.03%)	26 (44.83%)	0.2
3	7 (8.05%)	2 (6.90%)	5 (8.62%)	
4	0 (0.00%)	0 (0.00%)	0 (0.00%)	
Her-2 status				
Positive	11 (12.64%)	2 (6.90%)	9 (15.52%)	0.323
Negative	76 (87.36%)	27 (93.10%)	49 (84.48%)	
Ki-67				
Positive	72 (82.76%)	22 (75.86%)	50 (86.21%)	0.229
Negative	15 (17.24%)	7 (24.14%)	8 (13.79%)	
ER				
Positive	67 (77.01%)	22 (75.86%)	45 (77.59%)	0.857
Negative	20 (22.99%)	7 (24.14%)	13 (22.41%)	
PR				
Positive	71 (81.61%)	25 (86.21%)	46 (79.31%)	0.434
Negative	16 (18.39%)	4 (13.79%)	12 (20.69%)	
Molecular typing (CSCO 2020)				
Luminal A	13 (14.94%)	5 (17.24%)	8 (13.79%)	
Luminal B	62 (71.26%)	21 (72.41%)	41 (70.69%)	0.494
Her-2 positive	1 (1.15%)	0 (0.00%)	1 (1.72%)	
Triple negative	11 (12.64%)	3 (10.34%)	8 (13.79%)	
Vessel carcinoma embolus				
Yes	17 (19.54%)	5 (17.24%)	12 (20.69%)	0.702
No	70 (80.46%)	24 (82.76%)	46 (79.31%)	
N stage (AJCC 8th)				
0	53 (60.92%)	19 (65.52%)	34 (58.62%)	
1	18 (20.69%)	5 (17.24%)	13 (22.41%)	0.549
2	14 (16.09%)	5 (17.24%)	9 (15.52%)	
3	2 (2.30%)	0 (0.00%)	2 (3.45%)	

Table 1. *Cont.*

Category	Breast Carcinoma with Neuroendocrine Features (n = 87) (%)	NEN Group (n = 29) (%)	IBC with Neuroendocrine Differentiation (n = 58) (%)	*p* Value
Skin or chest wall invasion				—
Yes	0 (0.00%)	0 (0.00%)	0 (0.00%)	
No	87 (100.00%)	29 (100.00%)	58 (100.00%)	
Distant metastasis				0.55
Yes	2 (2.30%)	0 (0.00%)	2 (3.45%)	
No	85 (97.70%)	29 (100.00%)	56 (96.55%)	
Stage (AJCC 8th)				0.577
I	26 (29.88%)	11 (37.93%)	15 (25.86%)	
II	43 (49.43%)	11 (37.93%)	32 (55.17%)	
III	16 (18.39%)	7 (24.14%)	9 (15.52%)	
IV	2 (2.30%)	0 (0.00%)	2 (3.45%)	

BMI: body mass index; ER: estrogen receptor; PR: progesterone receptor. Her-2: human epidermal growth factor receptor 2; IBC: invasive breast cancer; AJCC 8th: The eighth version of America joint committee on cancer staging system; CSCO 2020: The guidelines of Chinese Society of Clinical Oncology in 2020. [a] There were two male breast cancer patients in the study group.

3.2. Special Clinicopathological Features of Breast Carcinoma with Neuroendocrine Features Patients

A total of 22.99% of patients had clinical symptoms such as pain, nipple discharge, or both. About 21.84% of patients were complicated, with thyroid diseases such as thyroid nodule, diffuse thyroid disease, and thyroid cancer. A total of 34 patients with lymph node metastasis all had axillary lymph node metastasis, two cases also had supraclavicular lymph node metastasis, and one case had subclavian lymph node metastasis at the same time (Table 2).

Table 2. Supplement of part clinical features of breast carcinoma with neuroendocrine features patients.

Category	n = 87
Clinical symptoms	
Yes	20 (22.99%)
Pain	14
Nipple discharge	5
Nipple discharge with pain	1
Paraneoplastic symptoms	0
No	67 (77.01%)
Thyroid diseases	
Yes	19 (21.84%)
Thyroid nodule	15
Diffuse thyroid disease	3
Thyroid cancer	1
No	68 (78.16%)
Lymph node metastasis	
Yes	34 (39.08%) [a]
Axillary lymph node metastasis	34
Supraclavicular lymph node metastasis	2
Subclavian lymph node metastasis	1
No	53 (60.92%)

[a] One patient had axillary lymph node metastasis, supraclavicular lymph node metastasis, and subclavian lymph node metastasis at the same time. One patient had axillary lymph node metastasis and supraclavicular lymph node metastasis at the same time.

A total of 87 cases with breast carcinoma with neuroendocrine features were classified into 29 cases of NEN and 58 cases of IBC with neuroendocrine differentiation (Table 3). After H and E staining and immunohistochemical staining, some tumor cells showed typical positive synaptophysin (Syn) and Chromogranin (CgA) staining (Figure 1). There were also some cases that had CD56+ tumors and neuron-specific enolase (NSE)+ tumors (Table 3).

Table 3. Supplement of part pathological characteristics of breast carcinoma with neuroendocrine features patients.

Category	n = 87
Classification	
NEN	29 (33.34%)
NET	26
Simple type	21
Mixed type	5
NEC	3
Small cell carcinoma	1
Large cell carcinoma	1
Mixed type	1
IBC with neuroendocrine differentiation	58 (66.66%)
IBC of no special type	
Invasive ductal carcinoma	50
IBC of special type	
Solid papillary carcinoma	2
Invasive papillary carcinoma	1
Type B mucinous carcinoma	1
Mixed type	4
Neuroendocrine markers	
Syn	
0	3 (3.45%)
1+	76 (87.35%)
2+	4 (4.60%)
3+	4 (4.60%)
CgA	
0	10 (11.49%)
1+	11 (12.64%)
2+	25 (28.74%)
3+	2 (2.30%)
NK	39 (44.83%)
CD56	
0	11 (12.64%)
1+	26 (29.88%)
2+	42 (48.28%)
3+	1 (1.15%)
NK	7 (8.05%)
NSE	
0	1 (1.15%)
1+	7 (8.05%)
2+	2 (2.30%)
3+	1 (1.15%)
NK	76 (87.35%)
Ductal carcinoma in situ composition	
Yes	29 (33.33%)
No	58 (66.67%)

NEN: neuroendocrine neoplasm; NET: neuroendocrine tumor; NEC: neuroendocrine carcinoma; IBC: invasive breast cancer; Syn: synaptophysin; CgA: chromogranin A; NSE: neuron-specific enolase; NK: not known.

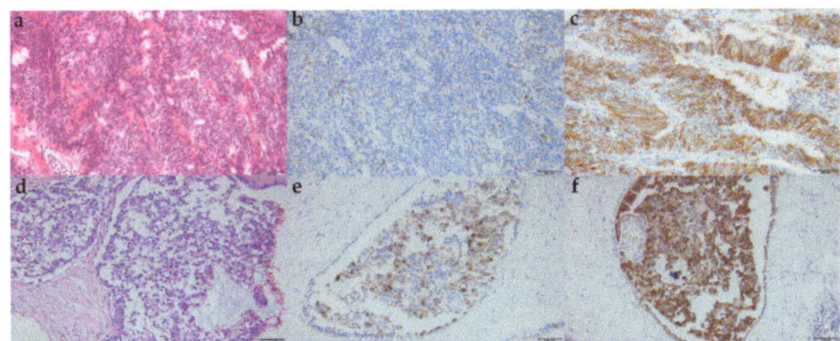

Figure 1. Morphological and immunohistochemical features. (**a**). H and E staining of NEC of the breast (×200); (**b**) positive immunostaining for chromogranin A of NEC of the breast (×200); (**c**) positive immunostaining for synaptophysin of NEC of the breast (×200); (**d**) H and E staining of no special type IBC with neuroendocrine differentiation (×200); (**e**) positive immunostaining for chromogranin A of no special type IBC with neuroendocrine differentiation (×200); (**f**) positive immunostaining for synaptophysin of no special type IBC with neuroendocrine differentiation (×200). NEC: neuroendocrine carcinoma; IBC: invasive breast cancer; H and E staining: Hematoxylin and Eosin staining.

3.3. Treatment and Follow-Up of Breast Carcinoma with Neuroendocrine Features Patients

Only 7 out of 87 patients received neoadjuvant chemotherapy. All patients received surgical treatment of the breast and/or axillary lymph node. There were 75 cases that underwent mastectomy, while the rest of the patients underwent breast-conserving surgery or nipple-areola complex-sparing mastectomy. As for adjuvant therapy, 81.61% of patients received chemotherapy, 31.03% of patients received radiotherapy, 73.56% of patients received endocrine therapy, and only 9.20% of patients received targeted therapy after surgery (Table 4).

Table 4. Treatment strategy of breast carcinoma with neuroendocrine features patients.

Category	n = 87
Neoadjuvant chemotherapy	
Yes	7 (8.05%)
No	80 (91.95%)
Surgery treatment	
Breast	
Mastectomy	75 (86.20%)
BCS	8 (9.20%)
NSM	3 (3.45%)
None [a]	1 (1.15%)
Axillary lymph node	
SLNB	19 (21.84%)
ALND	67 (77.01%)
None [b]	1 (1.15%)
Adjuvant therapy	
Chemotherapy	
Yes	71 (81.61%)
No	16 (18.39%)
Radiotherapy	
Yes	27 (31.03%)
No	60 (68.97%)
Endocrine therapy	
Yes	64 (73.56%)
No	23 (26.44)
Targeted therapy	
Yes	8 (9.20%)
No	79 (90.80%)

BCS: breast conserving surgery; NSM: nipple-areola complex sparing mastectomy; SLNB: sentinel lymph node biopsy; ALND: axillary lymph node dissection. [a] One patient had occult breast cancer, receiving ALND only. [b] One patient who was elderly received breast surgery only.

A total of 87 patients were followed up, and 2 patients dropped out. The median follow-up time was 57 (25–74) months. During the follow-up period, local recurrence involving the breast, axilla, or chest wall occurred in five cases. There were 12 cases recorded of distant metastasis including bone, lung, liver, brain, retroperitoneal lymph node, and contralateral axillary lymph node. Eight cases died, six of them died of breast cancer and two died of other causes (Table 5).

Table 5. Local recurrence, distant metastasis, and mortality of breast carcinoma with neuroendocrine features patients.

Category	n = 85 [a]
Local recurrence	5 (5.88%)
Breast	1
Chest wall	3
Axilla	1
Distant metastasis	12 (14.14%)
Bone	4
Lung	6
Liver	2
Brain	3
Retroperitoneal lymph node	1
Contralateral axillary lymph node	2
Death	8 (9.41%)
Breast cancer	6
Other causes	2
Follow-up time (months, IQR)	57 (25–74)

IQR: inter-quartile range. [a] There were 87 patients totally, and two patients dropped out.

3.4. Kaplan-Meier Survival Analysis of Breast Carcinoma with Neuroendocrine Features Patients

The five-year PFS, five-year DSS, and five-year OS of all of the patients were 81.19% (95%CI: 0.6964–0.8869), 91.53% (95%CI: 0.8022–0.9651), and 90.25% (95%CI: 0.7901–0.9564), respectively. No significant differences were found in the five-year PFS, five-year DSS, and five-year OS between NEN and IBC with neuroendocrine differentiation ($p > 0.05$) (Figures 2–4).

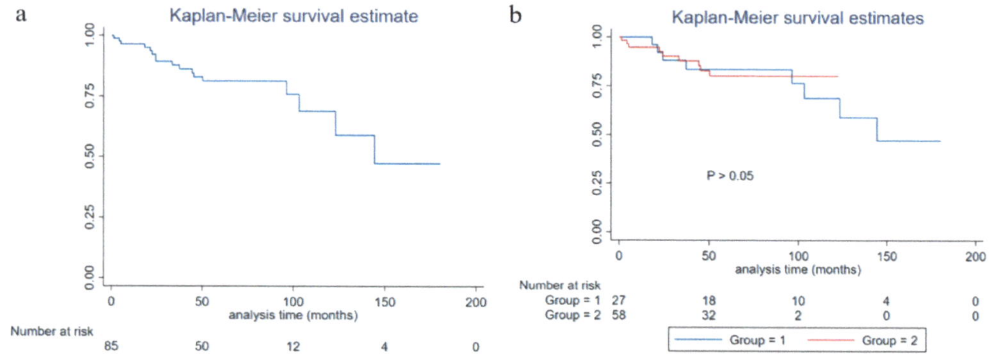

Figure 2. Progression-free survival of breast carcinoma with neuroendocrine features patients (group 1: neuroendocrine neoplasm patients; group 2: invasive breast cancer with neuroendocrine differentiation patients). (**a**) Progression-free survival of all breast carcinoma with neuroendocrine features patients; (**b**) progression-free survival of grouped breast carcinoma with neuroendocrine features patients.

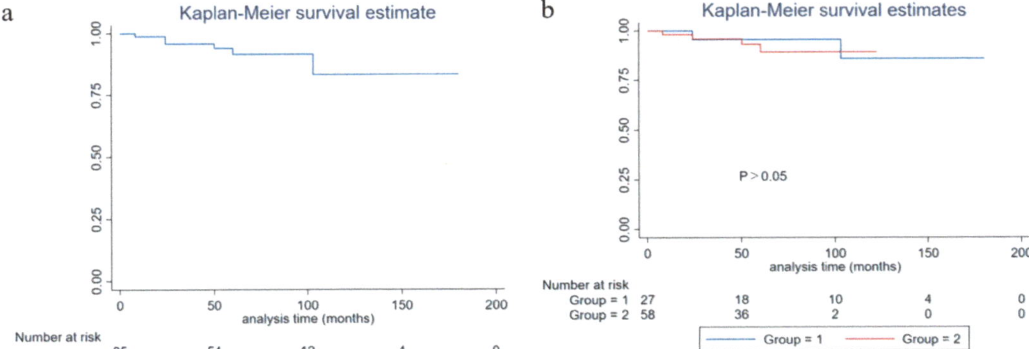

Figure 3. Disease-specific survival of breast carcinoma with neuroendocrine features patients (group 1: neuroendocrine neoplasm patients; group 2: invasive breast cancer with neuroendocrine differentiation patients). (**a**) Disease-specific survival of all breast carcinoma with neuroendocrine features patients; (**b**) disease-specific survival of grouped breast carcinoma with neuroendocrine features patients.

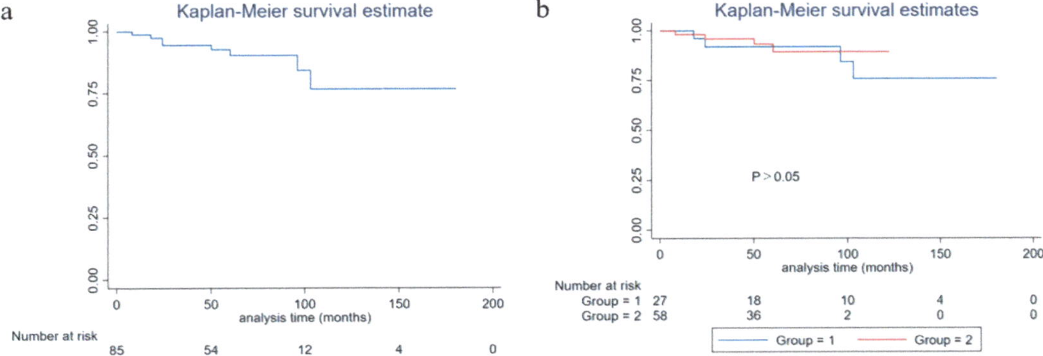

Figure 4. Overall survival of breast carcinoma with neuroendocrine features patients (group 1: neuroendocrine neoplasm patients; group 2: invasive breast cancer with neuroendocrine differentiation patients). (**a**) Overall survival of all breast carcinoma with neuroendocrine features patients; (**b**) overall survival of grouped breast carcinoma with neuroendocrine features patients.

3.5. Univariate Analysis of PFS in Breast Carcinoma with Neuroendocrine Features Patients

The results of the univariate analysis of PFS showed that smoking history, diabetes, Her-2 positive disease, N3 stage disease, distant metastasis, and targeted therapy were related to the progression-free survival of breast carcinoma in neuroendocrine features patients ($p = 0.038$, $p = 0.008$, $p = 0.003$, $p < 0.001$, $p < 0.001$; $p = 0.004$, respectively). The results of Her-2 status and family history slightly missed the margin of significance ($p = 0.097$ and $p = 0.092$, respectively) (Table 6).

Table 6. Univariate analysis of PFS in breast carcinoma with neuroendocrine features patients.

Variable	HR (95% CI)	*p* Value
Age (year)		
>60	1	
≤60	0.9670 (0.3566–2.6224)	0.947
Gender		
Female	1	
Male	1.6943 (0.2138–13.4258)	0.618
Laterality		
Right	1	
Left	0.4458 (0.1587–1.2522)	0.125
Smoking history		
No	1	
Yes	9.1939 (1.1251–75.1266)	0.038
Drinking history		
No	1	
Yes	—	1
BMI (kg/m^2)		
<24	1	
≥24	1.2394 (0.4535–3.3873)	0.676
Family history		
No	1	
Yes	5.8103 (0.7485–45.1052)	0.092
Menopausal status		
No	1	
Yes	0.7091 (0.2629–1.9125)	0.497
Hypertension		
No	1	
Yes	0.8888 (0.2882–2.7409)	0.837
Diabetes		
No	1	
Yes	4.8491 (1.5089–15.5837)	0.008
Hyperlipidemia		
No	1	
Yes	1.1692 (0.3771–3.6253)	0.787
T stage		
1	1	
2	1.6177 (0.5770–4.5348)	0.36
3	1.5048 (0.3172–7.1392)	0.607
Her-2 status		
Negative	1	
Positive	2.6226 (0.8390–8.1975)	0.097
Ki-67		
Negative	1	
Positive	1.1088 (0.3391–3.6256)	0.864
ER		
Positive	1	
Negative	1.6555 (0.5819–4.7100)	0.345
PR	1	
Positive	1.6112 (0.5242–4.9525)	
Negative		0.405
Molecular typing		
Luminal A	1	
Luminal B	1.7925 (0.3694–8.6984)	0.469
Her-2 positive	72.1912 (4.4462–1172.1380)	0.003
Triple negative	2.3484 (0.3846–14.3384)	0.355
Vessel carcinoma embolus		
No	1	
Yes	0.4462 (0.1013–1.9650)	0.286
N stage		
0	1	

Table 6. *Cont.*

Variable	HR (95% CI)	*p* Value
1	1.6375 (0.5346–5.0153)	0.388
2	0.9315 (0.1976–4.3919)	0.929
3	150.1349 (12.6995–1774.9190)	<0.001
Skin or chest wall invasion		
No	1	
Yes	—	—
Distant metastasis		
No	1	
Yes	133.1195 (11.7712–1505.4410)	<0.001
Stage		
I	1	
II	2.4426 (0.6494–9.1871)	0.186
III	1.5589 (0.3132–7.7588)	0.588
IV	243.9231 (17.2429–3450.6020)	<0.001
Classification		
IBC with neuroendocrine differentiation	1	
NEN	0.9300 (0.3073–2.8146)	0.898
Neoadjuvant chemotherapy		
Yes	1	
No	0.4378 (0.1241–1.5447)	0.199
Breast treatment		
Mastectomy	1	
BCS	1.1105 (0.2508–4.9179)	0.89
NSM	1.1265 (0.1455–8.7217)	0.909
Axillary lymph node treatment		
ALND	1	
SLNB	0.6931 (0.1972–2.4355)	0.568
Chemotherapy		
Yes	1	
No	2.1270 (0.7460–6.0646)	0.158
Radiotherapy		
Yes	1	
No	1.1213 (0.4000–3.1428)	0.828
Endocrine therapy		
Yes	1	
No	1.4339 (0.5213–3.9437)	0.485
Targeted therapy		
Yes	1	
No	0.1838 (0.0579–0.5840)	0.004

BMI: body mass index; ER: estrogen receptor; PR: progesterone receptor. Her-2: human epidermal growth factor receptor 2; NEN: neuroendocrine neoplasm; IBC: invasive breast cancer; BCS: breast conserving surgery; NSM: nipple-areola complex sparing mastectomy; SLNB: sentinel lymph node biopsy; ALND: axillary lymph node dissection; PFS: progression-free survival; HR: hazard ratio; CI: confidence interval.

3.6. Multivariate Analysis of PFS in Breast Carcinoma with Neuroendocrine Features Patients

The results of the multivariate analysis of PFS showed that diabetes and stage IV disease were related to the progression-free survival of breast carcinoma in neuroendocrine features patients ($p = 0.004$ and $p < 0.001$, respectively). The result of targeted therapy showed a barely detectable statistical significance ($p = 0.051$) (Table 7).

Table 7. Multivariate analysis of PFS in breast carcinoma in neuroendocrine features patients.

Variable	HR (95%CI)	*p* Value
Smoking history		
No	1	
Yes	—	1
Diabetes		
No	1	
Yes	7.2526 (1.8874–27.8691)	0.004
Molecular typing		
Luminal A	1	
Luminal B	0.8695 (0.1472–5.1356)	0.877
Her-2 positive	—	—
Triple negative	1.5148 (0.2420–9.4801)	0.657
Stage		
I	1	
II	1.9789 (0.5132–7.6316)	0.322
III	0.8796 (0.1515–5.1080)	0.886
IV	2.44×10^{19} (5.72×10^{17}–1.04×10^{21})	<0.001
Targeted therapy		
Yes	1	
No	1.7777 (0.0313–1.0085)	0.051

PFS: progression-free survival; HR: hazard ratio; CI: confidence interval.

4. Discussion

Breast carcinoma with neuroendocrine features is a group of heterogeneous tumors. Its definitions and diagnostic criteria have varied with the revisions of the WHO classification of breast tumors. As a result, the results of some studies on the clinicopathological characteristics of this disease have been controversial for a long time. Breast carcinoma with neuroendocrine features is a group of tumors that exhibit morphological features similar to those of neuroendocrine tumors of the gastrointestinal tract and lung [11]. Before 2003, there were no criteria for the definition and diagnosis of this disease. With the further study of breast carcinoma, the consensus on breast carcinoma with neuroendocrine features has gradually been formed. The third version of the WHO criteria of breast tumors defined it as >50% tumor cells with neuroendocrine differentiation confirmed by immunohistochemical staining [12]. From then on, it was recognized as single breast carcinoma entities named "neuroendocrine breast carcinomas". In 2012, the WHO classification used the category of "carcinoma with neuroendocrine features" and described this disease as tumors expressing neuroendocrine markers to any extent [13]. It included well-differentiated NET, poorly differentiated NEC, and carcinoma with neuroendocrine differentiation. In this version, small cell neuroendocrine carcinoma (SCNEC) was brought into the NEC group. The current WHO classification adopted the term "NEN", including well-differentiated (NET) and poorly-differentiated (NEC) tumors with predominant neuroendocrine differentiation [3]. The main distinction between the latest classification and the past version is that carcinoma with neuroendocrine differentiation without distinct or uniform enough neuroendocrine histological features and neuroendocrine marker expression is no longer classified as NET or NEC. In this version, large cell neuroendocrine carcinoma (LCNEC) was classified into NEC as well. All the criteria mentioned above are used for the classification of primary breast carcinoma with neuroendocrine features; however, before a diagnosis of primary NEN is made, the possibility of metastasis from other organs should be carefully ruled out. Immunohistochemistry staining is conducive to distinction between NEN derived from other organs from invasive mammary carcinoma with neuroendocrine features [14]. This study only discussed primary breast carcinoma with neuroendocrine features.

The results of the previous studies in different periods were different or even contradictory. Previous studies found that most patients were 50 years old or older [15,16], and that the clinical symptoms of this disease were mainly bloody nipple discharge [16], which was consistent with the results of this study. NEN can be divided into functional and non-functional tumors according to whether the tumor has hormone activity, and most NENs are non-functional. Functional NEN produces excessive hormones, leading to clinical symptoms such as diarrhea and facial flushing. Non-functional NENs do not produce enough hormones to cause these symptoms [17]. Paraneoplastic endocrine syndrome may occur in breast cancer with or without neuroendocrine differentiation [18]. There were few reports on paraneoplastic endocrine syndrome related to breast neuroendocrine tumors. One case with hyperprolactinemia was reported in the previous literature [19]. Studies have illustrated that patients often had ER/PR positive and Her-2 negative tumors [5,20], supporting this study. Another study found that neuroendocrine differentiation was more common in Luminal B breast cancer [21], which is consistent with the results of this study. However, among them, Her-2 positive patients were rare, only occasionally seen in case reports [22]. Research showed that breast carcinoma with neuroendocrine features was highly aggressive, with a high rate of local recurrence and distant metastasis, and a poor prognosis [5]. A study has also shown that its general clinical characteristics are not different from other breast cancers, and its biological behavior was not aggressive; on the contrary, it tended to be an independent good-prognosis subgroup [20].

Morphologically, the typical features of the lung/gastrointestinal tract NET, such as ribbons, cords, and rosettes, are not prominent in the breast NET; histological and immunohistochemical features of breast NEC are sometimes difficult to distinguish from lung NEC features [23]. CgA, Syn, NSE, and CD56 were neuroendocrine differentiation markers for breast carcinoma with neuroendocrine features [4]. It has been reported that only 23% of patients were detected with Syn+ and CgA+ at the same time [15]. The expression level of neuroendocrine markers in tumor tissues of patients in this study was also not high. There were significant differences in cytological characteristics between focal and diffuse neuroendocrine differentiated breast carcinomas [24]. However, when breast cancer of no special type with focal neuroendocrine differentiation was regarded as a separate entity, focal neuroendocrine differentiation had no obvious significance for its prognosis [15,25]. In this study, there was no difference between NEN and IBC with neuroendocrine differentiation in five-year OS, five-year DSS, and five-year PFS.

SCNEC was classified into NEC in 2012 and LCNEC was brought into NEC in 2019. A study indicated that approximately half of these patients had triple-negative breast cancer, with a 61.6% five-year DSS rate and 47.7% five-year OS rate [26]. Chemotherapy, surgery, and stage were predictive factors of prognosis. In this study, there was only one case of pure SCNEC and pure LCNEC, respectively, accounting for a relatively low proportion. The special type of breast cancer with neuroendocrine differentiation was rare. According to one study, half of the invasive solid papillary carcinomas were accompanied by neuroendocrine differentiation [7]. In this study, there were four cases of a special type of breast cancer with neuroendocrine differentiation, including two cases of solid papillary carcinoma, one case of invasive papillary carcinoma, and one case of type B mucinous carcinoma.

A study reported that the five-year OS and disease-free survival rates of HR positive/Her-2 negative breast cancer were 93.0% and 92.6%, respectively [27]. The PFS of breast carcinoma with neuroendocrine features was lower than that of the same molecular typing of breast cancer of no special type [28], which was consistent with the results of this study. The overall five-year PFS of patients in this study was 82.37% (95%CI: 0.7084–0.8966), five-year DSS was 91.53% (95%CI: 0.8022–0.9651), and five-year OS was 90.25% (95%CI: 0.7901–0.9564). The study of neoadjuvant therapy for this disease has been limited. In this study, there were seven patients receiving neoadjuvant chemotherapy, and none of them reached a pathological complete response. A study found that endocrine therapy or radiotherapy might improve the prognosis [5]. It was suggested that HR positive NEN patients receive endocrine therapy, especially those with SCNEC with recurrence and metastasis [29]. Endocrine therapy was also found to be effective for liver metastasis of breast cancer with neuroendocrine differentiation [6]. However, surgery was still the main treatment method, and the effect of chemotherapy on prognosis was still uncertain [16].

This study also analyzed the related factors of PFS. A study found that the prognostic factors of NEN of the breast were similar to those of gastrointestinal tract tumors [30], among which lymph node metastasis was an adverse factor of OS [5]. A previous study found that histological grade, pathological stage, ER status, and HER2 status were independent prognostic indicators of OS and disease-free survival [31]. The results of this study showed that diabetes and stage IV disease were related to the PFS of breast carcinoma with neuroendocrine features patients. The influence of diabetes on the PFS of this disease may be related to higher BMI, and there is a lack of relevant research at present.

Limitations

This study was a retrospective study, and the sample size was relatively insufficient. We failed to compare this disease with other types of breast cancer because of lacking a control group. In addition, this study did not describe its imaging characteristics. Because of the rarity of this disease, there was a gap between the length of follow-up time of the patients.

5. Conclusions

Breast carcinoma with neuroendocrine features is relatively rare compared with other types of breast cancer. In this study, it was illustrated that this disease tended to be HR+/Her-2- tumor. In addition, diabetes and stage IV were related to the PFS of patients. These results may provide evidence for the treatment and prognosis prediction of breast carcinoma with neuroendocrine features. Further studies such as large-sample randomized clinical trials are needed to validate the theoretical value and practical significance of these findings and improve understanding of this disease.

Author Contributions: Conceptualization, Y.Q. and J.W.; Methodology, L.Z. (Li Zhu); Investigation, X.H. and Y.D.; Data Collection, L.Z. (Liping Zhang) and B.B.; Writing—Original Draft Preparation, Y.Q. and Y.C.; Writing—Review & Editing, J.W.; Supervision, Y.D. and J.W. All authors have read and agreed to the published version of the manuscript.

Funding: This research was funded by the National Health Commission Capacity Building and Continuing Education Center (grant number: GWJJ2021100303).

Institutional Review Board Statement: All procedures performed in this study involving human participants were in accordance with the Declaration of Helsinki (as revised in 2013). The study was approved by ethics committee of the Chinese PLA General Hospital (NO.: S2022-746 and date of approval 11 January 2023).

Informed Consent Statement: Individual consent for this retrospective analysis was waived.

Data Availability Statement: The data that support the findings of this study are available from the corresponding authors, [J.W. and Y.C.], upon reasonable request.

Conflicts of Interest: The authors declare no conflict of interest.

References

1. Zhang, Y.; Geng, C.; Xu, B. Controversies and research progress on the diagnosis and treatment of breast neuroendocrine neoplasm. *Chin. J. Clin. Oncol.* **2022**, *49*, 968–972. (In Chinese) [CrossRef]
2. Wang, J.; Wei, B.; Albarracin, C.T.; Hu, J.; Abraham, S.C.; Wu, Y. Invasive neuroendocrine carcinoma of the breast: A population-based study from the surveillance, epidemiology and end results (SEER) database. *BMC Cancer* **2014**, *14*, 147. [CrossRef]
3. WHO Classification of Tumors Editorial Board. Breast Tumours. In *WHO Classification of Tumors*, 5th ed.; IARC: Lyon, France, 2019.
4. Rovera, F.; Lavazza, M.; La Rosa, S.; Fachinetti, A.; Chiappa, C.; Marelli, M.; Sessa, F.; Giardina, G.; Gueli, R.; Dionigi, G.; et al. Neuroendocrine breast cancer: Retrospective analysis of 96 patients and review of literature. *Int. J. Surg.* **2013**, *11*, S79–S83. [CrossRef]
5. Wei, B.; Ding, T.; Xing, Y.; Wei, W.; Tian, Z.; Tang, F.; Abraham, S.C.; Nayeemuddin, K.M.; Hunt, K.K.; Wu, Y. Invasive neuroendocrine carcinoma of the breast: A distinctive subtype of aggressive mammary carcinoma. *Cancer* **2010**, *116*, 4463–4473. [CrossRef]

6. Rischke, H.C.; Staib-Sebler, E.; Mose, S.; Adams, S.W.; Herrmann, G.; Böttcher, H.D.; Lorenz, M. Metastasiertes Mamma-Karzinom neuroendokriner Differenzierung–kombinierte Therapie mit Tamoxifen und dem Somatostatin-Analogon Octreotid [Metastatic breast carcinoma with neuroendocrine differentiation–its combined therapy with tamoxifen and the somatostatin analog octreotide]. *DMW-Dtsch. Med. Wochenschr.* **1999**, *124*, 182–186. [CrossRef]
7. Lin, X.; Matsumoto, Y.; Nakakimura, T.; Ono, K.; Umeoka, S.; Torii, M.; Yoshibayashi, H.; Toi, M. Invasive solid papillary carcinoma with neuroendocrine differentiation of the breast: A case report and literature review. *Surg. Case Rep.* **2020**, *6*, 143. [CrossRef]
8. Salemis, N.S. Primary neuroendocrine carcinoma of the breast: A rare presentation and review of the literature. *Intractable Rare Dis. Res.* **2020**, *9*, 233–246. [CrossRef]
9. Giuliano, A.E.; Connolly, J.L.; Edge, S.B.; Mittendorf, E.A.; Rugo, H.S.; Solin, L.J.; Weaver, D.L.; Winchester, D.J.; Hortobagyi, G.N. Breast Cancer-Major changes in the American Joint Committee on Cancer eighth edition cancer staging manual. *CA Cancer J. Clin.* **2017**, *67*, 290–303. [CrossRef]
10. Jiang, Z.; Song, E.; Wang, X.; Wang, H.; Wang, X.; Wu, J.; Yin, Y.; Zhang, Q.; Chen, J.; Che, W.; et al. Guidelines of Chinese Society of Clinical Oncology (CSCO) on Diagnosis and Treatment of Breast Cancer (2020 version). *Transl. Breast Cancer Res.* **2020**, *1*, 27. [CrossRef]
11. Rindi, G.; Klimstra, D.S.; Abedi-Ardekani, B.; Asa, S.L.; Bosman, F.T.; Brambilla, E.; Busam, K.J.; De Krijger, R.R.; Dietel, M.; El-Naggar, A.K.; et al. A common classification framework for neuroendocrine neoplasms: An International Agency for Research on Cancer (IARC) and World Health Organization (WHO) expert consensus proposal. *Mod. Pathol.* **2018**, *31*, 1770–1786. [CrossRef]
12. Tavassoli, F.A.; Devilee, P. *World Health Organization Classification of Tumours: Pathology and Genetics of Tumours of the Breast and Female Genital Organs*; IARC Press: Lyon, France, 2003.
13. Lakhani, S.R.; Ellis, I.O.; Schnitt, S.J.; Tan, P.H.; van de Vijver, M.J. *WHO Classification of Tumours of the Breast*, 4th ed.; IARC Press: Lyon, France, 2012.
14. Mohanty, S.K.; Kim, S.A.; DeLair, D.F.; Bose, S.; Laury, A.R.; Chopra, S.; Mertens, R.B.; Dhall, D. Comparison of metastatic neuroendocrine neoplasms to the breast and primary invasive mammary carcinomas with neuroendocrine differentiation. *Mod. Pathol.* **2016**, *29*, 788–798. [CrossRef]
15. Van Krimpen, C.; Elferink, A.; Broodman, C.; Hop, W.; Pronk, A.; Menke, M. The prognostic influence of neuroendocrine differentiation in breast cancer: Results of a long-term follow-up study. *Breast* **2004**, *13*, 329–333. [CrossRef] [PubMed]
16. Li, Y.; Du, F.; Zhu, W.; Xu, B. Neuroendocrine carcinoma of the breast: A review of 126 cases in China. *Chin. J. Cancer* **2017**, *36*, 1–4. [CrossRef] [PubMed]
17. Ozaki, Y.; Miura, S.; Oki, R.; Morikawa, T.; Uchino, K. Neuroendocrine Neoplasms of the Breast: The Latest WHO Classification and Review of the Literature. *Cancers* **2021**, *14*, 196. [CrossRef]
18. Dimitriadis, G.K.; Angelousi, A.; Weickert, M.O.; Randeva, H.S.; Kaltsas, G.; Grossman, A. Paraneoplastic endocrine syndromes. *Endocrine-Related Cancer* **2017**, *24*, R173–R190. [CrossRef]
19. Zhang, Q.; He, L.; Lv, W.; Wang, N. Neuroendocrine carcinoma of the breast with hyperprolactinemia: Report of two cases and a minireview. *Int. J. Clin. Exp. Pathol.* **2020**, *13*, 1457–1462.
20. Menendez, J.A.; Lopez-Bonet, E.; Alonso-Ruano, M.; Barraza, G.; Vázquez-Martín, A.; Bernadó, L. Solid neuroendocrine breast carcinomas: Incidence, clinico-pathological features and immunohistochemical profiling. *Oncol. Rep.* **1994**, *20*, 1369–1374. [CrossRef] [PubMed]
21. Wachter, D.L.; Hartmann, A.; Beckmann, M.W.; Fasching, P.A.; Hein, A.; Bayer, C.M.; Agaimy, A. Expression of Neuroendocrine Markers in Different Molecular Subtypes of Breast Carcinoma. *BioMed Res. Int.* **2014**, *2014*, 408459. [CrossRef]
22. Marijanović, I.; Kraljević, M.; Buhovac, T.; Križanac, D.K. Rare Human Epidermal Growth Factor Receptor 2 (HER-2)-Positive Neuroendocrine Carcinoma of the Breast: A Case Report with 9-Year Follow-up. *Am. J. Case Rep.* **2020**, *21*, e925895. [CrossRef]
23. Tsang, J.Y.; Tse, G.M. Breast cancer with neuroendocrine differentiation: An update based on the latest WHO classification. *Mod. Pathol.* **2021**, *34*, 1062–1073. [CrossRef]
24. Talu, C.K.; Guzelbey, B.; Hacihasanoglu, E.; Cakir, Y.; Nazli, M.A. The effect of the extent of neuroendocrine differentiation on cytopathological findings in primary neuroendocrine neoplasms of the breast. *J. Cytol.* **2021**, *38*, 216. [CrossRef]
25. Bogina, G.; Munari, E.; Brunelli, M.; Bortesi, L.; Marconi, M.; Sommaggio, M.; Lunardi, G.; Gori, S.; Massocco, A.; Pegoraro, M.C.; et al. Neuroendocrine differentiation in breast carcinoma: Clinicopathological features and outcome. *Histopathology* **2015**, *68*, 422–432. [CrossRef]
26. Zhu, J.; Wu, G.; Zhao, Y.; Yang, B.; Chen, Q.; Jiang, J.; Meng, Y.; Ji, S.; Gu, K. Epidemiology, Treatment and Prognosis Analysis of Small Cell Breast Carcinoma: A Population-Based Study. *Front. Endocrinol.* **2022**, *13*, 802339. [CrossRef]
27. Wang, Y.; Zhang, S.; Zhang, H.; Liang, L.; Xu, L.; Cheng, Y.J.; Duan, X.N.; Liu, Y.H.; Li, T. Clinicopathological features and prognosis of hormone receptor-positive/hum an epidermal growth factor receptor 2-negative breast cancer. *Beijing Da Xue Xue Bao Yi Xue Ban* **2022**, *54*, 853–862. (In Chinese)
28. Lavigne, M.; Menet, E.; Tille, J.-C.; Lae, M.; Fuhrmann, L.; Bonneau, C.; Deniziaut, G.; Melaabi, S.; Ng, C.C.K.; Marchiò, C.; et al. Comprehensive clinical and molecular analyses of neuroendocrine carcinomas of the breast. *Mod. Pathol.* **2018**, *31*, 68–82. [CrossRef]

29. Alkaied, H.; Harris, K.; Brenner, A.; Awasum, M.; Varma, S. Does Hormonal Therapy Have a Therapeutic Role in Metastatic Primary Small Cell Neuroendocrine Breast Carcinoma? Case Report and Literature Review. *Clin. Breast Cancer* **2012**, *12*, 226–230. [CrossRef]
30. Roininen, N.; Takala, S.; Haapasaari, K.-M.; Jukkola-Vuorinen, A.; Mattson, J.; Heikkilä, P.; Karihtala, P. Neuroendocrine Breast Carcinomas Share Prognostic Factors with Gastroenteropancreatic Neuroendocrine Tumors: A Putative Prognostic Role of Menin, p27, and SSTR-2A. *Oncology* **2018**, *96*, 147–155. [CrossRef]
31. Kwon, S.Y.; Bae, Y.K.; Gu, M.J.; Choi, J.E.; Kang, S.H.; Lee, S.J.; Kim, A.; Jung, H.R.; Kang, S.H.; Oh, H.K.; et al. Neuroendocrine differentiation correlates with hormone receptor expression and decreased survival in patients with invasive breast carcinoma. *Histopathology* **2013**, *64*, 647–659. [CrossRef]

life

Review

Breast Reconstruction following Mastectomy for Breast Cancer or Prophylactic Mastectomy: Therapeutic Options and Results

Laurentiu Simion [1,2], Ina Petrescu [3], Elena Chitoran [1,2,*], Vlad Rotaru [1,2], Ciprian Cirimbei [1,2,*], Sinziana-Octavia Ionescu [1,2], Daniela-Cristina Stefan [4], Dan Luca [1,2], Dana Lucia Stanculeanu [4,5], Adelina Silvana Gheorghe [4,5], Horia Doran [1,6] and Ioana Mihaela Dogaru [7,8]

[1] Department of General Surgery, "Carol Davila" University of Medicine and Pharmacy, 050474 Bucharest, Romania; laurentiu.simion@umfcd.ro (L.S.); rotaru.vlad@gmail.com (V.R.); sinziana.ionescu@umfcd.ro (S.-O.I.); luca_dan94@yahoo.com (D.L.); horia.doran@umfcd.ro (H.D.)

[2] General Surgery and Surgical Oncology Department I, Bucharest Institute of Oncology "Prof. Dr. Alexandru Trestioreanu", 022328 Bucharest, Romania

[3] Zetta Hospital, 020311 Bucharest, Romania; inapetrescu@yahoo.com

[4] Department of Oncology, "Carol Davila" University of Medicine and Pharmacy, 050474 Bucharest, Romania; cristina.stefan10@gmail.com (D.-C.S.); dana.stanculeanu@umfcd.ro (D.L.S.); adelina.silvana.gheorghe@gmail.com (A.S.G.)

[5] Oncology Department I, Bucharest Institute of Oncology "Prof. Dr. Alexandru Trestioreanu", 022328 Bucharest, Romania

[6] Surgical Clinic I, Clinical Hospital Dr. I. Cantacuzino, 030167 Bucharest, Romania

[7] Department of Plastic Surgery, "Carol Davila" University of Medicine and Pharmacy, 050474 Bucharest, Romania; ioana.m.dogaru@gmail.com

[8] Department of Plastic Surgery, Emergency University Hospital, 050098 Bucharest, Romania

* Correspondence: chitoran.elena@gmail.com (E.C.); ciprian.cirimbei@umfcd.ro (C.C.)

Citation: Simion, L.; Petrescu, I.; Chitoran, E.; Rotaru, V.; Cirimbei, C.; Ionescu, S.-O.; Stefan, D.-C.; Luca, D.; Stanculeanu, D.L.; Gheorghe, A.S.; et al. Breast Reconstruction following Mastectomy for Breast Cancer or Prophylactic Mastectomy: Therapeutic Options and Results. *Life* **2024**, *14*, 138. https://doi.org/10.3390/life14010138

Academic Editors: Taobo Hu, Mengping Long, Lei Wang and Riccardo Autelli

Received: 11 December 2023
Revised: 3 January 2024
Accepted: 12 January 2024
Published: 18 January 2024

Abstract: (1) Importance of problem: Breast cancer accounted for 685,000 deaths globally in 2020, and half of all cases occur in women with no specific risk factor besides gender and age group. During the last four decades, we have seen a 40% reduction in age-standardized breast cancer mortality and have also witnessed a reduction in the medium age at diagnosis, which in turn means that the number of mastectomies performed for younger women increased, raising the need for adequate breast reconstructive surgery. Advances in oncological treatment have made it possible to limit the extent of what represents radical surgery for breast cancer, yet in the past decade, we have seen a marked trend toward mastectomies in breast-conserving surgery-eligible patients. Prophylactic mastectomies have also registered an upward trend. This trend together with new uses for breast reconstruction like chest feminization in transgender patients has increased the need for breast reconstruction surgery. (2) Purpose: The purpose of this study is to analyze the types of reconstructive procedures, their indications, their limitations, their functional results, and their safety profiles when used during the integrated treatment plan of the oncologic patient. (3) Methods: We conducted an extensive literature review of the main reconstructive techniques, especially the autologous procedures; summarized the findings; and presented a few cases from our own experience for exemplification of the usage of breast reconstruction in oncologic patients. (4) Conclusions: Breast reconstruction has become a necessary step in the treatment of most breast cancers, and many reconstructive techniques are now routinely practiced. Microsurgical techniques are considered the "gold standard", but they are not accessible to all services, from a technical or financial point of view, so pediculated flaps remain the safe and reliable option, along with alloplastic procedures, to improve the quality of life of these patients.

Keywords: breast reconstruction; reconstruction following mastectomy; prophylactic mastectomy; chest feminization; transgender; implant reconstruction of breast; immediate reconstruction; delayed reconstruction; two-stage breast reconstruction; autologous breast reconstruction

1. Introduction

Breast cancer accounted for 685,000 deaths globally in 2020, and half of all cases occur in women with no specific risk factor besides gender and age group. During the last four decades, we have seen a 40% reduction in age-standardized breast cancer mortality [1] and have also witnessed a reduction in the medium age at diagnosis, which in turn means that the number of mastectomies performed for younger women increased, raising the need for adequate breast reconstructive surgery. Advances in oncological treatment have made it possible to limit the extent of what represents radical surgery for breast cancer, yet in the past decade, we have seen a marked trend toward mastectomies in breast-conserving surgery-eligible patients [2]. Prophylactic mastectomies have also registered an upward trend [3,4]. This trend together with new uses for breast reconstruction like chest feminization in transgender patients [5] has increased the need for breast reconstruction surgery.

Breast cancer is the most commonly diagnosed neoplasm in the female population in the world [6]. It is the leading cause of cancer-related death in women in most countries of the world, except in developed countries, where it ranks second after lung tumors. However, mortality has been steadily declining for over 30 years, with an average 5-year survival of 86% and 75% at 10 years [7]. This trend is attributed both to the increase in the effectiveness of oncological treatments and to early screenings and screening programs similar to those for other neoplastic diseases [8,9].

Breast reconstruction is an important component of breast cancer treatment. With the increase in life expectancy, it has become essential to ensure a good quality of life for patients, forcing a continuous evolution of surgical techniques. Breast reconstruction is necessary not only after performing a modified radical mastectomy, but also after conservative interventions on the breast that have not been accompanied by an optimal aesthetic effect. The need to complete the surgical treatment of breast cancer with breast reconstruction derives from the beneficial impact at the psychological level, respectively, for the body image, sexuality, and general quality of life of patients [10]. In recent years, the ever-increasing number of patients opting for prophylactic mastectomies due to a genetic predisposition for developing breast cancer or a family history of cancer [3,4,11–13] has given birth to a new type of integrated treatment plan in oncology. Changes in guidelines, prompting the genetic testing of BRCA mutations and the availability of those tests even in the absence of an oncologist's recommendation, have determined an increase in the number of women getting tested and then opting for a contralateral or bilateral prophylactic mastectomy. A good example for this trend is the Angelina Jolie effect on the Western population; following the known actress's double prophylactic mastectomy, there was a noticeable increase in the number of healthy women requesting this procedure and having it performed.

Although it is not the focus of this study, breast reconstruction has had another extremely important purpose in recent years, namely, for the chest feminization of male-to-female transgender patients. In combination with hormone and psychological therapy, breast enhancement is the most common physical modification in this populational subset [5,14–16], contributing to a reduction in the patient's dysphoria. For this purpose, all surgical reconstructive techniques used in patients with mastectomies can be employed.

This study reviews the main techniques, especially the autologous and mixed procedures, and investigates available data from the literature, indicating their indications and results.

2. Problem at Hand

2.1. Dimension of Problem

Breast cancer accounted for 685,000 deaths globally in 2020, and half of all cases occur in women with no specific risk factor. During last four decades, we have seen a 40% reduction in age-standardized breast cancer mortality [1], but we have also witnessed the reduction in the medium age at diagnosis, which in turn means that the number of mastectomies performed for younger women increased, raising the need for adequate reconstructive surgery. Breast cancer is the most prevalent form of cancer in the world, with a total of 7.8 million women alive in 2020 who were diagnosed with a form of this malady in the previous 5-year period [1]. Advances in oncological treatment, which have prolonged patients' survivorship after breast cancer, have also made it possible to limit the extent of what represents radical surgery for breast cancer, yet in the past decade, we have seen a marked trend toward mastectomy in breast-conserving surgery (BCS)-eligible patients [2]. Prophylactic mastectomy has also registered an upward trend [3,4]. This trend, together with new indications for breast reconstruction like chest feminization in transgender patients [5,14,16] or the need to resolve the asymmetry of the contralateral breast [17–19], have increased the need for breast reconstruction surgery.

2.2. Mastectomy vs. Breast-Conserving Surgery

Many breast cancer patients elect to have a radical mastectomy, rather than a conservative surgical procedure, even though they are perfect candidates for BCS, and some long-term studies have suggested a slightly more favorable outcome of lumpectomies associated with radiotherapy vs. mastectomies [20], maintaining the high frequency of mastectomies. Patients opting for mastectomies over BCS usually do not choose by taking into account the histology, localization, or aggressiveness of the tumor, but rather more subjective reasons like a lack of trust that BCS can offer the same likelihood of cure as a more extensive procedure [21] or fear of additional procedures. The surgeon's recommendation is a key factor in the decision-making process, but it is overshadowed by the patient's fear of cancer [22].

In recent years, an increase in mastectomy rates in early-stage breast cancer patients was observed. The reasons for which patients tend to select a more aggressive procedure when breast-conserving surgery is an option are unclear and include, besides a so-called "peace-of-mind" and a more laxed surveillance schedule, the easy access to reconstructive surgery and the patient's confidence in the aesthetic results of reconstructive techniques [23].

2.3. Lymphadenectomy and Sentinel-Lymph Node Biopsy

Radical surgery for breast cancer comprises the excision of the tumor (mastectomy or various breast-conserving techniques) and a procedure addressed to the axillary lymph nodes (inferior lymphadenectomy, extensive lymphadenectomy, or identification and excision of the sentinel lymph nodes using radioactive material or intravital dyes like Indocyanine green or Methylene Blue). The extension of the excision of the lymphatic tissue can influence the results of the reconstruction of the breast in both immediate and delayed settings by increasing the number of complications. Complete axillary lymph node dissection has a more pronounced effect when compared to that of a limited lymphadenectomy of a sentinel lymph node excision and is associated with a greater probability of implant loss independent of the associated radiotherapy [24,25]. There are studies that proved that the excision of each node increases the risk of reconstructive surgery complications by 4% [25]. The same study concluded that the removal of four or more lymph nodes can adversely affect the immediate reconstructive procedure by seroma formation or even implant loss [25]. However, the complications after immediate reconstruction of the breast are associated with the use of implants. For this reason, in patients requiring axillary lymph node dissection, the oncoplastic surgeon should offer the autologous methods of reconstruction [24].

Half of the patients with mastectomies for breast cancer elect to undergo reconstructive surgery [26,27] due to aesthetic considerations and an improved quality of life [28–31] through reduced body dysmorphia. Yet, following reconstruction, many patients experience sequelae like functional limitations of the upper limb (strength and mobility) and pain [31–33]. Axillary lymphadenectomy can cause neurological syndromes like pain, paresthesia, and limitations of mobility after reconstructive surgery. This effect can be reduced by preserving the sensitive nerves during the lymphadenectomy [34].

2.4. Impact of Radiotherapy in Surgical Options and Results

After reconstructive surgery, radiation therapy may affect the aspect of the operated breast, including the altered skin color and rigidity. It can also lead to capsular contraction, which mandates the removal of the implant. Patients undergoing radiation therapy after reconstructive surgery need to be advised about the possibility of additional corrective surgery [35].

2.5. Quality of Life following Surgery for Breast Cancer

After mastectomies, patients report a significant alteration of their quality of life (QoL) through a series of mechanisms: body dysmorphia affecting both emotional and sexual functioning, especially in younger patients [36]; pain and limited mobility in the ipsilateral upper limb; and psychological effects like negative emotions such as sadness, low mood, and dejection [37]. Although there is a significant reduction in the alteration of QoL following immediate breast reconstruction, many women tend to underestimate the impact of the mastectomy and to be overly optimistic about the impact of reconstructive surgery, and a significant proportion of them (up to 20% in some studies) come to regret breast reconstruction [36,38]. This particular aspect needs to be taken into account when discussing breast reconstruction with the patients in order to make sure they have realistic expectations.

When discussing QoL in patients who underwent mastectomies, we cannot leave out the problems caused by breast asymmetry, especially in large breasts, leading to alterations of the skeletal system (like scoliosis). Asymmetry of the breasts can occur even in patients with reconstructive surgery after their mastectomy if the procedure was unilateral and performed with implants. The remaining breast tends to be more ptotic, resulting in undesired aesthetic effects and causing the patient to request corrective surgery of the contralateral breast.

2.6. Oncologic Follow-Up and Results after Reconstructive Surgery

Breast reconstruction surgery is a safe procedure from the oncological point of view, regardless of using an autologous or implant-based method for reconstruction and regardless of an immediate or delayed timing of the procedure [39], and it does not increase the local or systemic recurrence rates nor disease-free and overall survival [40,41]. The type of reconstructive surgery after a mastectomy does not influence the recurrence rate independently of the aggressive histology of the tumor [42], lymphatic invasion, and the positive resection margins [43].

Another safety concern after breast reconstruction following a modified radical mastectomy is the possibility of the detection of recurrence, since autologous tissue below the skin flap may interfere with the detection of recurrent nodules, and fatty necrosis can confuse the diagnosis [44,45]. Recurrence after reconstructive surgery using implants may be challenging to detect beneath the implant [46]. A thicker skin flap (over 1.5 cm) may interfere with the detection of a palpable mass upon examination, and an extremely slim skin flap (under 0.5 cm), although conducive to an early clinical detection of recurrence, is more prone to necrosis of the flap. A delicate balance needs to be achieved, and in our opinion, a 1 cm thickness of the skin flap is optimal. Lastly, in order to minimize the risk of flap necrosis, techniques using Indocyanine green may be employed for assessing the perfusion of the flap used for breast reconstruction [47].

3. Breast Reconstruction

3.1. Timing of Breast Reconstruction—Immediate or Delayed

Breast reconstruction is classified by the type and time of surgery. Immediate reconstruction takes place at the same time as the mastectomy, and secondary (or delayed) reconstruction is performed from a few months to several years after the mastectomy. Currently, it is performed at least three months after the end of radiotherapy and generally at about a year after the mastectomy [48]. The two main types of reconstruction are with implants or autologous tissue; they can also be used together in mixed procedures.

Immediate breast reconstruction (Figure 1) has certain benefits over the secondary one, especially in terms of patient satisfaction, quality of life, and psychological status post-mastectomy [49]. These patients are relatively more protected from the psychological effects of mastectomies, and studies have shown a stable evolution of quality of life and satisfaction of this group compared to patients receiving delayed reconstruction [50,51]. In the latter case, the quality of life is significantly improved with the reconstructive procedures, with the results ultimately being equalized in the long run [52]. Also, after the immediate reconstruction, a more natural and aesthetic result is obtained, with the intervention usually being associated with a skin-sparing mastectomy which respects the inframammary groove and keeps the skin intact, proven safe from the oncological point of view by a series of studies, provided the correct selection of patients [53–56]. An important factor for selection is the appreciation of the thickness of the skin flap, this being correlated with the aesthetic results and with the possible postoperative complications [57]. Patients undergo fewer major surgeries and require fewer days of hospitalization, recovering faster postoperatively. From an oncological point of view, immediate reconstruction is considered safe and has been shown to not increase the risk of local recurrence compared to that of mastectomies without reconstruction [58,59]. At the same time, this technique does not change the effectiveness of adjuvant radiotherapy [60].

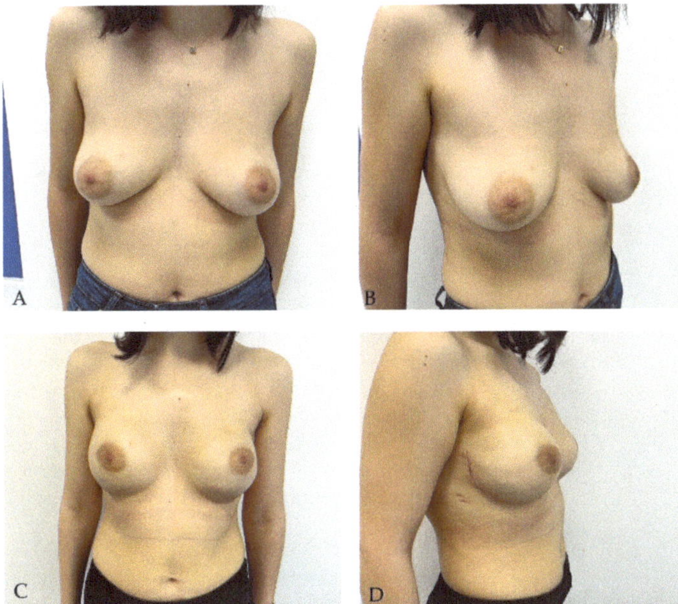

Figure 1. Patient with stage Ia invasive ductal carcinoma of left breast and BRCA positive status. She underwent bilateral subcutaneous mastectomy with left sentinel lymph node identification using Indocyanine green followed by immediate bilateral reconstruction with 350 cc round implants: (**A,B**) aspect before reconstructive surgery; (**C,D**) aspect at 3 months after reconstructive surgery.

Despite the many benefits of immediate reconstruction, many surgeons choose to postpone the operation for another time (Figure 2), often for reasons of oncological safety. Definite diagnoses of malignancies of radiologically detected breast tumors are made more and more frequently by guided biopsies, and the real extension of the tumor tissue can be evaluated macroscopically only intraoperatively, and microscopically only when examining the mastectomy piece. Thus, the subsequent therapeutic attitude is often decided intraoperatively [61]. However, in order to evaluate the quality and thickness of the skin flap intended for immediate breast reconstruction, mammography, breast ultrasound, and magnetic resonance imaging (MRI) can be used preoperatively to complete the clinical examination. The results of these investigations guide the decision on surgery for immediate breast reconstruction and have been shown to be true to intraoperative findings. The thickness of the flap is important in choosing the type of implant used, but also for avoiding postoperative complications such as skin necrosis [57].

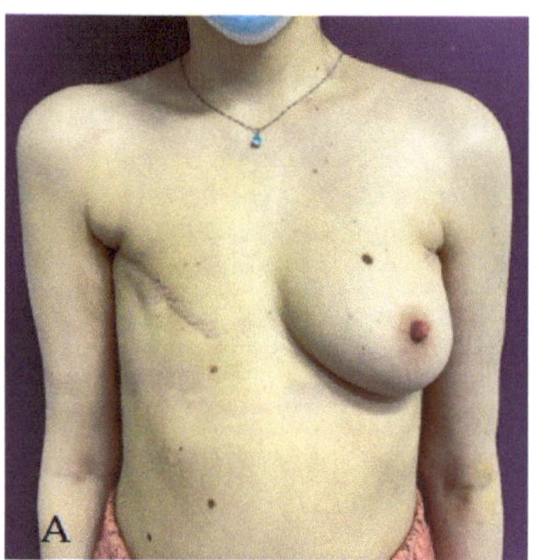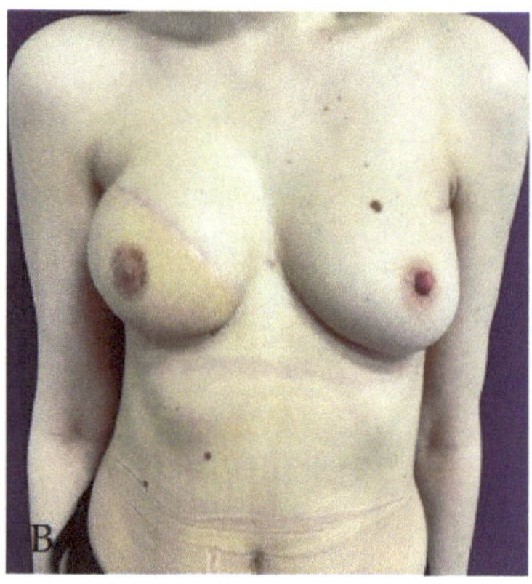

Figure 2. Patient with right radical mastectomy for breast cancer followed by radiotherapy; she underwent right breast delayed reconstruction using latissimus dorsi pediculated flap and a 225 cc round implant. (**A**) aspect before reconstructive surgery; (**B**) aspect at 3 months after reconstructive surgery.

The indications of radiotherapy, typically applied to patients at high risk of recurrence (>4 positive lymph nodes or positive resection margins), tend to increase, with studies proving its usefulness in patients with 1–3 positive lymph nodes [62–64]. Although radiotherapy does not contraindicate immediate reconstruction, the higher rate of complications, especially in implant-only reconstructions, is a second reason why in these patients, either a two-stage reconstruction or a delayed reconstruction is chosen [65]. After reconstructive surgery, radiation therapy may affect the aspect of the operated breast, including altered skin color and rigidity. It can also lead to capsular contraction which mandates the removal of implant. Patients undergoing radiation therapy after reconstructive surgery need to be advised about the possibility of additional corrective surgery [35].

Another contraindication of immediate reconstruction is any modifications of the flaps (tegument and subcutaneous tissue), namely, the presence of necrosis, inflammation, or signs of dermal neoplastic dissemination resulting in a large skin defect after the mastectomy.

3.2. Two-Stage Breast Reconstruction

In 2002, the technique of two-stage breast reconstruction was initially described for delayed reconstruction. Later, the technique was especially used to improve the results in cases associated with radiotherapy [66]. The ionizing radiation used on either the chest wall or the axilla irreversibly alters the tissues in the irradiated field, regardless of their nature. In the short term, erythema and scaling of the skin can appear, and in the long-term, severe fibrosis, telangiectasia, hyperpigmentation, and tissue atrophy [67]. Under these conditions, many surgeons prefer to place a tissue expander at the time of the mastectomy, preserved during radiotherapy, which aims to maintain both the shape and the skin needed for the final reconstruction [68]. The expander can be filled at the time of the intervention, or progressively, depending on the condition of the flaps and the center where the intervention is carried out [69]. It can be partially emptied before radiotherapy sessions in order to favor the alignment of the irradiation fields, but this step is not always necessary [70]. Subsequently—it is recommended no later than 3 months after the completion of radiotherapy—the second stage of reconstruction is performed, usually with autologous tissue. For patients who do not require adjuvant radiotherapy, the recommendation is that the second stage of reconstruction be performed no later than two weeks after the mastectomy [65].

3.3. Breast Reconstruction with Implant

Regarding the type of intervention, at present, it is estimated that 80% of breast reconstructions are performed with an implant [68]. This type of intervention is shorter and easier from a technical point of view, and postoperative recovery is faster.

In the long run, however, complications are more common than in cases of breast augmentation (30% at 5 years compared to 12% at 5 years) and are accentuated by the history of radiation therapy [69,70]. The main complications are capsular contracture, implant rupture, hematoma, and infections [71]. Implant reconstruction is associated with aesthetic complications like asymmetry, chest wall deformity, mispositioning or displacement, ptosis, wrinkling or rippling (wrinkling of the implant that can be felt or seen through the skin), skin rush, redness and bruising, and inflammation. The implant can suffer deflation, rupture, or extrusion. Many of these complications will require additional surgeries, a possibility that the patient should be informed about. Seroma, hematoma, delayed wound healing, infection, and necrosis of the skin/flap can also occur after implant breast reconstruction. These complications will require additional treatment and will most often delay adjuvant therapies with effects on the overall oncologic outcome. Following infection, hematoma formation, and seroma formation, capsular contraction can occur. Grade III and IV capsular contraction (hardening of the breast around the implant, causing painful tightening of the tissues) will require corrective surgery, but could occur again after the procedure. Implants are associated also with more exotic complications including other cancers; there have been reports of Breast Implant-Associated Anaplastic Large Cell Lymphoma (BIA-ALCL), squamous cell carcinoma, and mesenchymal tumors after breast reconstruction with implants.

However, implant reconstructions remain preferable for many surgeons because they avoid the complications of the donor areas and generate lower costs, and in the absence of radiotherapy or in a two-stage reconstruction, they are a simple solution with good aesthetic results. Acellular dermal matrices (ADMs) are used either as a first intervention to support the implant in the lower pole, not covered by the pectoralis major, or in reinterventions [72]. These biological materials are made from human, bovine, or porcine dermis processed to remove all cellular components—which can generate an immune response—and keeping the extracellular matrix containing mainly collagen (85%) along with proteoglycans, glycosaminoglycans, and elastin, arranged in a network, in the meshes of which the host cells are arranged [73]. This integration of the matrix provides good support for the breast prosthesis and a high-quality capsule, resulting in a natural appearance of the final reconstructed breast. The high costs are the main disadvantage and make dermal substitutes inaccessible on a large scale. With the increase in accessibility, it is estimated that the approach to implant reconstructions in a single stage will change significantly.

3.4. Autologous Breast Reconstruction Techniques

Autologous techniques are considered by many authors to be the gold standard in breast reconstruction. They consist in restoring the contour and volume of the mammary gland with the help of either rotating, pediculated flaps, which retain their vascular source, or micro-surgically freely transferred flaps from other areas of the body, most often from the abdomen. The intervention can be performed immediately or delayed, like the implant reconstruction. Moreover, if the volume provided by the flap is not sufficient, other techniques such as the free transfer of autologous fat (lipofilling) or placement of an implant may be associated [74].

a.   The advantages of flap reconstruction.

Flap reconstruction offers several advantages, including improving the quality of irradiated tissue by bringing healthy tissue into a scar area; the final appearance after reconstruction is a natural one that mimics, in time, the physiological ptosis of the contralateral breast, does not require reinterventions for replacement after a period of time, can be used in patients who do not want or do not tolerate an implant, and is the recommended type of reconstruction for the radio-treated patients [75].

b.   Types of transferred free flaps.

First described in 1989 by Koshima and Soeda [76], the freely transferred flap based on the inferior epigastric artery (DIEP) has long been the preferred alternative in autologous breast reconstruction in specialized centers [76]. Other free flaps that are described but rarely used in practice are the TRAM (transverse rectus abdominis myocutaneous flap), more commonly used in its pediculated version; TUG (transverse upper gracilis flap); SGAP (upper gluteal artery perforator flap); or IGAP (lower gluteal artery flap). Lower-limb flaps are indicated in selected cases, in the absence of a suitable abdominal donor area or in patients with previous interventions at this level [77].

c.   Latissimus dorsi pediculated flap.

First described by Tansini for covering chest wall defects in 1906, the latissimus dorsi pediculated flap began to be used in breast reconstruction after almost 70 years [78]. Until the middle of the twentieth century, the radical mastectomy technique described by Halsted recommended either grafting or secondhand healing of the resulting defect, strongly contraindicating any form of reconstruction, as it was considered to "hide possible recurrences and promote the spread of tumor cells" [79].

The evolution of oncological treatments, a better understanding of the pathology, and the increase of patients' life expectancy, together with the appearance of breast implants, changed the approach of these cases. Schneider and Botswick described in 1977 and 1978, respectively, the latissimus dorsi flap accompanied by the implant in restoring the physiological contour and ptosis of the breast after a mastectomy [80,81]. Subsequently, Papp and McCraw modified the flap, including subcutaneous adipose tissue overlying the muscle in order to achieve implant-free reconstruction [82].

Although it is no longer the gold standard in autologous reconstruction, the reliability and predictability of its anatomy still make it preferred by many surgeons for delayed reconstructions and also the preferred rescue option in the event of free-transfer flaps failure [83]. Currently, its primary uses are in patients who do not have sufficient reserves for a free flap; those with a personal history of abdominal interventions; or those with significant comorbidities such as obesity and diabetes or in smoking patients [84].

The most common complications are seromas in the donor area, usually easy to treat without further intervention. Associated with alloplastic procedures, capsular contracture has been described more frequently in association with implants and less frequently in two-stage reconstructions, when the implant is preceded by an expander. Rare complications are contour defects in the donor area, limited shoulder mobility, and decreased muscle strength in the arm and the scapula alatae [85].

In Figures 3–5, we present various immediate or delayed reconstructive techniques using the autologous or mixed procedures we employed for our patients.

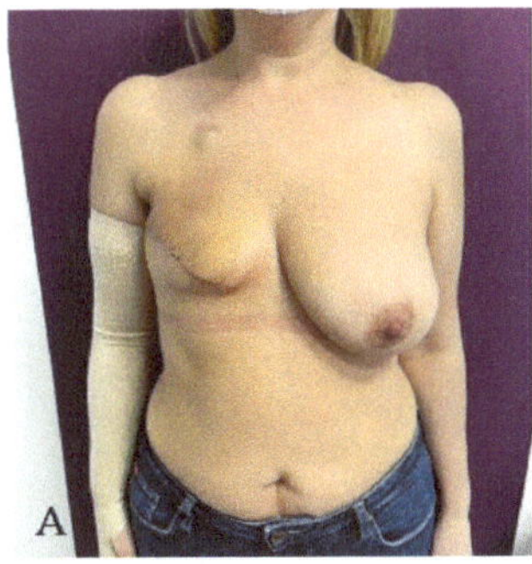 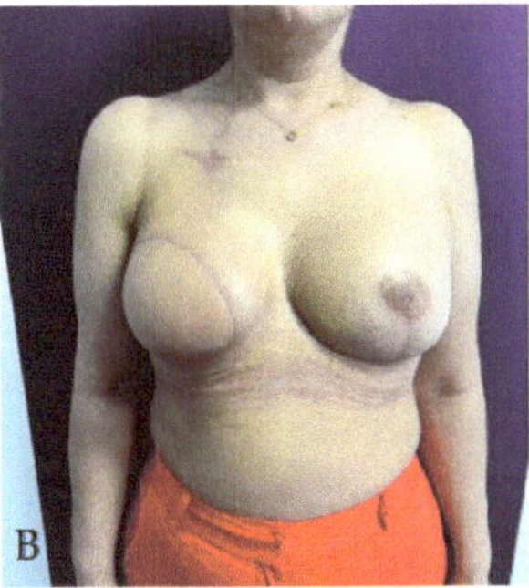

Figure 3. Patient with right radical mastectomy for breast cancer followed by radiotherapy; she underwent right breast delayed reconstruction using latissimus dorsi pediculated flap and a 320 cc round implant simultaneous with prophylactic left subcutaneous mastectomy (due to BRCA-positive status) with immediate reconstruction using a pediculated inferior dermoadipous flap and a 350 cc round implant: (**A**) aspect before reconstructive surgery; (**B**) aspect at 3 months after reconstructive surgery.

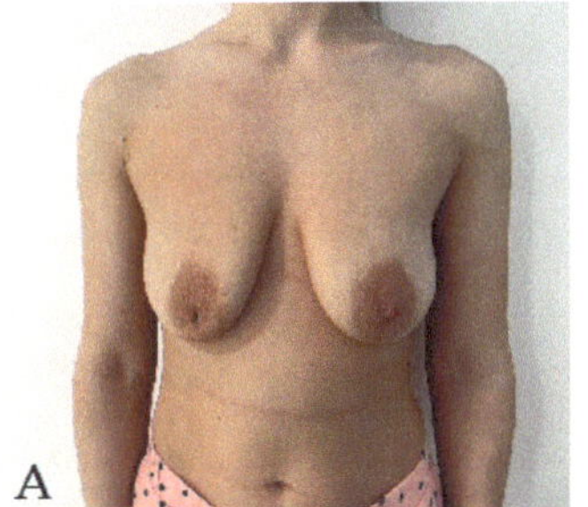 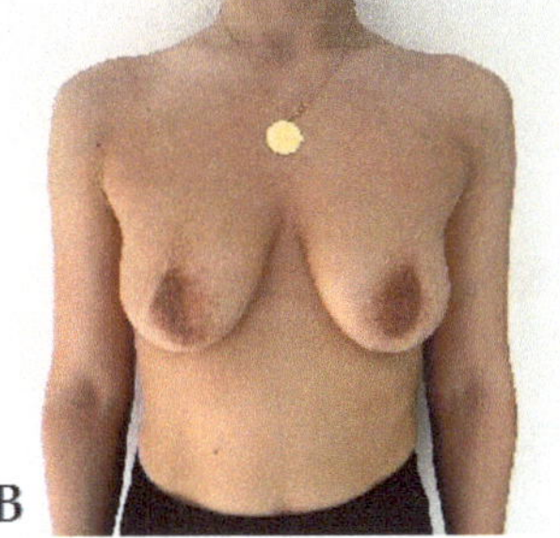 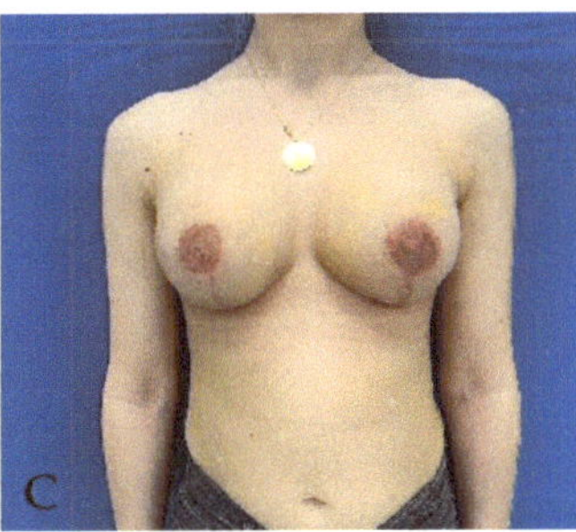

Figure 4. Bilateral prophylactic mastectomy in patient with BRCA-positive status and heavy family history of breast cancer. Immediate reconstruction using 325 cc round implant and pediculated inferior dermoadipous flap followed by nipple–areola complex graft: (**A**) aspect before reconstructive surgery; (**B**) aspect at 3 months after reconstructive surgery; (**C**) final aspect at 32 months after reconstructive surgery.

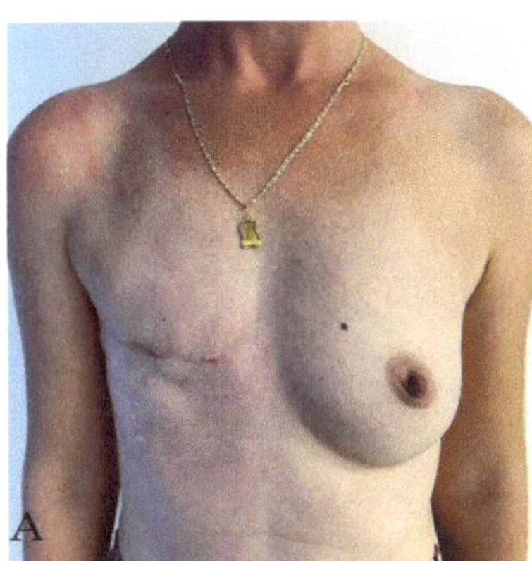 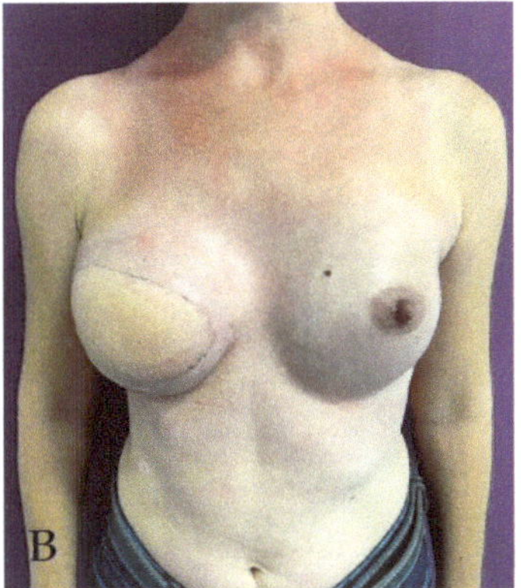

Figure 5. Patient with right radical mastectomy for breast cancer; she underwent right breast delayed reconstruction using latissimus dorsi pediculated flap and a 275 cc round implant simultaneous with prophylactic left subcutaneous mastectomy (due to BRCA-positive status) with immediate reconstruction using a 325 cc round implant: (**A**) aspect before reconstructive surgery; (**B**) aspect at 3 months after reconstructive surgery.

4. Discussion

Breast surgery has rapidly evolved in parallel with oncological treatments; while in 1970, the safety of reconstruction after a mastectomy was still questioned, today it is suggested for most patients who want it, so today, the task of choosing the most appropriate technique for each case is on the shoulders of the surgeon. With all the options available, the surgeon chooses the right technique by taking into account his own experience and preference; available resources and factors related to the patient, such as the breast size to be reconstructed, skin quality, type of mastectomy indicated, disease stage, adjuvant treatments, surgical history, and the general condition of the patient; and last but not least, her preference.

The contraindications of reconstruction are relatively few, limited to patients with a precarious general condition which do not allow an elective intervention as well as cases with a definite unfavorable life prognosis, which do not justify additional interventions. Also, patients with unrealistic expectations about the end result or who do not accept postoperative scars are not good candidates for reconstruction [86]. Age is no longer considered a contraindication to either the procedure itself or the choice of surgical technique, although, for reasons beyond the general condition and possible associated diseases, techniques involving freely transferred flaps are not usually recommended in patients over 65 years [48].

The selection of oncological surgery, tumorectomy, or mastectomy, as the case may be, contributes significantly to the end result. The decision on whether or not to preserve the mammary gland in early cases is still a matter of debate. A study by Veronesi et al. [87] following the development of 700 women with tumors < 2 cm for 20 years showed that breast preservation interventions (tumorectomies/lumpectomies) do not change the long-term survival when compared to mastectomies, although the local recurrence rate is higher in the first situation; Morrow and co-workers [88] also showed that for stages 0-II, a third of patients end up requiring a mastectomy. The American Society of Breast Surgery has recommended breast preservation whenever possible, along with the association with adjuvant oncological treatments such as chemo- and radiotherapy [89]. However, more recent data from the United States show an increase in the preference for mastectomies, especially prophylactic, in patients with and without BRCA 1/2 mutations [90].

The long-term benefit of this radical gesture has been demonstrated in cases with the presence of mutations, or in familial cases, in studies such as that performed by Boughey et al. [91], which followed a group of 385 women with a family history and stage I or II tumors and found that after 17 years, survival was significantly improved in patients with bilateral mastectomies. Meanwhile, another study by the same author [92] showed that bilateral mastectomy increases hospitalization costs and the number of on-call visits in the first 2 years, recommending that these data be explained to patients before making a decision.

Hoskin et al. [93] conducted a study in the USA on 3195 women operated on for breast tumors over a period of 5 years, between 2009 and 2014. Of the patients who required mastectomies, the proportion of patients who opted for immediate reconstruction increased by 31%. The percentage of prophylactic bilateral mastectomies with immediate reconstruction increased by 20%, while for the same intervention without reconstruction, the percentage decreased by 10%, from 22 to 12%.

Complications after intervention are not significantly different between tumoral and healthy breasts, but in the case of bilateral procedures, the complication rate increases significantly compared to that of unilateral ones, from 6.3% to 10.6%, according to some authors [94], respectively, and from 4.2% to 7.6%, according to other studies [95], this aspect being one of the main criticisms of this trend.

Statistics on the incidence of breast cancer in Romania are limited. The existence of a national patient record that would include, among other things, the stage at the time of diagnosis would contribute to the understanding of epidemiology and would facilitate a unified, multidisciplinary approach and faster access of patients to treatments. From the experience of oncological surgery centers, many patients with breast cancer who present for treatment are detected to be in advanced stages locally, with larger tumors and often with clinical or radiological lymph node involvement. This situation significantly changes the surgical indications and, implicitly, the reconstructive options. Although surgical excision is sometimes possible primarily through the radical mastectomy technique, patients usually receive neoadjuvant chemotherapy. Given the stage of the disease, reconstruction in such cases is most often delayed until the completion of oncological treatments [96]. However, the evolution in the diagnosis and treatment of breast cancer has led to the development of oncoplastic surgery that not only allows for the preservation of the breast, but also obtains better aesthetic results in oncological safety conditions [97].

A number of studies have evaluated the safety of immediate breast reconstruction in neoadjuvant-treated patients with favorable results. A meta-analysis conducted in 2020 by Varghese et al. [98] evaluated 17 observational studies, comprising 3429 cases, and revealed that it does not increase the risks of perioperative complications such as hematoma, seroma, or difficult wound healing and does not delay adjuvant treatment. The study instead showed a lower rate of complications in younger patients, as well as a higher rate of complications in patients who smoke or have a high body mass index. Also, patients with large breasts (>600 g) had a higher complication rate. Neoadjuvant chemotherapy slightly increases the risk of complications related to implants or expanders and insignificant risks related to autologous procedures, as noted by the same authors.

The effect of adjuvant chemotherapy on the results of reconstruction is difficult to estimate, as most patients also benefit from radiation therapy during treatment. One study showed a relative risk of liponecrosis of 4.8 in cases where immediate reconstruction with a free flap was performed [99].

Radiotherapy can significantly affect the postoperative outcomes, especially in alloplastic procedures. Reconstruction using autologous procedures is safer, with a lower rate of complications. El-Sawabi [100] performed a meta-analysis on complications after breast reconstruction in irradiated patients and showed that autologous procedures are associated with a lower rate of post-procedural complications (wound healing, seromas, hematomas, infections, and reinterventions) when compared to implant-based reconstruction (30.9% vs. 41.3%) [100]. Failure of the intervention occurred in 16.8% of alloplastic procedures and only 1.6% of autologous ones. When radiotherapy was performed on the temporary device, the complication rate was higher than when it was performed on the permanent implant (18.8% and 14.4%, respectively).

Among the autologous procedures, the latissimus dorsi myocutaneous flap has long been the basic choice for reconstruction, associated or not with an implant. Almost any patient can benefit from this technique due to the reliability and versatility of this flap. The main controversies are related to the transferred volume, the aesthetic result, and the secondary functional deficit of the shoulder and arm.

As early as 1986, Russel and colleagues [101] observed that although there is a decrease in scapular girdle muscle function immediately postoperatively and this effect may be more evident in athletic or elderly patients, this deficit does not have a significant impact on daily activities—except in athletes, skiers, swimmers, and climbers—and fades in about 6 months due to the development of synergistic musculature. The muscle strength has been showed to be comparable to preoperative levels in 3 months (2015 study by Yang) [102]. However, this can also be associated with the neurologic alterations following the sectioning of the sensitive nerves during axillary lymphadenectomy; this can be mitigated by the utilization of a modified surgical technique for lymphadenectomy which preserves the intercostobrachial nerve and the third and fourth intercostals [34].

On the subject of aesthetic results of the reconstruction after a mastectomy, Lindegren [98] conducted a study on secondary autologous-type reconstructions with 70 irradiated patients comparing the perceptions of both patients and surgeons of the aesthetic results after using the latissimus dorsi flap or DIEP. Although the surgeons favored the DIEP due to the natural shape and volume of the reconstructed breast, the patients were more satisfied with the latissimus dorsi flap reconstruction. This result was unexpected for the authors, which they hypothesized was correlated to higher satisfaction in the latissimus flap with the scar of the donor area [103]. Another study had the opposite results in a larger group of patients but a small percentage of irradiated patients [104].

The appropriate volume for larger breasts can be recreated either by a combination with the implant, by serial lipofilling sessions, or by changing the skin palette to include more subcutaneous adipose tissue [105].

Breast reconstruction has become a necessary step in the treatment of most breast cancers, and many reconstructive techniques are now routinely practiced. Microsurgical techniques are considered the "gold standard", but they are not accessible to all services, from a technical or financial point of view, so pediculated flaps remain the safe and reliable option, along with mixed and alloplastic procedures, to improve the quality of life of these patients.

Author Contributions: Conceptualization, I.M.D. and L.S.; methodology, I.M.D. and I.P.; validation, S.-O.I. and C.C.; data curation, H.D., C.C., D.L.S., A.S.G. and D.L.; writing—original draft preparation, I.M.D., E.C. and V.R.; writing—review and editing, E.C. and V.R.; supervision, L.S. and D.-C.S.; project administration, L.S. All authors have read and agreed to the published version of the manuscript.

Funding: Publication of this paper was supported by the University of Medicine and Pharmacy Carol Davila, through the institutional program Publish not Perish.

Institutional Review Board Statement: The study was conducted in accordance with the Declaration of Helsinki and approved by the Institutional Ethics Committee of Bucharest Oncological Institute "Prof. Dr. Alexandru Trestioreanu" (approval number 9676 from 31 July 2023).

Informed Consent Statement: Informed consent was obtained from all subjects involved in the study. All the photos included come from the authors' collection (I.P.), and the consent of the respective patients was specifically obtained to use these photos for scientific purposes.

Data Availability Statement: Not applicable.

Acknowledgments: Publication of this paper was supported by the University of Medicine and Pharmacy *Carol Davila*, through the institutional program *Publish not Perish*. We acknowledge the instrumental efforts of Rotaru Vlad to the writing and development of this manuscript, his contribution being equal to those of the first, last and corresponding authors.

Conflicts of Interest: The authors declare no conflicts of interest.

References

1. WHO. Statistics on Breast Cancer. Available online: https://www.who.int/news-room/fact-sheets/detail/breast-cancer (accessed on 22 July 2023).
2. Kummerow, K.L.; Du, L.; Penson, D.F.; Shyr, Y.; Hooks, M.A. Nationwide Trends in Mastectomy for Early-Stage Breast Cancer. *JAMA Surg.* **2015**, *150*, 9–16. [CrossRef] [PubMed]
3. Tuttle, T.M.; Habermann, E.B.; Grund, E.H.; Morris, T.J.; Virnig, B.A. Increasing Use of Contralateral Prophylactic Mastectomy for Breast Cancer Patients: A Trend toward More Aggressive Surgical Treatment. *J. Clin. Oncol.* **2007**, *25*, 5203–5209. [CrossRef] [PubMed]
4. Soran, A.; Kamali Polat, A.; Johnson, R.; McGuire, K.P. Increasing Trend of Contralateral Prophylactic Mastectomy: What Are the Factors behind This Phenomenon? *Surgeon* **2014**, *12*, 316–322. [CrossRef] [PubMed]
5. Patel, H.; Samaha, Y.; Ives, G.; Lee, T.-Y.; Cui, X.; Ray, E. Chest Feminization in Male-to-Female Transgender Patients: A Review of Options. *Transgender Health* **2021**, *6*, 244–255. [CrossRef] [PubMed]
6. Ferlay, J.; Steliarova-Foucher, E.; Lortet-Tieulent, J.; Rosso, S.; Coebergh, J.W.W.; Comber, H.; Forman, D.; Bray, F. Cancer Incidence and Mortality Patterns in Europe: Estimates for 40 Countries in 2012. *Eur. J. Cancer* **2013**, *49*, 1374–1403. [CrossRef]
7. Sancho-Garnier, H.; Colonna, M. Epidémiologie Des Cancers Du Sein Breast Cancer Epidemiology. *Presse Medicale* **2019**, *48*, 1076–1084. [CrossRef]
8. Simion, L.; Rotaru, V.; Cirimbei, C.; Gales, L.; Stefan, D.-C.; Ionescu, S.-O.; Luca, D.; Doran, H.; Chitoran, E. Inequities in Screening and HPV Vaccination Programs and Their Impact on Cervical Cancer Statistics in Romania. *Diagnostics* **2023**, *13*, 2776. [CrossRef]
9. Simion, L.; Chitoran, E.; Cirimbei, C.; Stefan, D.-C.; Neicu, A.; Tanase, B.; Ionescu, S.O.; Luca, D.C.; Gales, L.; Gheorghe, A.S.; et al. A Decade of Therapeutic Challenges in Synchronous Gynecological Cancers from the Bucharest Oncological Institute. *Diagnostics* **2023**, *13*, 2069. [CrossRef]

10. UpToDate. Overview of Breast Reconstruction. Available online: https://www.uptodate.com/contents/overview-of-breast-reconstruction?search=Overview%20of%20breast%20reconstruction&source=search_result&selectedTitle=1~150&usage_type=default&display_rank=1 (accessed on 21 July 2023).

11. Angelos, P.; Bedrosian, I.; Euhus, D.M.; Herrmann, V.M.; Katz, S.J.; Pusic, A. Prophylactic Mastectomy: Challenging Considerations for the Surgeon. *Ann. Surg. Oncol.* **2015**, *22*, 3208–3212. [CrossRef]

12. Dragun, A.E.; Pan, J.; Riley, E.C.; Kruse, B.; Wilson, M.R.; Rai, S.; Jain, D. Increasing Use of Elective Mastectomy and Contralateral Prophylactic Surgery among Breast Conservation Candidates: A 14-Year Report from a Comprehensive Cancer Center. *Am. J. Clin. Oncol.* **2013**, *36*, 375–380. [CrossRef]

13. Metcalfe, K.A.; Eisen, A.; Poll, A.; Candib, A.; McCready, D.; Cil, T.; Wright, F.; Demsky, R.; Mancuso, T.; Sun, P.; et al. Frequency of Contralateral Prophylactic Mastectomy in Breast Cancer Patients with a Negative BRCA1 and BRCA2 Rapid Genetic Test Result. *Ann. Surg. Oncol.* **2021**, *28*, 4967–4973. [CrossRef] [PubMed]

14. Miller, T.J.; Wilson, S.C.; Massie, J.P.; Morrison, S.D.; Satterwhite, T. Breast Augmentation in Male-to-Female Transgender Patients: Technical Considerations and Outcomes. *JPRAS Open* **2019**, *21*, 63–74. [CrossRef] [PubMed]

15. Bekeny, J.C.; Zolper, E.G.; Fan, K.L.; Del Corral, G. Breast Augmentation for Transfeminine Patients: Methods, Complications, and Outcomes. *Gland Surg.* **2020**, *9*, 788–796. [CrossRef] [PubMed]

16. Bekeny, J.C.; Zolper, E.G.; Manrique, O.J.; Fan, K.L.; Corral, G. Del Breast Augmentation in the Transgender Patient: Narrative Review of Current Techniques and Complications. *Ann. Transl. Med.* **2021**, *9*, 611. [CrossRef] [PubMed]

17. Khan, U.D. Preoperative Planning and Breast Implant Selection for Volume Difference Management in Asymmetrical Breasts. *Plast. Aesthet. Res.* **2017**, *4*, 108–115. [CrossRef]

18. Patlazhan, G.; Shkolnaya, O.; Torubarov, I.; Gomes, M. Our 10 Years' Experience in Breast Asymmetry Correction. *Aesthetic Plast. Surg.* **2020**, *44*, 706–715. [CrossRef]

19. Peterson, B.; Alajmi, H.; Ladak, A.; Samargandi, O.A. Breast Equalization Augmentation: The Use of Ultrasonic Assisted Liposuction for Correction of Primary Breast Asymmetry with Bilateral Augmentation. *Aesthetic Plast. Surg.* **2022**, *46*, 667–674. [CrossRef]

20. Christiansen, P.; Carstensen, S.L.; Ejlertsen, B.; Kroman, N.; Offersen, B.; Bodilsen, A.; Jensen, M.B. Breast Conserving Surgery versus Mastectomy: Overall and Relative Survival—A Population Based Study by the Danish Breast Cancer Cooperative Group (DBCG). *Acta Oncol.* **2018**, *57*, 19–25. [CrossRef]

21. Benedict, S.; Cole, D.J.; Baron, L.; Baron, P. Factors Influencing Choice Between Mastectomy and Lumpectomy for Women in the Carolinas. *J. Surg. Oncol.* **2001**, *76*, 6–12. [CrossRef]

22. Nold, R.J.; Beamer, R.L.; Helmer, S.D.; McBoyle, M.F. Factors Influencing a Woman's Choice to Undergo Breast-Conserving Surgery versus Modified Radical Mastectomy. *Am. J. Surg.* **2000**, *180*, 413–418. [CrossRef]

23. Farhangkhoee, H.; Matros, E.; Disa, J. Trends and Concepts in Post-Mastectomy Breast Reconstruction. *J. Surg. Oncol.* **2016**, *113*, 891–894. [CrossRef] [PubMed]

24. Wang, F.; Peled, A.W.; Chin, R.; Fowble, B.; Alvarado, M.; Ewing, C.; Esserman, L.; Foster, R.; Sbitany, H. The Impact of Radiation Therapy, Lymph Node Dissection, and Hormonal Therapy on Outcomes of Tissue Expander-Implant Exchange in Prosthetic Breast Reconstruction. *Plast. Reconstr. Surg.* **2016**, *137*, 1–9. [CrossRef]

25. Verma, R.; Klein, G.; Dagum, A.; Khan, S.; Bui, D.T. The Effect of Axillary Lymph Node Sampling during Mastectomy on Immediate Alloplastic Breast Reconstruction Complications. *Plast. Reconstr. Surg. Glob. Open* **2019**, *7*, E2224. [CrossRef] [PubMed]

26. Morrow, M.; Li, Y.; Alderman, A.K.; Jagsi, R.; Hamilton, A.S.; Graff, J.J.; Hawley, S.T.; Katz, S.J. Access to Breast Reconstruction After Mastectomy and Patient Perspectives on Reconstruction Decision Making. *JAMA Surg* **2014**, *149*, 1015–1021. [CrossRef]

27. Steffen, L.E.; Johnson, A.; Levine, B.J.; Mayer, D.K.; Avis, N.E. Met and Unmet Expectations for Breast Reconstruction in Early Posttreatment Breast Cancer Survivors. *Plast. Surg. Nurs.* **2017**, *37*, 146–153. [CrossRef] [PubMed]

28. Wilkins, E.G.; Cederna, P.S.; Lowery, J.C.; Davis, J.A.; Myra Kim, H.; Roth, R.S.; Goldfarb, S.; Izenberg, P.H.; Houin, H.P.; Shaheen, K.W.; et al. Prospective Analysis of Psychosocial Outcomes in Breast Reconstruction: One-Year Postoperative Results from the Michigan Breast Reconstruction Outcome Study. *Plast. Reconstr. Surg.* **2000**, *106*, 1014–1025. [CrossRef]

29. Rubino, C.; Figus, A.; Lorettu, L.; Sechi, G. Post-Mastectomy Reconstruction: A Comparative Analysis on Psychosocial and Psychopathological Outcomes. *J. Plast. Reconstr. Aesthetic Surg.* **2007**, *60*, 509–518. [CrossRef]

30. Helms, R.L.; O'Hea, E.L.; Corso, M. Body Image Issues in Women with Breast Cancer. *Psychol. Health Med.* **2008**, *13*, 313–325. [CrossRef]

31. Reddy, K.G.; Strassle, P.D.; McGuire, K.P. Role of Age, Tumor Grade, and Radiation Therapy on Immediate Postmastectomy Breast Reconstruction. *Clin. Breast Cancer* **2018**, *18*, 313–319. [CrossRef]

32. Patel, A.U.; Day, S.J.; Pencek, M.; Roussel, L.O.; Christiano, J.G.; Punekar, I.R.; Koltz, P.F.; Langstein, H.N. Functional Return after Implant-Based Breast Reconstruction: A Prospective Study of Objective and Patient-Reported Outcomes. *J. Plast. Reconstr. Aesthet. Surg.* **2020**, *73*, 850–855. [CrossRef]

33. Myung, Y.; Choi, B.; Kwon, H.; Heo, C.Y.; Kim, E.K.; Kang, E.; Jeong, J.H.; Yang, E.J. Quantitative Analysis of Shoulder Function and Strength after Breast Reconstruction: A Retrospective Cohort Study. *Medicine* **2018**, *97*, e10979. [CrossRef]

34. Rotaru, V.; Chitoran, E.; Cirimbei, C.; Cirimbei, S.; Simion, L. Preservation of Sensory Nerves During Axillary Lymphadenectomy. In Proceedings of the 35th Balkan Medical Week, Athens, Greece, 25–27 September 2018. Available online: https://www.webofscience.com/wos/woscc/full-record/WOS:000471903700045 (accessed on 20 July 2023).

35. Doherty, C.; Mcclure, J.A.; Baxter, N.N.; Brackstone, M. Complications from Postmastectomy Radiation Therapy in Patients Undergoing Immediate Breast Reconstruction: A Population-Based Study. *Adv. Radiat. Oncol.* **2023**, *8*, 101104. [CrossRef] [PubMed]

36. Fortunato, L.; Loreti, A.; Cortese, G.; Spallone, D.; Toto, V.; Cavaliere, F.; Farina, M.; La Pinta, M.; Manna, E.; Detto, L.; et al. Regret and Quality of Life After Mastectomy with or Without Reconstruction. *Clin. Breast Cancer* **2021**, *21*, 162–169. [CrossRef] [PubMed]

37. Kuliński, W.; Kosno, M. Quality of Life in Women after Mastectomy. Clinical and Social Study. *Wiad Lek* **2021**, *74*, 429–435. [CrossRef]

38. Lee, C.N.H.; Pignone, M.P.; Deal, A.M.; Blizard, L.; Hunt, C.; Huh, R.; Liu, Y.J.; Ubel, P.A. Accuracy of Predictions of Patients with Breast Cancer of Future Well-Being after Immediate Breast Reconstruction. *JAMA Surg.* **2018**, *153*, e176112. [CrossRef] [PubMed]

39. Qin, Q.; Tan, Q.; Lian, B.; Mo, Q.; Huang, Z.; Wei, C. Postoperative Outcomes of Breast Reconstruction after Mastectomy A Retrospective Study. *Medicine* **2018**, *97*, e9766. [CrossRef] [PubMed]

40. De La Cruz, L.; Moody, A.M.; Tappy, E.E.; Blankenship, S.A.; Hecht, E.M. Overall Survival, Disease-Free Survival, Local Recurrence, and Nipple–Areolar Recurrence in the Setting of Nipple-Sparing Mastectomy: A Meta-Analysis and Systematic Review. *Ann. Surg. Oncol.* **2015**, *22*, 3241–3249. [CrossRef]

41. Petit, J.Y.; Gentilini, O.; Rotmensz, N.; Rey, P.; Rietjens, M.; Garusi, C.; Botteri, E.; De Lorenzi, F.; Martella, S.; Bosco, R.; et al. Oncological Results of Immediate Breast Reconstruction: Long Term Follow-up of a Large Series at a Single Institution. *Breast Cancer Res. Treat.* **2008**, *112*, 545–549. [CrossRef]

42. Ha, J.H.; Hong, K.Y.; Lee, H.B.; Moon, H.G.; Han, W.; Noh, D.Y.; Lim, J.; Yoon, S.; Chang, H.; Jin, U.S. Oncologic Outcomes after Immediate Breast Reconstruction Following Mastectomy: Comparison of Implant and Flap Using Propensity Score Matching. *BMC Cancer* **2020**, *20*, 78. [CrossRef]

43. Fujihara, M.; Yamasaki, R.; Ito, M.; Shien, T.; Maeda, R.; Kin, T.; Ueno, A.; Kajiwara, Y.; Kawasaki, K.; Ichimura, K.; et al. Risk Factors of Local Recurrence Following Implant-Based Breast Reconstruction in Breast Cancer Patients. *BMC Womens Health* **2021**, *21*, 147. [CrossRef]

44. Mammography in the Evaluation of Masses in Breasts Reconstructed with TRAM Flaps. Annals of Plastic Surgery. Available online: https://journals.lww.com/annalsplasticsurgery/Abstract/1998/09000/Mammography_in_the_Evaluation_of_Masses_in_Breasts.1.aspx (accessed on 22 July 2023).

45. Hsu, W.; Sheen-Chen, S.M.; Eng, H.L.; Ko, S.F. Mammographic Microcalcification in an Autogenously Reconstructed Breast Simulating Recurrent Carcinoma. *Tumori J.* **2018**, *94*, 574–576. [CrossRef]

46. Juanpere, S.; Perez, E.; Huc, O.; Motos, N.; Pont, J.; Pedraza, S. Imaging of Breast Implants—A Pictorial Review. *Insights Imaging* **2011**, *2*, 653–670. [CrossRef] [PubMed]

47. Simion, L.; Ionescu, S.; Chitoran, E.; Rotaru, V.; Cirimbei, C.; Madge, O.-L.; Nicolescu, A.C.; Tanase, B.; Dicu-Andreescu, I.-G.; Dinu, D.M.; et al. Indocyanine Green (ICG) and Colorectal Surgery: A Literature Review on Qualitative and Quantitative Methods of Usage. *Medicina* **2023**, *59*, 1530. [CrossRef] [PubMed]

48. Mathelin, C.; Bruant-Rodier, C. Indications for Breast Reconstruction after Mastectomy According to the Oncological Situation. *Ann. Chir. Plast. Esthet.* **2018**, *63*, 580–584. [CrossRef] [PubMed]

49. Atisha, D.; Alderman, A.K.; Lowery, J.C.; Kuhn, L.E.; Davis, J.; Wilkins, E.G. Prospective Analysis of Long-Term Psychosocial Outcomes in Breast Reconstruction: Two-Year Postoperative Results from the Michigan Breast Reconstruction Outcomes Study. *Ann. Surg.* **2008**, *247*, 1019–1028. [CrossRef] [PubMed]

50. Platt, J.; Zhong, T. Patient-Centered Breast Reconstruction Based on Health-Related Quality-of-Life Evidence. *Clin. Plast. Surg.* **2018**, *45*, 137–143. [PubMed]

51. Zhong, T.; McCarthy, C.; Min, S.; Zhang, J.; Beber, B.; Pusic, A.L.; Hofer, S.O.P. Patient Satisfaction and Health-Related Quality of Life after Autologous Tissue Breast Reconstruction: A Prospective Analysis of Early Postoperative Outcomes. *Cancer* **2012**, *118*, 1701–1709. [CrossRef] [PubMed]

52. Zhong, T.; Hu, J.; Bagher, S.; Vo, A.; O'Neill, A.C.; Butler, K.; Novak, C.B.; Hofer, S.O.P.; Metcalfe, K.A. A Comparison of Psychological Response, Body Image, Sexuality, and Quality of Life between Immediate and Delayed Autologous Tissue Breast Reconstruction: A Prospective Long-Term Outcome Study. *Plast. Reconstr. Surg.* **2016**, *138*, 772–780. [CrossRef]

53. Simmons, R.M.; Fish, S.K.; Gayle, L.; La Trenta, G.S.; Swistel, A.; Christos, P.; Osborne, M.P. Local and Distant Recurrence Rates in Skin-Sparing Mastectomies Compared With Non-Skin-Sparing Mastectomies. *Ann. Surg. Oncol.* **1999**, *6*, 676–681. [CrossRef]

54. Toth, B.A.; Lappert, P. Modified Skin Incisions for Mastectomy: The Need for Plastic Surgical Input in Preoperative Planning. *Plast. Reconstr. Surg.* **1991**, *87*, 1048–1053. [CrossRef]

55. Kroll, S.; Khoo, A.; Singletary, E.; Ames, F.; Wang, B.G.; Reece, G.; Miller, M.; Evans, G.; Robb, G. Local Recurrence Risk after Skin-Sparing and Conventional Mastectomy: A 6-Year Follow-Up. *Plast. Reconstr. Surg.* **1999**, *104*, 421–425. [CrossRef] [PubMed]

56. Galimberti, V.; Vicini, E.; Corso, G.; Morigi, C.; Fontana, S.; Sacchini, V.; Veronesi, P. Nipple-Sparing and Skin-Sparing Mastectomy: Review of Aims, Oncological Safety and Contraindications. *Breast* **2017**, *34*, S82–S84. [CrossRef] [PubMed]

57. Radu, M.; Bordea, C.; Noditi, A.; Blidaru, A. Assessment of Mastectomy Skin Flaps for Immediate Implant-Based Breast Reconstruction. *J. Med. Life* **2018**, *11*, 137–145. [PubMed]

58. Huang, C.J.; Hou, M.F.; Lin, S.D.; Chuang, H.Y.; Huang, M.Y.; Fu, O.Y.; Lian, S.L. Comparison of Local Recurrence and Distant Metastases between Breast Cancer Patients after Postmastectomy Radiotherapy with and without Immediate TRAM Flap Reconstruction. *Plast. Reconstr. Surg.* **2006**, *118*, 1079–1086. [CrossRef] [PubMed]

59. Gieni, M.; Avram, R.; Dickson, L.; Farrokhyar, F.; Lovrics, P.; Faidi, S.; Sne, N. Local Breast Cancer Recurrence after Mastectomy and Immediate Breast Reconstruction for Invasive Cancer: A Meta-Analysis. *Breast* **2012**, *21*, 230–236. [CrossRef]

60. Strålman, K.; Mollerup, C.L.; Kristoffersen, U.S.; Elberg, J.J. Long-Term Outcome after Mastectomy with Immediate Breast Reconstruction. *Acta Oncol.* **2008**, *47*, 704–708. [CrossRef]

61. Kronowitz, S.J. Delayed-Immediate Breast Reconstruction: Technical and Timing Considerations. *Plast. Reconstr. Surg.* **2010**, *125*, 463–474. [CrossRef]

62. Gradishar, W.; Salerno, K.E. NCCN guidelines update: Breast cancer. *J. Natl. Compr. Cancer Netw.* **2016**, *14*, 641–644.

63. Harris, E.E.R.; Freilich, J.; Lin, H.Y.; Chuong, M.; Acs, G. The Impact of the Size of Nodal Metastases on Recurrence Risk in Breast Cancer Patients with 1-3 Positive Axillary Nodes after Mastectomy. *Int. J. Radiat. Oncol. Biol. Phys.* **2013**, *85*, 609–614. [CrossRef]

64. McBride, A.; Allen, P.; Woodward, W.; Kim, M.; Kuerer, H.M.; Drinka, E.K.; Sahin, A.; Strom, E.A.; Buzdar, A.; Valero, V.; et al. Locoregional Recurrence Risk for Patients with T1,2 Breast Cancer with 1-3 Positive Lymph Nodes Treated with Mastectomy and Systemic Treatment. *Int. J. Radiat. Oncol. Biol. Phys.* **2014**, *89*, 392–398. [CrossRef]

65. Kronowitz, S.J.; Hunt, K.K.; Kuerer, H.M.; Babiera, G.; McNeese, M.D.; Buchholz, T.A.; Strom, E.A.; Robb, G.L. Delayed-Immediate Breast Reconstruction. *Plast. Reconstr. Surg.* **2004**, *113*, 1617–1628. [CrossRef]

66. See, M.S.F.; Farhadi, J. Radiation Therapy and Immediate Breast Reconstruction: Novel Approaches and Evidence Base for Radiation Effects on the Reconstructed Breast. *Clin. Plast. Surg.* **2018**, *45*, 13–24. [CrossRef] [PubMed]

67. Bellini, E.; Pesce, M.; Santi, P.L.; Raposio, E. Two-Stage Tissue-Expander Breast Reconstruction: A Focus on the Surgical Technique. *Biomed. Res. Int.* **2017**, *2017*, 1791546.

68. Cemal, Y.; Albornoz, C.R.; Disa, J.J.; Mccarthy, C.M.; Mehrara, B.J.; Pusic, A.L.; Cordeiro, P.G.; Matros, E. A Paradigm Shift in U.S. Breast Reconstruction: Part 2. the Influence of Changing Mastectomy Patterns on Reconstructive Rate and Method. *Plast. Reconstr. Surg.* **2013**, *131*, 320e–326e. [CrossRef] [PubMed]

69. Sherine, E.G.; Woods, J.E.; O'Fallon, M.; Beard, M.; Kurland, L.T.; Melton, J.L. Complications Leading to Surgery after Breast Implantation. *N. Engl. J. Med.* **1997**, *336*, 677–682. [CrossRef]

70. Percec, I.; Bucky, L.P. Successful Prosthetic Breast Reconstruction after Radiation Therapy. *Ann. Plast. Surg.* **2008**, *60*, 527–531. [CrossRef]

71. Thamm, O.C.; Andree, C. Immediate Versus Delayed Breast Reconstruction: Evolving Concepts and Evidence Base. *Clin. Plast. Surg.* **2018**, *45*, 119–127. [CrossRef]

72. Nahabedian, M.Y. Acellular Dermal Matrices in Primary Breast Reconstruction. *Plast. Reconstr. Surg.* **2012**, *130*, 44S–53S. [CrossRef]

73. Uitto, J.; Olsen, D.R.; Fazio, M.J. Extracellular Matrix of the Skin: 50 Years of Progress. *J. Investig. Dermatol.* **1989**, *92*, S61–S77. [CrossRef]

74. Ho, A.Y.; Hu, Z.I.; Mehrara, B.J.; Wilkins, E.G. Radiotherapy in the Setting of Breast Reconstruction: Types, Techniques, and Timing. *Lancet Oncol.* **2017**, *18*, e742–e753. [CrossRef]

75. Chen, V.W.; Lin, A.; Hoang, D.; Carey, J. Trends in Breast Reconstruction Techniques at a Large Safety Net Hospital: A 10-Year Institutional Review. *Ann. Breast Surg.* **2018**, *2*, 14. [CrossRef]

76. Koshima, I.; Soeda, S. Inferior Epigastric Artery Skin Flaps without Rectus Abdominis Muscle. *J. Plast. Surg.* **1989**, *42*, 645–648. [CrossRef] [PubMed]

77. Healy, C.; Allen, R.J. The Evolution of Perforator Flap Breast Reconstruction: Twenty Years after the First DIEP Flap. *J. Reconstr. Microsurg.* **2014**, *30*, 121–126. [CrossRef] [PubMed]

78. Maxwell, P. Iginio Tansini and the Origin of the Latissimus Dorsi musculocutaneous Flap. *Plast. Reconstr. Surg.* **1989**, *65*, 686–692. [CrossRef]

79. Champaneria, M.C.; Wong, W.W.; Hill, M.E.; Gupta, S.C. The Evolution of Breast Reconstruction: A Historical Perspective. *World J. Surg.* **2012**, *36*, 730–742. [CrossRef]

80. Bostwick, J.; Vasconez, L.O.; Jurkiewicz, M.J. Breast Reconstruction after a Radical Mastectomy. *Plast. Reconstr. Surg.* **1978**, *61*, 682–693. [CrossRef]

81. Schneider, W.J.; Hill, H.L.; Brown, R.G. Latissimus dorsi myocutaneous flap for breast reconstruction. *Br. J. Plast. Surg.* **1977**, *30*, 277–281. [CrossRef]

82. Papp, C.; McCraw, J.B. Autogenous Latissimus Breast Reconstruction. *Clin. Plast. Surg.* **1998**, *25*, 261–266.

83. DeLong, M.R.; Tandon, V.J.; Rudkin, G.H.; Da Lio, A.L. Latissimus Dorsi Flap Breast Reconstruction—A Nationwide Inpatient Sample Review. *Ann. Plast. Surg.* **2017**, *78*, S185–S188. [CrossRef]

84. Spear, S.L.; Clemens, M.W. Latissimus Dorsi Flap Breast Reconstruction. In *Plastic Surgery*, 3rd ed.; Saunders (Elsevier): Philadelphia, PA, USA, 2012; pp. 370–392.

85. Hardwicke, J.T.; Prinsloo, D.J. An Analysis of 277 Consecutive Latissimus Dorsi Breast Reconstructions: A Focus on Capsular Contracture. *Plast. Reconstr. Surg.* **2011**, *128*, 63–70. [CrossRef]

86. Warren, A.G.; Morris, D.J.; Houlihan, M.J.; Slavin, S.A. Breast Reconstruction in a Changing Breast Cancer Treatment Paradigm. *Plast. Reconstr. Surg.* **2008**, *121*, 1116–1126. [CrossRef] [PubMed]

87. Veronesi, U.; Cascinelli, N.; Mariani, L.; Greco, M.; Saccozzi, R.; Luini, A.; Aguilar, M.; Marubini, E. Twenty-Year Follow-up of a Randomized Study Comparing Breast-Conserving Surgery with Radical Mastectomy for Early Breast Cancer. *N. Engl. J. Med.* **2002**, *347*, 1227–1232. [CrossRef] [PubMed]

88. Morrow, M.; Jagsi, R.; Alderman, A.K.; Griggs, J.J.; Hawley, S.T.; Hamilton, A.S.; Graff, J.J.; Katz, S.J. Surgeon Recommendations and Receipt of Mastectomy for Treatment of Breast Cancer. *JAMA J. Am. Med. Assoc.* **2009**, *302*, 1551. [CrossRef]

89. Boughey, J.C.; Attai, D.J.; Chen, S.L.; Cody, H.S.; Dietz, J.R.; Feldman, S.M.; Greenberg, C.C.; Kass, R.B.; Landercasper, J.; Lemaine, V.; et al. Contralateral Prophylactic Mastectomy (CPM) Consensus Statement from the American Society of Breast Surgeons: Data on CPM Outcomes and Risks. *Ann. Surg. Oncol.* **2016**, *23*, 3100. [CrossRef]

90. Panchal, H.; Matros, E. Current Trends in Postmastectomy Breast Reconstruction. *Plast. Reconstr. Surg.* **2017**, *140*, 7S–13S. [CrossRef]

91. Boughey, J.C.; Hoskin, T.L.; Degnim, A.C.; Sellers, T.A.; Johnson, J.L.; Kasner, M.J.; Hartmann, L.C.; Frost, M.H. Contralateral Prophylactic Mastectomy Is Associated with a Survival Advantage in High-Risk Women with a Personal History of Breast Cancer. *Ann. Surg. Oncol.* **2010**, *17*, 2702. [CrossRef]

92. Boughey, J.C.; Schilz, S.R.; Van Houten, H.K.; Zhu, L.; Habermann, E.B.; Lemaine, V. Contralateral Prophylactic Mastectomy with Immediate Breast Reconstruction Increases Healthcare Utilization and Cost. *Ann. Surg. Oncol.* **2017**, *24*, 2957–2964. [CrossRef]

93. Hoskin, T.L.; Hieken, T.J.; Degnim, A.C.; Jakub, J.W.; Jacobson, S.R.; Boughey, J.C. Use of Immediate Breast Reconstruction and Choice for Contralateral Prophylactic Mastectomy. *Surgery* **2016**, *159*, 1199–1209. [CrossRef]

94. Erdahl, L.; Shah, A.R.; Boughey, J.C.; Hieken, T.J.; Hoskin, T.L.; Degnim, A.C. Comparison of Side-Specific Complications Af-Ter Contralateral Prophylactic Mastectomy vs Treatment Mastectomy for Unilateral Breast Cancer. *Ann. Surg. Oncol.* **2014**, *37*, 46–47.

95. Osman, F.; Saleh, F.; Jackson, T.D.; Corrigan, M.A.; Cil, T. Increased Postoperative Complications in Bilateral Mastectomy Patients Compared to Unilateral Mastectomy: An Analysis of the NSQIP Database. *Ann. Surg. Oncol.* **2013**, *20*, 3212–3217. [CrossRef]

96. Jones, C.; Lancaster, R. Evolution of Operative Technique for Mastectomy. *Surg. Clin. N. Am.* **2018**, *98*, 835–844. [CrossRef]

97. Blidaru, A.; Bordea, C.I.; Ichim, E.; El Houcheimi, B.; Matei Purge, I.; Noditi, A.; Sterie, I.; Gherghe, M.; Radu, M. Breast Cancer Surgery in Images. *Chirurgia* **2017**, *112*, 486–492. [CrossRef]

98. Varghese, J.; Gohari, S.S.; Rizki, H.; Faheem, I.; Langridge, B.; Kümmel, S.; Johnson, L.; Schmid, P. A Systematic Review and Meta-Analysis on the Effect of Neoadjuvant Chemotherapy on Complications Following Immediate Breast Reconstruction. *Breast* **2021**, *55*, 55–62. [CrossRef]

99. Li, L.; Chen, Y.; Chen, J.; Chen, J.; Yang, B.; Li, J.; Huang, X.; Shen, Z.; Shao, Z.; Yu, P.; et al. Adjuvant Chemotherapy Increases the Prevalence of Fat Necrosis in Immediate Free Abdominal Flap Breast Reconstruction. *J. Plast. Reconstr. Aesthet. Surg.* **2014**, *67*, 461–467. [CrossRef]

100. El-Sabawi, B.; Sosin, M.; Carey, J.N.; Nahabedian, M.Y.; Patel, K.M. Breast Reconstruction and Adjuvant Therapy: A Systematic Review of Surgical Outcomes. *J. Surg. Oncol.* **2015**, *112*, 458–464. [CrossRef]

101. Russell, R.C.; Pribaz, J.; Zook, E.G.; Leighton, W.D.; Eriksson, E.; Smith, C.J. Functional Evaluation of Latissimus Dorsi Donor Site. *Plast. Reconstr. Surg.* **1986**, *78*, 336–344. [CrossRef]

102. Yang, J.D.; Huh, J.S.; Min, Y.S.; Kim, H.J.; Park, H.Y.; Jung, T. Du Physical and Functional Ability Recovery Patterns and Quality of Life after Immediate Autologous Latissimus Dorsi Breast Reconstruction: A 1-Year Prospective Observational Study. *Plast. Reconstr. Surg.* **2015**, *136*, 1146–1154. [CrossRef]

103. Lindegren, A.; Halle, M.; Docherty Skogh, A.C.; Edsander-Nord, A. Postmastectomy Breast Reconstruction in the Irradiated Breast: A Comparative Study of DIEP and Latissimus Dorsi Flap Outcome. *Plast. Reconstr. Surg.* **2012**, *130*, 10–18. [CrossRef]

104. Yueh, J.H.; Slavin, S.A.; Adesiyun, T.; Nyame, T.T.; Gautam, S.; Morris, D.J.; Tobias, A.M.; Lee, B.T. Patient Satisfaction in Postmastectomy Breast Reconstruction: A Comparative Evaluation of DIEP, TRAM, Latissimus Flap, and Implant Techniques. *Plast. Reconstr. Surg.* **2010**, *125*, 1585–1595. [CrossRef]

105. Zhu, L.; Mohan, A.T.; Vijayasekaran, A.; Hou, C.; Sur, Y.J.; Morsy, M.; Saint-Cyr, M. Maximizing the Volume of Latissimus Dorsi Flap in Autologous Breast Reconstruction with Simultaneous Multisite Fat Grafting. *Aesthet. Surg. J.* **2016**, *36*, 169–178. [CrossRef]

Article

Identification of Hub Genes and Biological Mechanisms Associated with Non-Alcoholic Fatty Liver Disease and Triple-Negative Breast Cancer

Jingjin Zhu [1,2], Ningning Min [1,2], Wenye Gong [2,3], Yizhu Chen [2,3] and Xiru Li [2,*]

[1] School of Medicine, Nankai University, Tianjin 300071, China; 2120211513@mail.nankai.edu.cn (J.Z.); 2120201347@mail.nankai.edu.cn (N.M.)
[2] Department of General Surgery, The First Medical Center of Chinese PLA General Hospital, Beijing 100853, China; gongwenye1998@163.com (W.G.); chenyimizhuzi@163.com (Y.C.)
[3] Medical School of Chinese PLA, Beijing 100853, China
[*] Correspondence: 2468li@sina.com; Tel.: +86-010-6693-8371

Abstract: The relationship between non-alcoholic fatty liver disease (NAFLD) and triple-negative breast cancer (TNBC) has been widely recognized, but the underlying mechanisms are still unknown. The objective of this study was to identify the hub genes associated with NAFLD and TNBC, and to explore the potential co-pathogenesis and prognostic linkage of these two diseases. We used GEO, TCGA, STRING, ssGSEA, and Rstudio to investigate the common differentially expressed genes (DEGs), conduct functional and signaling pathway enrichment analyses, and determine prognostic value between TNBC and NAFLD. GO and KEGG enrichment analyses of the common DEGs showed that they were enriched in leukocyte aggregation, migration and adhesion, apoptosis regulation, and the PPAR signaling pathway. Fourteen candidate hub genes most likely to mediate NAFLD and TNBC occurrence were identified and validation results in a new cohort showed that ITGB2, RAC2, ITGAM, and CYBA were upregulated in both diseases. A univariate Cox analysis suggested that high expression levels of ITGB2, RAC2, ITGAM, and CXCL10 were associated with a good prognosis in TNBC. Immune infiltration analysis of TNBC samples showed that NCF2, ICAM1, and CXCL10 were significantly associated with activated CD8 T cells and activated CD4 T cells. NCF2, CXCL10, and CYBB were correlated with regulatory T cells and myeloid-derived suppressor cells. This study demonstrated that the redox reactions regulated by the NADPH oxidase (NOX) subunit genes and the transport and activation of immune cells regulated by integrins may play a central role in the co-occurrence trend of NAFLD and TNBC. Additionally, ITGB2, RAC2, and ITGAM were upregulated in both diseases and were prognostic protective factors of TNBC; they may be potential therapeutic targets for treatment of TNBC patients with NAFLD, but further experimental studies are still needed.

Keywords: triple-negative breast cancer (TNBC); non-alcoholic fatty liver disease (NAFLD); non-alcoholic steatohepatitis (NASH); bioinformatics analysis; hub genes; prognostic value

Citation: Zhu, J.; Min, N.; Gong, W.; Chen, Y.; Li, X. Identification of Hub Genes and Biological Mechanisms Associated with Non-Alcoholic Fatty Liver Disease and Triple-Negative Breast Cancer. *Life* **2023**, *13*, 998. https://doi.org/10.3390/life13040998

Academic Editor: Riccardo Autelli

Received: 2 February 2023
Revised: 20 March 2023
Accepted: 6 April 2023
Published: 12 April 2023

1. Introduction

Breast cancer is the most commonly diagnosed cancer and the leading cause of cancer death among women [1]. Triple-negative breast cancer (TNBC), defined by estrogen receptor (ER)-negative, progesterone receptor (PR)-negative, and human epidermal growth factor receptor-2 (HER2)-negative histological presentation, accounts for approximately 20% of all breast cancer cases [2,3]. In contrast to other breast cancer types, TNBC has a more aggressive expression profile (high p53 and Ki67 and low Bcl-2 expression), large tumor size, and high histological grade, and is associated with an increased risk of early relapse and poor prognosis [4]. Many potential risk factors for TNBC have been reported, including non-modifiable factors such as age, sex, race, genetic mutations, breast tissue density, and family history of breast disease, and modifiable factors such as diet, lifestyle, obesity, and hormone replacement therapy [5–7]. In particular, the increasing proportion of obesity worldwide has led to a sharp rise in patients with metabolic syndrome and an increased risk of certain malignancies [8,9]. It has been shown that metabolic syndrome is positively associated with breast cancer and significantly associated with TNBC [10].

Non-alcoholic fatty liver disease (NAFLD) is a common chronic liver disease intimately related to metabolic syndrome and abdominal obesity [11–13]. It is becoming increasingly evident that NAFLD is not only linked to an increased risk of liver-related mortality or morbidity, but also associated with extrahepatic complications such as cardiovascular disease, chronic kidney disease, pulmonary insufficiency, and extrahepatic malignancies [14–16]. A retrospective study by Nseir et al. [17] indicated that NAFLD is associated with breast cancer independent of known risk factors.

In addition, breast cancer patients often develop non-alcoholic fatty liver disease during the course of disease. Bilici et al. [18] reported a prevalence of NAFLD as high as 63% and 72% in newly diagnosed and systematically treated breast cancer patients, respectively. It has also been reported that long-term selective estrogen receptor modulator (SERM) administration may increase the risk of NAFLD development, with fatty liver reported in 48.5% of tamoxifen-treated and 50.2% of toremifene-treated breast cancer patients at 60 months [19]. It is of particular note that patients who present with NAFLD have been reported to have longer disease-free survival (DFS) [20,21], but the underlying mechanisms associated with improved clinical outcomes have not been thoroughly investigated.

Recently, advances in sequencing technology and bioinformatics have made it possible to explore the pathogenesis of diseases and the interactions between different diseases at the gene level, which is expected to shed new light on the pathogenesis, diagnosis, and treatment of diseases [22–24]. Here, we investigated the common differentially expressed genes (DEGs) of NAFLD and TNBC from public RNA-sequencing databases and identified 14 candidate hub genes most likely to mediate NAFLD and TNBC occurrence. Next, the biological functional pathways of the hub genes were estimated to explore the underlying mechanisms of both diseases. Finally, validation and prognostic analysis were performed in a new cohort of TNBC patients.

2. Materials and Methods

2.1. Study Design and Data Collection

Three microarray datasets, GSE63067 and GSE48452 of NAFLD, and GSE38959 of TNBC, were collected from the GEO (http://www.ncbi.nlm.nih.gov/geo/, accessed on 1 November 2022) database. The nature of the three microarray datasets from the GEO database is summarized in Table 1. Non-alcoholic steatohepatitis (NASH) is a stage of NAFLD that is usually associated with a worse prognosis [25] and has a higher prevalence and more advanced stage of neoplasms compared to steatosis [26]; thus, the NASH samples were selected as the representatives for analysis in this study. The GSE63067 dataset included the gene expression profiles of 18 samples, of which 9 were NASH patients and 7 were controls, and the GSE38959 dataset included 47 samples, of which 30 were TNBC patients and 13 were controls. The GSE48452 dataset consisted of human liver biopsy samples taken at different phases from control to NASH; 14 controls and 18 NASH samples were used in this study.

RNAseq profiling in the form of fragments per kilobase million (FPKM) and clinicopathological breast cancer data were obtained from the TCGA (https://portal.gdc.cancer.gov, accessed on 1 November 2022) database (TCGA-BRCA cohort). TNBC samples were selected according to the status of ER, PR, and HER2 by referring to the method of Craven et al. [27]. One sample with an unknown ID was excluded, and samples from 132 patients were finally selected for analysis.

Table 1. The nature of the three microarray datasets from the GEO database.

Series	Country	Status	Platforms	Type of Samples	Numbers
GSE63067	Sweden	Public on 7 November 2014	GPL570	non-alcoholic steatohepatitis steatosis healthy	9 2 7
GSE48452	Germany	Public on 8 August 2013	GPL11532	non-alcoholic steatohepatitis steatosis healthy obese control	18 14 27 14
GSE38959	Japan	Public on 21 December 2012	GPL4133	triple-negative breast cancer normal mammary ductal cells normal human vital organs including heart, lung, liver, and kidney	30 13 4

2.2. Differentially Expressed Gene (DEG) Selection

DEGs were extracted and analyzed separately using the R package "limma". The fold changes (FCs) were calculated for individual gene expression levels. Genes meeting specific cut-off criteria of p-value < 0.05 and $|\log FC| >$ with [mean $|\log FC|$) + 2 × sd($|\log FC|$)] were defined as DEGs. The overlapping DEGs between NAFLD and TNBC were delineated using the R package "ggVennDiagram". These common DEGs with consistent upregulation or downregulation trends were retained for subsequent analysis.

2.3. Functional Classification and Pathway Enrichment of DEGs

The above overlapping DEGs were submitted to Gene Ontology (GO) functional enrichment analysis, which consisted of biological process (BP), cellular component (CC), and molecular function (MF) analyses, and to Kyoto Encyclopedia of Genes and Genomes (KEGG) signaling pathway enrichment analysis using the R package "cluster Profiler". The enriched GO terms and KEGG pathways with an adjusted p-value < 0.1 were selected.

2.4. Protein–Protein Interaction (PPI) Establishment and Hub Gene Identification

To further explore the interactions among the common genes obtained as described above, the Search Tool for the Retrieval of Interacting Genes (STRING) (http://string-db.org/, accessed on 12 November 2022) was used for PPI network construction. Subsequently, Cytoscape software was used to visualize the PPI network. The Cytoscape plug-in Minimal Common Oncology Data Elements (MCODE, http://apps.cytoscape.org/apps/mcode, accessed on 12 November 2022) was used to screen out key protein expression molecules and multiple topological analysis algorithms in the cytoHubba plug-in (http://hub.iis. sinica.edu.tw/cytohubba/, accessed on 12 November 2022), such as MCC, MNC, Degree, and EPC, were used to screen the hub genes in the PPI network.

2.5. Hub Gene Expression Validation and Prognostic Analysis

The expression levels of the identified hub genes were validated in 132 TNBC samples and 113 controls from the TCGA cohort, and 18 NASH samples and 14 controls from the GEO database. The Wilcoxon test was used to compare the data between the two groups, and a two-sided p-value < 0.05 was considered significant. According to the median expression level for each gene, the TNBC samples were divided into high- or low-expression groups. Survival analysis was performed using univariate and multivariate Cox regression hazard analysis, providing hazard ratios (HRs) and 95% confidence intervals (CIs), and survival curves were derived using Kaplan–Meier (KM) survival analysis with log-rank tests for comparison.

2.6. Immune Infiltration Analysis

The ssGSEA (single-sample gene set enrichment analysis) algorithm is a rank-based method that defines a score representing the degree of absolute enrichment of a particular gene set in each sample [28,29]. The ssGSEA score utilized immune-cell-marker-associated gene sets (http://cis.hku.hk/TISIDB/data/download/CellReports.txt, accessed on 20 November 2022) to quantify the infiltration of immune cells in TNBC tissue and determine the level of immune infiltration in each sample. Pearson's correlation analysis was used to reveal the relationships between hub genes and immune cells.

3. Results

3.1. DEG Identification in NASH and TNBC

In the NASH and control groups in the GSE63067 dataset, there were 498 up-DEGs and 163 down-DEGs screened with a logFC threshold of 0.472 (Figure 1A). In the TNBC and control groups in the GSE38959 dataset, there were 453 up-DEGs and 422 down-DEGs screened with a logFC threshold of 1.834 (Figure 1B). A Venn diagram was used to determine the intersection and 42 common DEGs were identified (Figure 1C). After excluding genes with opposite expression trends, 27 DEGs with the same expression trends were found, including 19 common upregulated genes and 8 common downregulated genes (Table S1).

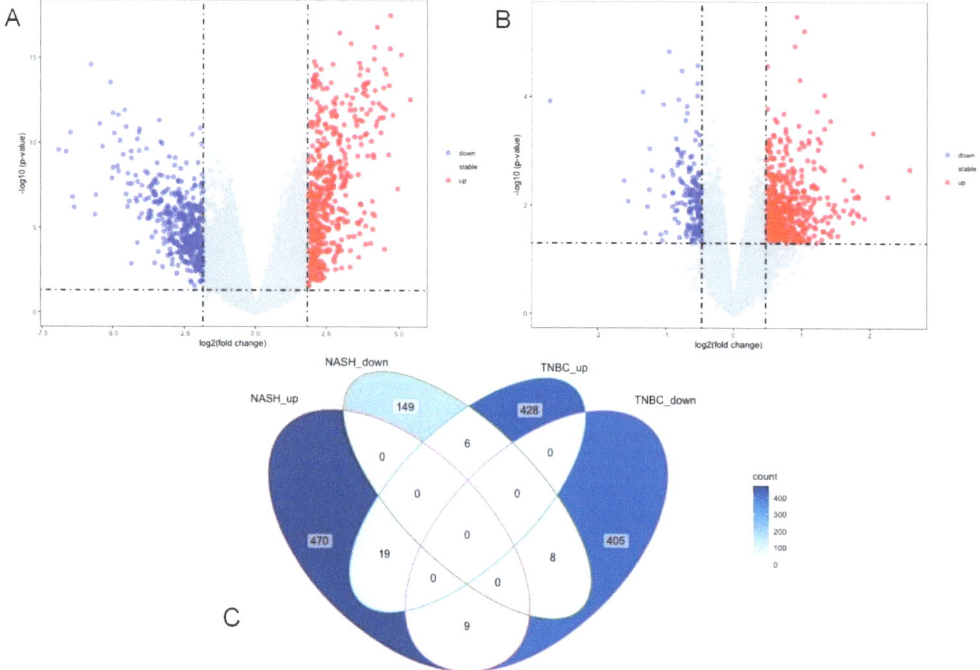

Figure 1. Characterization of the DEGs in NASH and TNBC. (**A**) Volcano map of DEGs between NASH samples and normal samples in GSE63067; (**B**) volcano map of DEGs between TNBC samples and normal samples in GSE38959; (**C**) Venn diagram of the common DEGS between the two upregulation and two downregulation modules in NAFLD and TNBC. DEGs, differentially expressed genes; NASH, non-alcoholic steatohepatitis; TNBC, triple-negative breast cancer.

3.2. GO and KEGG Enrichment Pathway Analysis of DEGs

To better understand the biological functions of the identified DEGs, GO and KEGG pathway enrichment analyses were performed. After screening with the threshold of adjusted $p < 0.1$, significantly enriched GO terms and KEGG terms were selected.

As shown in Figure 2, in the BP category, DEGs were mainly enriched in maintenance of location, leukocyte cell–cell adhesion, leukocyte aggregation, leukocyte migration involved in inflammatory response, protein nitrosylation, peptidyl-cysteine S-nitrosylation, and regulation of apoptotic signaling pathway. In the CC category, DEGs were principally associated with secretory granule lumen, cytoplasmic vesicle lumen, vesicle lumen, immunological synapse, collagen-containing extracellular matrix, and nuclear inner membrane. The analysis of the MF category indicated that DEGs were enriched in toll-like receptor binding, fatty acid derivative binding, fatty acid binding, monocarboxylic acid binding, microtubule binding, integrin binding, and tubulin binding. Furthermore, two KEGG pathways with significant enrichment were the peroxisome proliferator-activated receptor (PPAR) signaling pathway and the IL-17 signaling pathway.

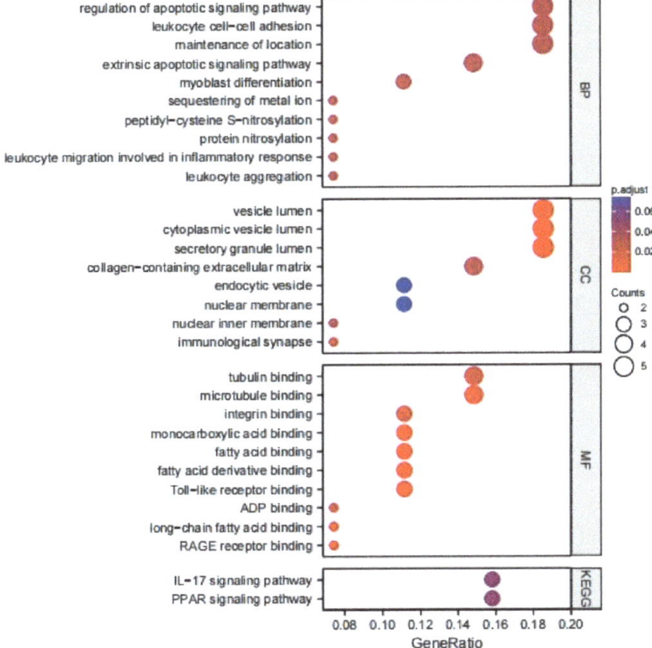

Figure 2. GO and KEGG analysis results of the common DEGs. GO, Gene Ontology; BP, biological process; CC, cellular component; MF, molecular function; KEGG, Kyoto Encyclopedia of Genes and Genomes.

3.3. PPI Network Construction and Hub Gene Identification

To determine the interactions among the DEGs and identify hub genes, the PPI network of the DEGs was generated using STRING. With the aim of preventing important hub genes being missed, we modified the PPI settings to have a minimum required interaction score of medium confidence (0.400) and a maximum number of interactions of no more than 20 interactors to increase the maximum number of interactions and the number of proteins directly related to the input proteins. Then, a PPI with 47 nodes and 122 edges, with a PPI enrichment p value $< 1.0 \times 10^{-16}$, was obtained and imported into Cytoscape software v3.9.1 for visualization (Figure 3A, Table S2).

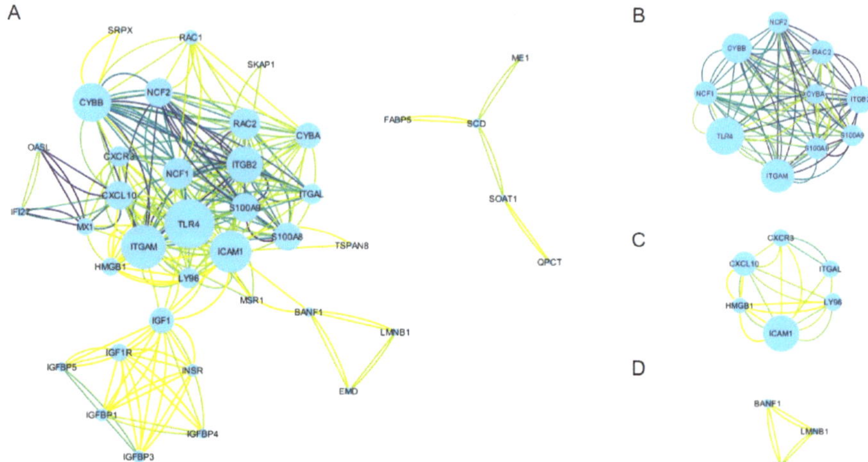

Figure 3. Construction of the PPI network and module analysis. (**A**) The visualization results of the PPI network of the common DEGs obtained from Cytoscape software v3.9.1; (**B–D**) three crucial clustering modules extracted by MCODE. PPI, protein–protein interaction; DEG, differentially expressed genes; STRING, Search Tool for the Retrieval of Interacting Genes; MCODE, Minimal Common Oncology Data Elements.

The MCODE plug-in was used to conduct module analysis to detect crucial clustering modules. Three modules were retrieved from the PPI network. The criteria were set as follows: Degree Cutoff = 2, Node Score Cutoff = 0.2, K-Core = 2, and Max. Depth = 100. Module 1 included 10 nodes and 84 edges with a cluster score (density times the number of members) of 9.333. Module 2 and module 3 had 6 nodes and 20 edges and 3 nodes and 6 edges, respectively, and the scores were 4.000 and 3.000, respectively (Figure 3B–D).

The CytoHubba plug-in was used to identify hub genes. Based on the MCC, MNC, Degree, and EPC algorithms, the top 15 important hub genes in the PPI networks were predicted. The intersection of these 15 genes from the four algorithms revealed 14 candidate hub genes: TLR4, CYBB, NCF1, NCF2, S100A8, S100A9, ITGB2, RAC2, ITGAM, CYBA, ICAM1, CXCL10, CXCR3, and ITGAL.

Combined with the logFC values of the hub genes in the GSE38959 dataset, the GO and KEGG enrichment pathways were analyzed. The top 10 GO terms and KEGG pathways are shown in Figure 4. In the BP category, nine hub genes including CYBB, NCF1, NCF2, ITGB2, RAC2, ITGAM, CYBA, ICAM1, and TLR4 were enriched in reactive oxygen species metabolic process. Furthermore, S100A8, S100A9, ITGB2, RAC2, ITGAM, ICAM1, CXCL10, and CXCR3 were enriched in leukocyte migration. Enriched CC and MF were related to redox reactions like NADPH oxidase complex, superoxide-generating NADPH oxidase activity, and oxidoreductase activity. KEGG enrichment analyses showed that leukocyte transendothelial migration was highly correlated with these genes.

3.4. Hub Gene Expression Validation and Prognostic Analysis

Validation was performed in the TCGA-BRCA cohort for TNBC and the GSE48452 dataset for NASH. For TNBC, the differences in the expression levels of all hub genes between normal tissues and TNBC samples were statistically significant (Figure 5A). Compared with normal tissues, 13 hub genes were upregulated, including CYBB, NCF1, NCF2, S100A8, S100A9, ITGB2, RAC2, ITGAM, CYBA, ICAM1, CXCL10, CXCR3, and ITGAL, and TLR4 was downregulated. For NASH, the expressions of hub genes ITGB2, RAC2, ITGAM, and CYBA were upregulated, and the changes in other genes' expressions were not statistically significant (Figure 5B).

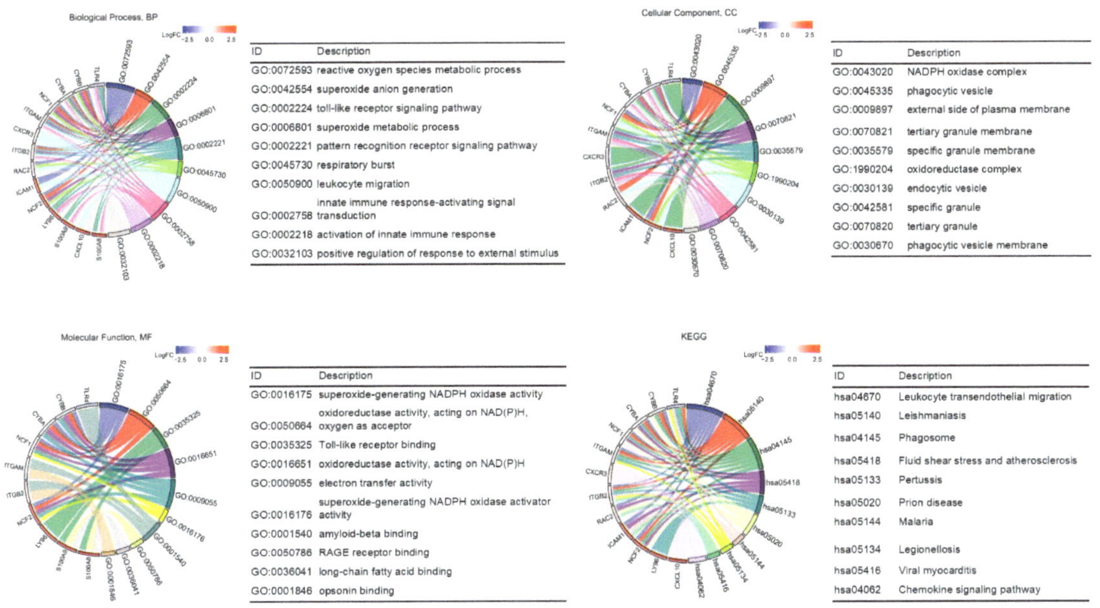

Figure 4. Functional enrichment analysis results of the hub genes. NADPH, nicotinamide adenine dinucleotide phosphate; RAGE, receptor for advanced glycation endproducts.

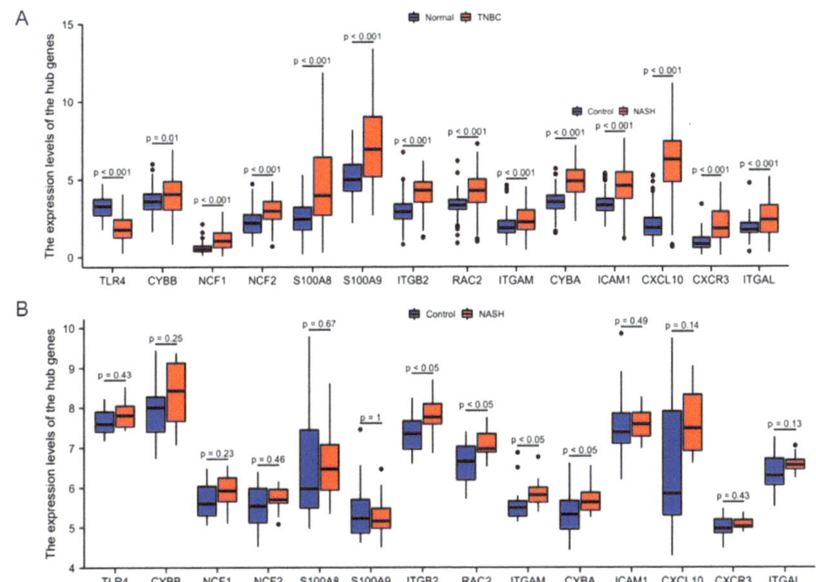

Figure 5. Comparison of hub genes between normal tissues and TNBC samples in TCGA-BRCA cohort (**A**) and control group and NASH samples in the GSE48452 dataset (**B**). TNBC, triple-negative breast cancer; TCGA, The Cancer Genome Atlas; BRCA, breast cancer; NASH, non-alcoholic steato-hepatitis; TNBC, triple-negative breast cancer.

To evaluate the clinical relevance of hub gene expression, the TNBC samples were divided into a high-expression group and a low-expression group according to the median expression level of each gene for prognostic analysis. Univariate Cox analysis suggested that high expression of ITGB2, RAC2, ITGAM, and CXCL10 was associated with better overall survival (OS) (Table 2). The corresponding KM survival curves are shown in Figure 6. In addition, two prognostic factors associated with worse OS were identified in terms of clinicopathological features, namely black or African American and Asian ethnicity, and N stage. All variables significant upon univariate Cox regression analysis ($p \leq 0.05$) were subjected to multivariate Cox regression analysis, and it was found that the N stage was an independent risk factor for overall survival and the remaining factors were not significant.

Table 2. The Cox regression analysis results of the hub genes and clinicopathological variables in the TCGA-BRCA group.

Factor	Univariate Cox Regression Analysis		Multivariate Cox Regression Analysis	
	HR (95%CI)	*p*-Value	HR (95%CI)	*p*-Value
TLR4	0.640 (0.237–1.728)	0.378		
CYBB	0.509 (0.183–1.411)	0.194		
NCF1	0.431 (0.155–1.200)	0.107		
NCF2	0.622 (0.228–1.694)	0.352		
S100A8	0.810 (0.295–2.226)	0.683		
S100A9	0.523 (0.181–1.507)	0.230		
ITGB2	0.157 (0.044–0.556)	0.004	0.213 (0.033–1.376)	0.104
RAC2	0.341 (0.118–0.984)	0.047	1.067 (0.282–4.036)	0.924
ITGAM	0.282 (0.094–0.842)	0.023	1.392 (0.318–6.100)	0.661
CYBA	1.750 (0.632–4.847)	0.282		
ICAM1	0.676 (0.250–1.828)	0.440		
CXCL10	0.244 (0.082–0.725)	0.011	0.430 (0.108–1.718)	0.232
CXCR3	0.419 (0.152–1.156)	0.093		
ITGAL	0.413 (0.148–1.151)	0.091		
Age	0.773 (0.249–2.406)	0.657		
Race	2.830 (1.019–7.860)	0.046	2.090 (0.631–6.922)	0.227
T stage				
(T2 vs. T1, T3/T4 vs. T1)	1.717 (0.471–6.255) 4.427 (0.858–22.838)	0.194		
N stage				
(N1/N2/N3 vs. N0)	5.641 (1.815–17.534)	0.003	4.681 (1.452–15.089)	0.010

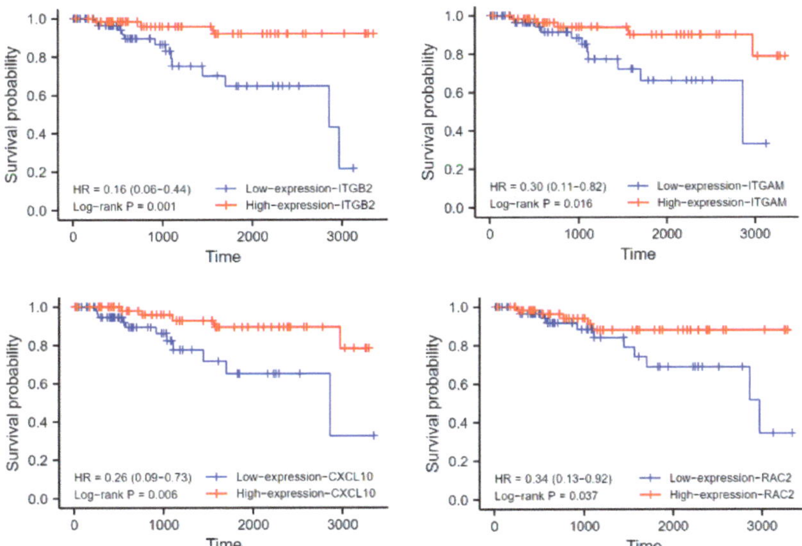

Figure 6. Kaplan–Meier survival curves of association between the expression levels of hub genes and the OS of TNBC patients (group cutoff = median). OS, overall survival; TNBC, triple-negative breast cancer.

3.5. Association between the Hub Genes and Immune Infiltration

Figure 7 shows the relationships between 14 hub genes and 27 immune cells (results for central memory CD4 T cells unavailable) according to the results of ssGSEA analysis. For TNBC samples from the GSE38959 dataset, NCF2, ICAM1, and CXCL10 were significantly associated with activated CD8 T cells (NCF2, r = 0.764, $p = 2.42 \times 10^{-9}$; ICAM1, r = 0.705, $p = 1.32 \times 10^{-7}$; CXCL10, r = 0.802, $p = 1.03 \times 10^{-10}$) and activated CD4 T cells (NCF2, r = 0.715, $p = 7.16 \times 10^{-8}$; ICAM1, r = 0.804, $p = 8.70 \times 10^{-11}$; CXCL10, r = 0.785, $p = 4.72 \times 10^{-10}$). Specifically, NCF2, CXCL10, and CYBB were correlated with regulatory T cells (NCF2, r = 0.773, $p = 1.22 \times 10^{-9}$; CXCL10, r = 0.730, $p = 2.86 \times 10^{-8}$; CYBB, r = 0.718, $p = 6.02 \times 10^{-8}$), and myeloid-derived suppressor cells (MDSCs) (NCF2, r = 0.743, $p = 1.19 \times 10^{-8}$; CXCL10, r = 0.79, $p = 4.24 \times 10^{-10}$; CYBB, r = 0.733, $p = 2.29 \times 10^{-8}$). In addition, ITGB2 was associated with activated B cells (r = 0.766, $p = 2.22 \times 10^{-9}$) and immature B cells (r = 0.733, $p = 2.36 \times 10^{-8}$).

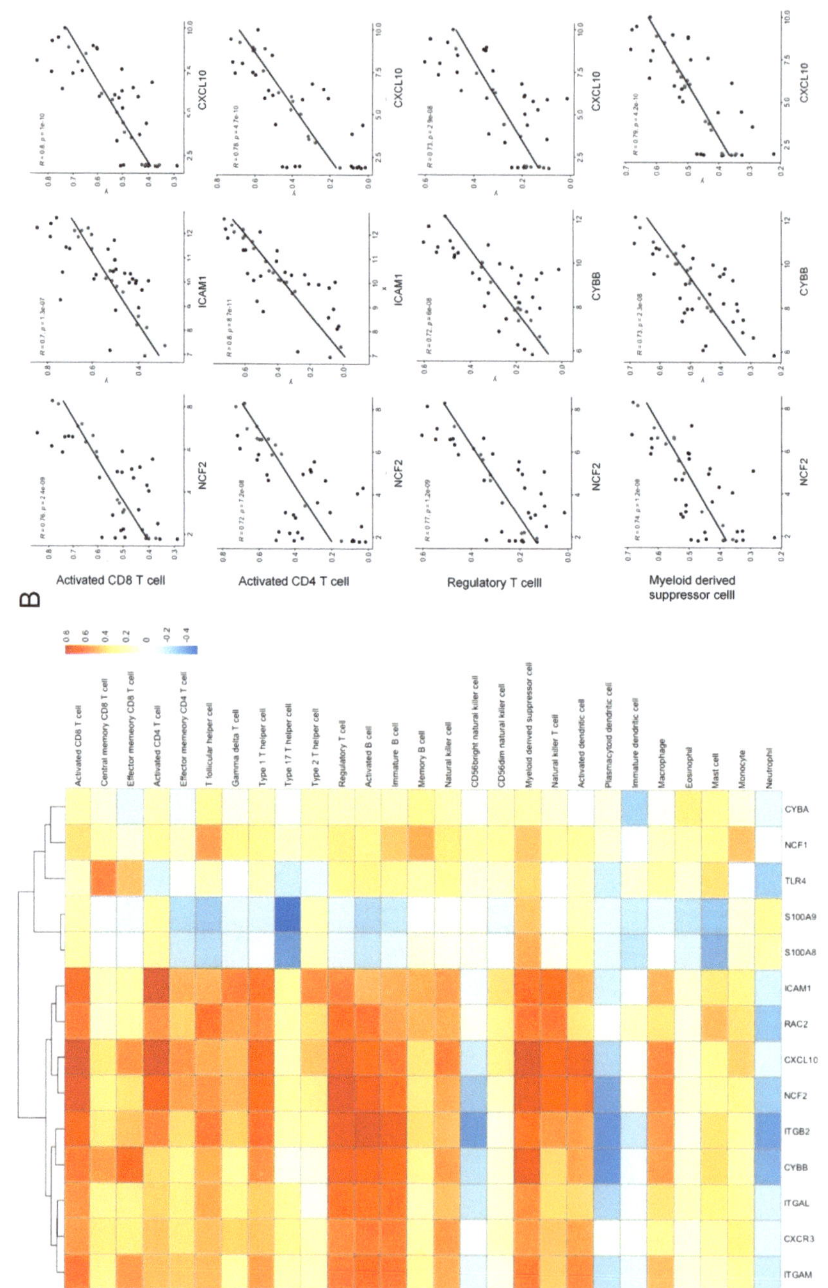

Figure 7. Associations between the hub genes and immune infiltration in the GSE38959 group (**A**) and genes significantly associated with activated CD8 T cells, activated CD4 T cells, regulatory T cells and myeloid-derived suppressor cells (**B**). Red: positive correlation; blue: negative correlation.

4. Discussion

In recent years, the relationship between NAFLD and breast cancer has become a research hotspot, and an increasing number of studies have confirmed the correlation. Some studies have reported that breast cancer is a common extrahepatic complication of NAFLD [16,30]. Simultaneously, it also has been suggested that patients with breast cancer, especially those receiving endocrine therapy, present an increased risk of NAFLD [31,32]. Based on these combined results, NAFLD may be related to the occurrence and progression of breast cancer. In addition, it has been proposed that liver metastasis in the diagnosis of fatty liver patients with breast cancer is significantly lower than that of patients with normal liver histology, further revealing the correlation between the two events in clinical practice [33]. However, these studies have been mostly observational, and the mechanism connecting NAFLD and TNBC remains unclear to date. Therefore, exploring the molecular mechanisms to enable early identification and intervention is undoubtedly of great clinical significance.

In this study, we explored common DEGs of NASH and TNBC datasets in public databases through bioinformatics analysis and observed the biological processes and signaling pathways in which they jointly participate. GO and KEGG enrichment analyses of the common DEGs showed enrichment in leukocyte aggregation, migration and adhesion, apoptosis regulation, and the PPAR signaling pathway, suggesting that TNBC in NAFLD patients was likely due to enhanced leukocyte recruitment in the inflammatory response and abnormal apoptosis. Interestingly, the PPAR signaling pathway not only controls the expression of genes encoding proteins of lipid metabolism, but is also involved in anti-cancer responses [34]. One of the mechanisms by which PPARs act to control cancer progression is to affect the NF-κB signaling pathway, or its upstream pathways, such as the Toll-like receptor 4 (TLR4) signaling pathway [35,36]. PPAR γ agonists have been found to induce apoptosis in TNBC cells and inhibit melanoma progression in mice [37,38].

A total of 14 candidate hub genes most likely to mediate NASH and TNBC occurrence were identified, including TLR4, CYBB, NCF1, NCF2, S100A8, S100A9, ITGB2, RAC2, ITGAM, CYBA, ICAM1, CXCL10, CXCR3, and ITGAL.

CYBB, CYBA, NCF1, NCF2, and RAC2 are NADPH oxidase (NOX) subunit genes and are associated with inflammation and fibrosis in multiple organs, such as the liver [39,40], lungs [41], and kidneys [42], as well as with various types of cancer [43]. NOX can produce reactive oxygen species (ROS) that cause changes in cellular redox status, leading to chronic liver injury and fibrosis, which is critical for alcoholic steatohepatitis and NASH [44,45]. The analysis of the TCGA cohort showed that NOX-related genes were expressed more highly in tumor cells than in normal tissues of the same tissue origin, which suggested that the abnormal expression and regulation of NOX may be related to tumorigenesis and the increase of ROS in tumor cells [46], which probably contributes to the increased susceptibility of TNBC patients to NAFLD compared to the healthy population. In addition, RAC2 was strongly associated with OS in patients. RAC2 is a 21 kDa RAS superfamily of GTPases that stabilize the cytoskeleton structure of actin [47,48]. Chen et al. [49] found the high expression of RAC2 can inhibit the proliferation of breast cancer cells.

ITGAM, ITGB2, and ITGAL are involved in the most common integrins expressed on leukocytes, including Mac-1 (αMβ2 or CD11b/CD18) and leukocyte function-related antigen 1 (LFA-1 or αLβ2) [50,51]. Activated integrins play a crucial role in trafficking immune cells into tissues, activating and promoting the proliferation of effector cells, and inducing the formation of immune synapses between cells [52,53]. Clinically, Mac-1 expression is increased in patients with metabolic syndrome [54]. It has also been confirmed that Mac-1 is required for pro-inflammatory gene expression by macrophages in adipose tissue inflammation and is related to recruiting monocytes from bone marrow and inducing them to transform into M1-like macrophages (pro-inflammatory and usually anti-tumor) to express cytotoxic factors to engulf and destroy tumor cells [55–57]. Rojas et al. [51] demonstrated that an integrin marker composed of ITGA4, ITGB2, ITGAX, ITGB7, ITGAM, ITGAL, and ITGA8 had the potential to recognize basal-like breast cancers with immune-infiltrating and favorable prognosis. Our results are similar to this finding, with ITGAM and ITGB2 highly expressed in both diseases and associated with a better prognosis in TNBC. Further analysis of immune infiltration showed a positive correlation between integrin genes and activated B cells and immature B cells.

CXCL10 and its homologous receptor CXCR3 are critical in the development of specific features of the NAFLD phenotype, wherein they are mainly involved in the induction of inflammation, regulation of adipogenesis and oxidative stress, and other related processes [58,59]. In the process of tumor progression, studies have shown that CXCL10 has a dual role. It can not only promote tumor progression by increasing cell proliferation and metastasis, but also exert an anti-malignancy function by inhibiting angiogenesis and influencing the tumor microenvironment [60–62]. Sun et al. [63] found that CXCL10 expression was significantly upregulated in mice with melanoma and that CXCL10 promoted the proliferation of monocyte-like MDSCs, leading to an immunosuppressive microenvironment. On the other hand, it has also been demonstrated that tumor-cell-derived CXCL9/CXCL10 regulates the recruitment of T cells in various tumors [64–66]. Our study suggested that CXCL10 is positively correlated with MDCSs and activated T cells, and TNBC patients with high CXCL10 expression obtained a better prognosis. Therefore, in the context of NAFLD, CXCL10 may play an anti-tumor role in TNBC, but more in-depth experimental research is still needed.

Although many studies have linked metabolic syndrome to the development of cancer and poor prognosis, it may be a symptom of a general metabolic disorder. Our study explored the relationship between NAFLD and TNBC at the genetic level for the first time, and found that the hub genes ITGB2, RAC2, and ITGAM were upregulated in both diseases and were prognostic protective factors in TNBC. This is inconsistent with our understanding of risk factors such as obesity, a high-fat diet, and NAFLD that promote the occurrence and progression of breast cancer. Therefore, further experimental studies will be of great significance and are expected to find new targets for diagnosis, prognostic assessment, and treatment of TNBC.

The study had several limitations. First, although the role of these genes has been elucidated in multiple studies, the key pathways and hub genes identified have not been validated in experiments. Second, due to the lack of a dataset, the validation of the hub gene was performed in patients with only NAFLD or TNBC, but not in patients with NAFLD combined with TNBC. Third, the relationship between the hub genes and the prognosis of TNBC patients needs to be confirmed by prospective clinical studies.

5. Conclusions

In conclusion, this study explored the hub genes of NAFLD and TNBC and illustrated the possible mechanisms for the co-occurrence trend of these two diseases. Redox reactions regulated by the NOX subunit genes and the transport and activation of immune cells regulated by integrins may play a central role in the development of NAFLD and TNBC. Additionally, the expressions of ITGB2, RAC2, ITGAM, and CXCL10 were significantly correlated with a good prognosis in TNBC and may be potential therapeutic targets for the development of gene therapies for TNBC patients with NAFLD.

Supplementary Materials: The following supporting information can be downloaded at: https://www.mdpi.com/article/10.3390/life13040998/s1; Table S1. The common DEGs; Table S2. The details of the PPI obtained from STRING database analysis.

Author Contributions: Conceptualization, J.Z. and X.L.; methodology, J.Z.; software, J.Z.; validation, N.M., W.G. and Y.C.; formal analysis, J.Z.; investigation, J.Z.; resources, J.Z.; data curation, J.Z.; writing—original draft preparation, J.Z.; writing—review and editing, X.L.; visualization, J.Z.; supervision, X.L.; project administration, X.L. All authors have read and agreed to the published version of the manuscript.

Funding: This research received no external funding.

Institutional Review Board Statement: Not applicable.

Informed Consent Statement: Not applicable.

Data Availability Statement: Publicly available datasets were analyzed in this study. The data can be found here: http://www.ncbi.nlm.nih.gov/geo/ (accessed on 1 November 2022), and https://portal.gdc.cancer.gov (accessed on 1 November 2022).

Acknowledgments: We thank Jiahao Han (The State Key Laboratory Breeding Base of Basic Science of Stomatology and Key Laboratory for Oral Biomedicine of Ministry of Education, School and Hospital of Stomatology, Wuhan University) for his valuable insights into this work.

Conflicts of Interest: The authors declare no conflict of interest.

References

1. Sung, H.; Ferlay, J.; Siegel, R.L.; Laversanne, M.; Soerjomataram, I.; Jemal, A.; Bray, F. Global Cancer Statistics 2020: GLOBOCAN Estimates of Incidence and Mortality Worldwide for 36 Cancers in 185 Countries. *CA Cancer J. Clin.* **2021**, *71*, 209–249. [CrossRef]

2. Billar, J.A.; Dueck, A.C.; Stucky, C.C.; Gray, R.J.; Wasif, N.; Northfelt, D.W.; McCullough, A.E.; Pockaj, B.A. Triple-negative breast cancers: Unique clinical presentations and outcomes. *Ann. Surg. Oncol.* **2010**, *17* (Suppl. 3), 384–390. [CrossRef]

3. Yin, L.; Duan, J.J.; Bian, X.W.; Yu, S.C. Triple-negative breast cancer molecular subtyping and treatment progress. *Breast Cancer Res.* **2020**, *22*, 61. [CrossRef]

4. Nofech-Mozes, S.; Trudeau, M.; Kahn, H.K.; Dent, R.; Rawlinson, E.; Sun, P.; Narod, S.A.; Hanna, W.M. Patterns of recurrence in the basal and non-basal subtypes of triple-negative breast cancers. *Breast Cancer Res. Treat* **2009**, *118*, 131–137. [CrossRef]

5. Almansour, N.M. Triple-Negative Breast Cancer: A Brief Review About Epidemiology, Risk Factors, Signaling Pathways, Treatment and Role of Artificial Intelligence. *Front. Mol. Biosci.* **2022**, *9*, 836417. [CrossRef]

6. Cornejo-Moreno, B.A.; Uribe-Escamilla, D.; Salamanca-Gómez, F. Breast cancer genes: Looking for BRACA's lost brother. *Isr. Med. Assoc. J.* **2014**, *16*, 787–792. [PubMed]

7. Toro, A.L.; Costantino, N.S.; Shriver, C.D.; Ellsworth, D.L.; Ellsworth, R.E. Effect of obesity on molecular characteristics of invasive breast tumors: Gene expression analysis in a large cohort of female patients. *BMC Obes.* **2016**, *3*, 22. [CrossRef]

8. Wang, C.H.; Lundh, M.; Fu, A.; Kriszt, R.; Huang, T.L.; Lynes, M.D.; Leiria, L.O.; Shamsi, F.; Darcy, J.; Greenwood, B.P.; et al. CRISPR-engineered human brown-like adipocytes prevent diet-induced obesity and ameliorate metabolic syndrome in mice. *Sci. Transl. Med.* **2020**, *12*, eaaz8664. [CrossRef]

9. Kazak, L.; Chouchani, E.T.; Lu, G.Z.; Jedrychowski, M.P.; Bare, C.J.; Mina, A.I.; Kumari, M.; Zhang, S.; Vuckovic, I.; Laznik-Bogoslavski, D.; et al. Genetic Depletion of Adipocyte Creatine Metabolism Inhibits Diet-Induced Thermogenesis and Drives Obesity. *Cell Metab.* **2017**, *26*, 693. [CrossRef] [PubMed]

10. Akinyemiju, T.; Oyekunle, T.; Salako, O.; Gupta, A.; Alatise, O.; Ogun, G.; Adeniyi, A.; Deveaux, A.; Hall, A.; Ayandipo, O.; et al. Metabolic Syndrome and Risk of Breast Cancer by Molecular Subtype: Analysis of the MEND Study. *Clin. Breast Cancer* **2022**, *22*, e463–e472. [CrossRef] [PubMed]

11. Chalasani, N.; Younossi, Z.; Lavine, J.E.; Diehl, A.M.; Brunt, E.M.; Cusi, K.; Charlton, M.; Sanyal, A.J. The diagnosis and management of non-alcoholic fatty liver disease: Practice guideline by the American Gastroenterological Association, American Association for the Study of Liver Diseases, and American College of Gastroenterology. *Gastroenterology* **2012**, *142*, 1592–1609. [CrossRef]

12. Lomonaco, R.; Sunny, N.E.; Bril, F.; Cusi, K. Nonalcoholic fatty liver disease: Current issues and novel treatment approaches. *Drugs* **2013**, *73*, 1–14. [CrossRef]

13. Younossi, Z.M.; Koenig, A.B.; Abdelatif, D.; Fazel, Y.; Henry, L.; Wymer, M. Global epidemiology of nonalcoholic fatty liver disease-Meta-analytic assessment of prevalence, incidence, and outcomes. *Hepatology* **2016**, *64*, 73–84. [CrossRef] [PubMed]

14. Wei, Z.; Ren, Z.; Hu, S.; Gao, Y.; Sun, R.; Lv, S.; Yang, G.; Yu, Z.; Kan, Q. Development and validation of a simple risk model to predict major cancers for patients with nonalcoholic fatty liver disease. *Cancer Med.* **2020**, *9*, 1254–1262. [CrossRef] [PubMed]

15. Simon, T.G.; Roelstraete, B.; Sharma, R.; Khalili, H.; Hagström, H.; Ludvigsson, J.F. Cancer Risk in Patients With Biopsy-ConfirMed. Nonalcoholic Fatty Liver Disease: A Population-Based Cohort Study. *Hepatology* **2021**, *74*, 2410–2423. [CrossRef] [PubMed]

16. Hong, C.; Yan, Y.; Su, L.; Chen, D.; Zhang, C. Development of a risk-stratification scoring system for predicting risk of breast cancer based on non-alcoholic fatty liver disease, non-alcoholic fatty pancreas disease, and uric acid. *Open Med. (Wars)* **2022**, *17*, 619–625. [CrossRef]

17. Nseir, W.; Abu-Rahmeh, Z.; Tsipis, A.; Mograbi, J.; Mahamid, M. Relationship between Non-Alcoholic Fatty Liver Disease and Breast Cancer. *Isr. Med. Assoc. J.* **2017**, *19*, 242–245.

18. Bilici, A.; Ozguroglu, M.; Mihmanli, I.; Turna, H.; Adaletli, I. A case-control study of non-alcoholic fatty liver disease in breast cancer. *Med. Oncol.* **2007**, *24*, 367–371. [CrossRef]

19. Yang, Y.J.; Kim, K.M.; An, J.H.; Lee, D.B.; Shim, J.H.; Lim, Y.S.; Lee, H.C.; Lee, Y.S.; Ahn, J.H.; Jung, K.H.; et al. Clinical significance of fatty liver disease induced by tamoxifen and toremifene in breast cancer patients. *Breast* **2016**, *28*, 67–72. [CrossRef]

20. Taroeno-Hariadi, K.W.; Putra, Y.R.; Choridah, L.; Widodo, I.; Hardianti, M.S.; Aryandono, T. Fatty Liver in Hormone Receptor-Positive Breast Cancer and Its Impact on Patient's Survival. *J. Breast Cancer* **2021**, *24*, 417–427. [CrossRef]

21. Zheng, Q.; Xu, F.; Nie, M.; Xia, W.; Qin, T.; Qin, G.; An, X.; Xue, C.; Peng, R.; Yuan, Z.; et al. Selective Estrogen Receptor Modulator-Associated Nonalcoholic Fatty Liver Disease Improved Survival in Patients With Breast Cancer: A Retrospective Cohort Analysis. *Medicine (Baltim.)* **2015**, *94*, e1718. [CrossRef] [PubMed]
22. Li, H.; Yu, L.; Zhang, X.; Shang, J.; Duan, X. Exploring the molecular mechanisms and shared gene signatuRes. between rheumatoid arthritis and diffuse large B cell lymphoma. *Front. Immunol.* **2022**, *13*, 1036239. [CrossRef] [PubMed]
23. Hu, Y.; Zeng, N.; Ge, Y.; Wang, D.; Qin, X.; Zhang, W.; Jiang, F.; Liu, Y. Identification of the Shared Gene SignatuRes. and Biological Mechanism in Type 2 Diabetes and Pancreatic Cancer. *Front. Endocrinol. (Lausanne)* **2022**, *13*, 847760. [CrossRef] [PubMed]
24. Chen, Y.; Ma, L.; Ge, Z.; Pan, Y.; Xie, L. Key Genes Associated With Non-Alcoholic Fatty Liver Disease and Polycystic Ovary Syndrome. *Front. Mol. Biosci.* **2022**, *9*, 888194. [CrossRef]
25. Cotter, T.G.; Rinella, M. Nonalcoholic Fatty Liver Disease 2020: The State of the Disease. *Gastroenterology* **2020**, *158*, 1851–1864. [CrossRef] [PubMed]
26. Wong, V.W.; Wong, G.L.; Tsang, S.W.; Fan, T.; Chu, W.C.; Woo, J.; Chan, A.W.; Choi, P.C.; Chim, A.M.; Lau, J.Y.; et al. High prevalence of colorectal neoplasm in patients with non-alcoholic steatohepatitis. *Gut* **2011**, *60*, 829–836. [CrossRef]
27. Craven, K.E.; Gökmen-Polar, Y.; Badve, S.S. CIBERSORT analysis of TCGA and METABRIC identifies subgroups with better outcomes in triple negative breast cancer. *Sci. Rep.* **2021**, *11*, 4691. [CrossRef]
28. Finotello, F.; Trajanoski, Z. Quantifying tumor-infiltrating immune cells from transcriptomics data. *Cancer Immunol. Immunother.* **2018**, *67*, 1031–1040. [CrossRef]
29. Bindea, G.; Mlecnik, B.; Tosolini, M.; Kirilovsky, A.; Waldner, M.; Obenauf, A.C.; Angell, H.; Fredriksen, T.; Lafontaine, L.; Berger, A.; et al. Spatiotemporal dynamics of intratumoral immune cells reveal the immune landscape in human cancer. *Immunity* **2013**, *39*, 782–795. [CrossRef]
30. Sanna, C.; Rosso, C.; Marietti, M.; Bugianesi, E. Non-Alcoholic Fatty Liver Disease and Extra-Hepatic Cancers. *Int. J. Mol. Sci.* **2016**, *17*, 717. [CrossRef] [PubMed]
31. Pan, H.J.; Chang, H.T.; Lee, C.H. Association between tamoxifen treatment and the development of different stages of nonalcoholic fatty liver disease among breast cancer patients. *J. Formos. Med. Assoc.* **2016**, *115*, 411–417. [CrossRef]
32. Yan, M.; Wang, J.; Xuan, Q.; Dong, T.; He, J.; Zhang, Q. The Relationship Between Tamoxifen-associated Nonalcoholic Fatty Liver Disease and the Prognosis of Patients With Early-stage Breast Cancer. *Clin. Breast Cancer* **2017**, *17*, 195–203. [CrossRef]
33. Wu, W.; Chen, J.; Ye, W.; Li, X.; Zhang, J. Fatty liver decreases the risk of liver metastasis in patients with breast cancer: A two-center cohort study. *Breast Cancer Res. Treat* **2017**, *166*, 289–297. [CrossRef] [PubMed]
34. Chinetti, G.; Fruchart, J.C.; Staels, B. Peroxisome proliferator-activated receptors (PPARs): Nuclear receptors at the crossroads between lipid metabolism and inflammation. *Inflamm. Res.* **2000**, *49*, 497–505. [CrossRef] [PubMed]
35. Michalik, L.; Wahli, W. PPARs Mediate Lipid Signaling in Inflammation and Cancer. *PPAR Res.* **2008**, *2008*, 134059. [CrossRef] [PubMed]
36. Zúñiga, J.; Cancino, M.; Medina, F.; Varela, P.; Vargas, R.; Tapia, G.; Videla, L.A.; Fernández, V. N-3 PUFA supplementation triggers PPAR-α activation and PPAR-α/NF-κB interaction: Anti-inflammatory implications in liver ischemia-reperfusion injury. *PLoS ONE* **2011**, *6*, e28502. [CrossRef]
37. Dana, N.; Haghjooy Javanmard, S.; Vaseghi, G. The effect of fenofibrate, a PPARα activator on toll-like receptor-4 signal transduction in melanoma both in vitro and in vivo. *Clin. Transl. Oncol.* **2020**, *22*, 486–494. [CrossRef] [PubMed]
38. Li, T.; Zhang, Q.; Zhang, J.; Yang, G.; Shao, Z.; Luo, J.; Fan, M.; Ni, C.; Wu, Z.; Hu, X. Fenofibrate induces apoptosis of triple-negative breast cancer cells via activation of NF-κB pathway. *BMC Cancer* **2014**, *14*, 96. [CrossRef]
39. Crosas-Molist, E.; Fabregat, I. Role of NADPH oxidases in the redox biology of liver fibrosis. *Redox Biol.* **2015**, *6*, 106–111. [CrossRef]
40. García-Ruiz, I.; Blanes Ruiz, N.; Rada, P.; Pardo, V.; Ruiz, L.; Blas-García, A.; Valdecantos, M.P.; Grau Sanz, M.; Solís Herruzo, J.A.; Valverde, Á.M. Protein tyrosine phosphatase 1b deficiency protects against hepatic fibrosis by modulating nadph oxidases. *Redox Biol.* **2019**, *26*, 101263. [CrossRef]
41. Hecker, L.; Vittal, R.; Jones, T.; Jagirdar, R.; Luckhardt, T.R.; Horowitz, J.C.; Pennathur, S.; Martinez, F.J.; Thannickal, V.J. NADPH oxidase-4 mediates myofibroblast activation and fibrogenic responses to lung injury. *Nat. Med.* **2009**, *15*, 1077–1081. [CrossRef]
42. Bondi, C.D.; Manickam, N.; Lee, D.Y.; Block, K.; Gorin, Y.; Abboud, H.E.; Barnes, J.L. NAD(P)H oxidase mediates TGF-beta1-induced activation of kidney myofibroblasts. *J. Am. Soc. Nephrol.* **2010**, *21*, 93–102. [CrossRef] [PubMed]
43. Panday, A.; Sahoo, M.K.; Osorio, D.; Batra, S. NADPH oxidases: An overview from structure to innate immunity-associated pathologies. *Cell Mol. Immunol.* **2015**, *12*, 5–23. [CrossRef] [PubMed]
44. Matuz-Mares, D.; Vázquez-Meza, H.; Vilchis-Landeros, M.M. NOX as a Therapeutic Target in Liver Disease. *Antioxidants* **2022**, *11*, 2038. [CrossRef]
45. Ma, Y.; Lee, G.; Heo, S.Y.; Roh, Y.S. Oxidative Stress Is a Key Modulator in the Development of Nonalcoholic Fatty Liver Disease. *Antioxidants* **2021**, *11*, 91. [CrossRef]
46. Bedard, K.; Krause, K.H. The NOX family of ROS-generating NADPH oxidases: Physiology and pathophysiology. *Physiol. Rev.* **2007**, *87*, 245–313. [CrossRef]
47. Arrington, M.E.; Temple, B.; Schaefer, A.; Campbell, S.L. The molecular basis for immune dysregulation by the hyperactivated E62K mutant of the GTPase RAC2. *J. Biol. Chem.* **2020**, *295*, 12130–12142. [CrossRef]

48. Watanabe, M.; Terasawa, M.; Miyano, K.; Yanagihara, T.; Uruno, T.; Sanematsu, F.; Nishikimi, A.; Côté, J.F.; Sumimoto, H.; Fukui, Y. DOCK2 and DOCK5 act additively in neutrophils to regulate chemotaxis, superoxide production, and extracellular trap formation. *J. Immunol.* **2014**, *193*, 5660–5667. [CrossRef] [PubMed]
49. Chen, Q.; Jun, H.; Yang, C.; Yang, F.; Xu, Y. The Pyroptosis-Related Risk Genes APOBEC3D, TNFRSF14, and RAC2 Were Used to Evaluate Prognosis and as Tumor Suppressor Genes in Breast Cancer. *J. Oncol.* **2022**, *2022*, 3625790. [CrossRef]
50. Abram, C.L.; Lowell, C.A. The ins and outs of leukocyte integrin signaling. *Annu. Rev. Immunol.* **2009**, *27*, 339–362. [CrossRef] [PubMed]
51. Rojas, K.; Baliu-Piqué, M.; Manzano, A.; Saiz-Ladera, C.; García-Barberán, V.; Cimas, F.J.; Pérez-Segura, P.; Pandiella, A.; Győrffy, B.; Ocana, A. In silico transcriptomic mapping of integrins and immune activation in Basal-like and HER2+ breast cancer. *Cell Oncol. (Dordr.)* **2021**, *44*, 569–580. [CrossRef]
52. Rose, D.M.; Liu, S.; Woodside, D.G.; Han, J.; Schlaepfer, D.D.; Ginsberg, M.H. Paxillin binding to the alpha 4 integrin subunit stimulates LFA-1 (integrin alpha L beta 2)-dependent T cell migration by augmenting the activation of focal adhesion kinase/proline-rich tyrosine kinase-2. *J. Immunol.* **2003**, *170*, 5912–5918. [CrossRef]
53. Harjunpää, H.; Llort Asens, M.; Guenther, C.; Fagerholm, S.C. Cell Adhesion Molecules and Their Roles and Regulation in the Immune and Tumor Microenvironment. *Front. Immunol.* **2019**, *10*, 1078. [CrossRef] [PubMed]
54. Jialal, I.; Adams-Huet, B.; Devaraj, S. Monocyte cell adhesion molecule receptors in nascent metabolic syndrome. *Clin. BioChem.* **2016**, *49*, 505–507. [CrossRef] [PubMed]
55. Winer, S.; Winer, D.A. The adaptive immune system as a fundamental regulator of adipose tissue inflammation and insulin resistance. *Immunol. Cell Biol.* **2012**, *90*, 755–762. [CrossRef] [PubMed]
56. Osborn, O.; Olefsky, J.M. The cellular and signaling networks linking the immune system and metabolism in disease. *Nat. Med.* **2012**, *18*, 363–374. [CrossRef] [PubMed]
57. Lumeng, C.N.; Bodzin, J.L.; Saltiel, A.R. Obesity induces a phenotypic switch in adipose tissue macrophage polarization. *J. Clin. Investig.* **2007**, *117*, 175–184. [CrossRef] [PubMed]
58. Xu, Z.; Zhang, X.; Lau, J.; Yu, J. C-X-C motif chemokine 10 in non-alcoholic steatohepatitis: Role as a pro-inflammatory factor and clinical implication. *Expert Rev. Mol. Med.* **2016**, *18*, e16. [CrossRef]
59. Zhang, X.; Han, J.; Man, K.; Li, X.; Du, J.; Chu, E.S.; Go, M.Y.; Sung, J.J.; Yu, J. CXC chemokine receptor 3 promotes steatohepatitis in mice through mediating inflammatory cytokines, macrophages and autophagy. *J. Hepatol.* **2016**, *64*, 160–170. [CrossRef]
60. Wu, X.; Sun, A.; Yu, W.; Hong, C.; Liu, Z. CXCL10 mediates breast cancer tamoxifen resistance and promotes estrogen-dependent and independent proliferation. *Mol. Cell Endocrinol.* **2020**, *512*, 110866. [CrossRef]
61. Zhou, H.; Wu, J.; Wang, T.; Zhang, X.; Liu, D. CXCL10/CXCR3 axis promotes the invasion of gastric cancer via PI3K/AKT pathway-dependent MMPs production. *BioMed. Pharmacother.* **2016**, *82*, 479–488. [CrossRef] [PubMed]
62. Fujita, M.; Zhu, X.; Ueda, R.; Sasaki, K.; Kohanbash, G.; Kastenhuber, E.R.; McDonald, H.A.; Gibson, G.A.; Watkins, S.C.; Muthuswamy, R.; et al. Effective immunotherapy against murine gliomas using type 1 polarizing dendritic cells—significant roles of CXCL10. *Cancer Res.* **2009**, *69*, 1587–1595. [CrossRef] [PubMed]
63. Sun, Y.; Mo, Y.; Jiang, S.; Shang, C.; Feng, Y.; Zeng, X. CXC chemokine ligand-10 promotes the accumulation of monocyte-like myeloid-derived suppressor cells by activating p38 MAPK signaling under tumor conditions. *Cancer Sci.* **2022**, *114*, 142–151. [CrossRef]
64. Mowat, C.; Mosley, S.R.; Namdar, A.; Schiller, D.; Baker, K. Anti-tumor immunity in mismatch repair-deficient colorectal cancers requiRes. type I IFN-driven CCL5 and CXCL10. *J. Exp. Med.* **2021**, *218*, e20210108. [CrossRef] [PubMed]
65. Reschke, R.; Gajewski, T.F. CXCL9 and CXCL10 bring the heat to tumors. *Sci. Immunol.* **2022**, *7*, eabq6509. [CrossRef]
66. Hoch, T.; Schulz, D.; Eling, N.; Gómez, J.M.; Levesque, M.P.; Bodenmiller, B. Multiplexed imaging mass cytometry of the chemokine milieus in melanoma characterizes featuRes. of the response to immunotherapy. *Sci. Immunol.* **2022**, *7*, eabk1692. [CrossRef]

Article

In Silico Analysis of Publicly Available Transcriptomic Data for the Identification of Triple-Negative Breast Cancer-Specific Biomarkers

Rachid Kaddoura, Fatma Alqutami, Mohamed Asbaita and Mahmood Hachim *

College of Medicine, Mohammed Bin Rashid University of Medicine and Health Sciences, Dubai P.O. Box 505055, United Arab Emirates
* Correspondence: mahmood.almashhadani@mbru.ac.ae

Abstract: Background: Breast cancer is the most common type of cancer among women and is classified into multiple subtypes. Triple-negative breast cancer (TNBC) is the most aggressive subtype, with high mortality rates and limited treatment options such as chemotherapy and radiation. Due to the heterogeneity and complexity of TNBC, there is a lack of reliable biomarkers that can be used to aid in the early diagnosis and prognosis of TNBC in a non-invasive screening method. Aim: This study aims to use in silico methods to identify potential biomarkers for TNBC screening and diagnosis, as well as potential therapeutic markers. Methods: Publicly available transcriptomic data of breast cancer patients published in the NCBI's GEO database were used in this analysis. Data were analyzed with the online tool GEO2R to identify differentially expressed genes (DEGs). Genes that were differentially expressed in more than 50% of the datasets were selected for further analysis. Metascape, Kaplan-Meier plotter, cBioPortal, and the online tool TIMER were used for functional pathway analysis to identify the biological role and functional pathways associated with these genes. Breast Cancer Gene-Expression Miner v4.7 was used to validify the obtained results in a larger cohort of datasets. Results: A total of 34 genes were identified as differentially expressed in more than half of the datasets. The DEG GATA3 had the highest degree of regulation, and it plays a role in regulating other genes. The estrogen-dependent pathway was the most enriched pathway, involving four crucial genes, including GATA3. The gene FOXA1 was consistently down-regulated in TNBC in all datasets. Conclusions: The shortlisted 34 DEGs will aid clinicians in diagnosing TNBC more accurately as well as developing targeted therapies to improve patient prognosis. In vitro and in vivo studies are further recommended to validate the results of the current study.

Keywords: triple-negative breast cancer; in silico analysis; differentially expressed gene; biomarkers; GATA 3; FOXA1; tumor microenvironment

Citation: Kaddoura, R.; Alqutami, F.; Asbaita, M.; Hachim, M. In Silico Analysis of Publicly Available Transcriptomic Data for the Identification of Triple-Negative Breast Cancer-Specific Biomarkers. *Life* **2023**, *13*, 422. https://doi.org/10.3390/life13020422

Academic Editors: Taobo Hu, Mengping Long, Lei Wang and Riccardo Autelli

Received: 2 January 2023
Revised: 28 January 2023
Accepted: 29 January 2023
Published: 2 February 2023

1. Introduction

Breast cancer is one of the most common types of cancer amongst women, with a very complex pathophysiology and 2.3 million newly identified cases globally in the year 2020, and a total of 7.8 million diagnoses by the end of that year [1]. Breast cancer is most commonly classified based on the molecular subtypes, which are dependent on the molecular profiles of the estrogen receptor (ER), progesterone receptor (PR), and human epidermal growth factor recpetor (HER2) [2]. Of the different molecular subtypes, the triple-negative breast cancer (TNBC) subtype is negative for all these receptors, accounts for 15–25% of the cases, and is considered to be the most aggressive subtype [3]. In the United States, TNBC has been found to yield a low five-year survival rate, of 8–16%, in comparison to the other molecular sybtypes [3].There has been a gradual increase in the incidence of breast cancer annually, with management of the disease being dependent on enhancing the outcome and survival of patients through early detection and diagnosis [4].

Current diagnostic procedures include imaging and immunohistochemistry, which aid in subtyping and classifying the disease for enhancing treatment options [5]. Recent technical developments in the transcriptomic and genomic profiling of tumors have shifted the traditional clinicopathological classification into an advanced classification based on subtyping, which demonstrated prognostic and therapeutic features [6]. Furthermore, the introduction of a minimally invasive procedures, such as liquid biopsies, can potentially increase the rate of early diagnosis as opposed to a more demanding and less appealing option—the solid biopsy. In a study that involved newly diagnosed patients, for example, the "predictive value" of plasma ddPCR using liquid biopsy for both primary EGFR mutation and KRAS mutation was 100 percent, meaning that patients who tested positive for either mutation carried said mutation in their tumor [7]. This screening accuracy, paired with the minimally invasive nature of liquid biopsies, could aid in introducing screening tests as a more common procedure, especially for situations such as of ruling out triple-negative breast cancers [7].

There is a clear distinction in the protein expression levels between the molecular subtypes of breast cancer—Luminal A, Luminal B, and HER2-enriched breast cancer—which is not present in TNBC. This lack of a precise molecular mechanism to explain TNBC limits treatment plans to the likes of chemotherapy, with an ambiguity in the levels of protein expression that are detrimental to TNBC diagnosis. Due to the disease complexity and heterogeneity, TNBC cannot be treated as a single entity, and there is no single biomarker that can be used for diagnosis, making it difficult for early recognition and prognosis [8,9]. To date, no clinical tools have been identified to easily assess whether the patient will respond to standard breast cancer treatment or have resistant de novo mutations in TNBC subtypes [9]. Therefore, there has been an increase in the drive to obtain reliable and accurate biomarkers to aid in the early detection and prognosis of TNBC, which is the motivation behind the conduction of this study.

Furthermore, in recent years, many immune cells have been found in the tumor microenvironment, each playing a different role. These different immune cells can be used as either biomarkers for tumor classification or potential therapeutic targets. For example, recruitment of tumor-associated macrophages is a potential target for tumor treatments in breast cancer [10]. Similarly, the proportion of immune cells in the tumor microenvironment can not only predict but also explain a patient's outcome and prognosis [11].

A comprehensive understanding of the molecular changes in TNBC might identify new players that can explain the pathogenesis and serve as potential and reliable markers, which is another incentive of this study. Output omics databases and patient datasets that are publicly available, as used in this study, are an excellent source for identifying such markers.

Breast cancer patients' expression profiles were re-analyzed after grouping them into TNBC and non-TNBC groups within their respective datasets. The aim of this study was to identify consistently differentially expressed genes (DEGs)—genes found in more than 50% of datasets—and their pathways, as well as potential patient impact. The shortlisted genes will aid clinicians in diagnosing TNBC more accurately as well as developing targeted therapies to improve patient prognosis.

2. Materials and Methods

2.1. Publicly Available Breast Cancer Transcriptomic Datasets

In order to identify consistently differentially expressed genes specific to TNBC compared to other types of breast cancer, we explored the publicly available transcriptomics data repository of the National Center for Biotechnology Information (NCBI), the Gene Expression Omnibus (GEO) (https://www.ncbi.nlm.nih.gov/geo/, accessed on 2 January 2023)—a genomic data repository—for datasets of patients with breast cancer. For consistency, we selected publicly available datasets which can be analyzed using GEO2R, a built-in platform within NCBI GEO, to carry out differential gene expression analysis on microarray data. This platform utilizes the computer language R and the limma statistical package to carry out various statistical calculations, such as the empirical Bayes statistics, to identify genes that are differentially expressed between different patient groups.

The inclusion criteria for the datasets were: human sample sources, data type was expression profiling by microarray, and datasets had breast cancer patients with TNBC patients included. A total of nine datasets (n = 1027; TNBC n = 207) were used for analysis (Table 1). Patients of each respective dataset were grouped into two groups: a TNBC group and non-TNBC group. Figure 1 illustrates a simplified flowchart of the re-analysis process.

Table 1. List of breast cancer studies used in this analysis from NCBI GEO database. TNBC: triple-negative breast cancer.

GEO Accession Number	Study Title	Samples	PMID
GSE2741	Breast Tumor's study	TNBC = 3 Non-TNBC = 8 Total samples = 11	16230372
GSE45255	Expression Profiles of Breast Tumors from Singapore and Europe	TNBC = 15 Non-TNBC = 124 Total samples = 139	23618380
GSE30682	Search for a gene-expression signature of breast cancer local recurrence in young women	TNBC = 58 Non-TNBC = 285 Total samples = 343	22271875
GSE36295	Transcriptomic analysis of breast cancer	TNBC = 11 Non-TNBC = 27 Total samples = 38	27177292
GSE19615	Integrated genomic and function characterization of the 8q22 gain	TNBC = 28 Non-TNBC = 87 Total samples = 115	20098429
GSE37751	Molecular Profiles of Human Breast Cancer and Their Association with Tumor Subtypes and Disease Prognosis (Affymetrix)	TNBC = 14 Non-TNBC = 47 Total samples = 61	30501643
GSE97177	Genome-wide multi-omics profiling reveals extensive genetic complexity in 8p11-p12 amplified breast carcinomas [expression]	TNBC = 9 Non-TNBC = 44 Total samples = 53	29844878
GSE18864	Tumor expression data from neoadjuvant trial of cisplatin monotherapy in triple-negative breast cancer patients	TNBC = 38 Non-TNBC = 46 Total samples = 84	20100965
GSE40115	Classifications within Molecular Subtypes Enables Identification of BRCA1/BRCA2 Mutation Carriers by RNA Tumor Profiling	TNBC = 31 Non-TNBC = 152 Total samples = 182	23704984

2.2. Identification of Differentially Expressed Genes

Each dataset was processed individually to identify DEGs using the GEO2R online tool (https://www.ncbi.nlm.nih.gov/geo/geo2r/, accessed on 1 October 2021). Samples were assigned to groups based on their subtype and analyzed using the standardized parameters of the tool. These standardized parameters include automated log2 transformation of non-transformed data, empirical Bayes method of calculation through the limma statistical package, and adjustment of p value using the default Benjamini and Hochberg (false discovery rate) method. p-value < 0.05 was used to indicate statistical significance.

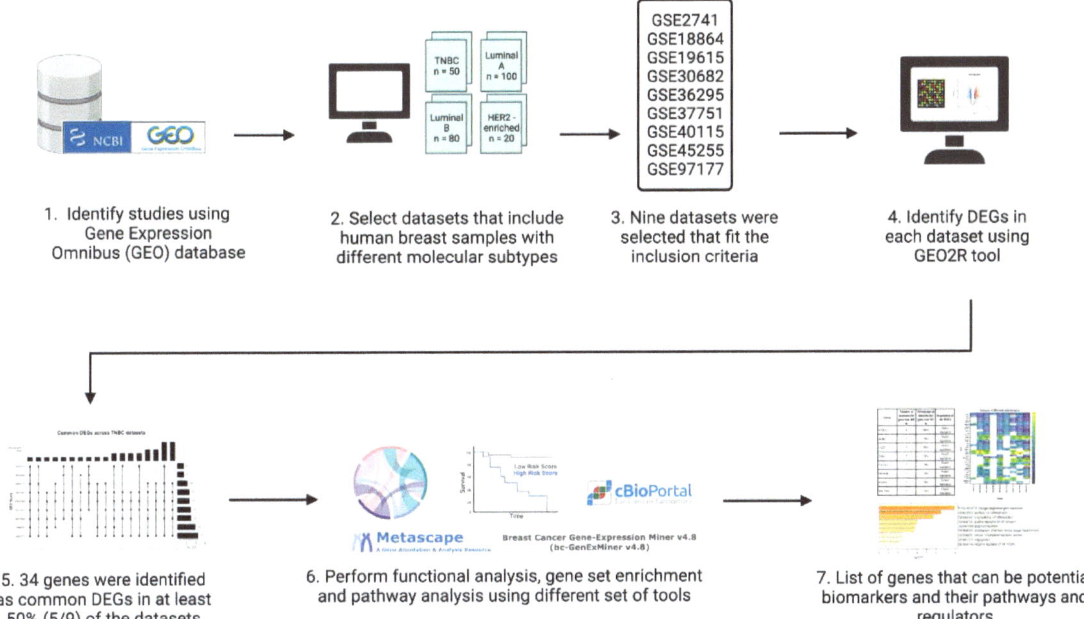

1. Identify studies using Gene Expression Omnibus (GEO) database

2. Select datasets that include human breast samples with different molecular subtypes

3. Nine datasets were selected that fit the inclusion criteria

4. Identify DEGs in each dataset using GEO2R tool

5. 34 genes were identified as common DEGs in at least 50% (5/9) of the datasets

6. Perform functional analysis, gene set enrichment and pathway analysis using different set of tools

7. List of genes that can be potential biomarkers and their pathways and regulators

Figure 1. Schematic representation and summary of the re-analysis process of the nine publicly available datasets retrieved from the GEO database. Publicly available data were identified from the NCBI GEO database and analyzed using the GEO2R online tool. The common differentially expressed genes (DEGs) in all datasets were identified and then further analyzed using Metascape, TIMER, cBioPortal, and Breast Cancer Gene-Expression Miner v4.7. Created with Biorender.com accessed on 1 October 2021.

To identify consistent DEGs across the nine datasets, all DEGs were intersected using the 'ComplexUpset' and 'UpSetR' libraries and functions in R studio (R version 4.2.2). Genes that were found to be common in more than 50% of the datasets (5/9 datasets or more) were selected for further analysis.

2.3. Gene Ontology and Pathway Analysis

The online database Metascape (http://metascape.org, accessed on 1 October 2021) was used to identify the biological role and functional pathways associated with the common DEGs. Metascape combines a variety of functions including gene annotation, functional enrichment, and membership in over 40 independent databases within a single integrated portal [12]. This tool was used to highlight the significance of the potential connectivity network of our genes and those needed for consideration in order to understand the full biological process [13]. Additionally, this tool streamlines different analysis types instead of searching each database individually. Outcomes include enriched pathways, top transcriptional factors, gene regulators, and protein–protein interactions.

2.4. Observing DEG Expression in Patients

To evaluate the expression of the identified DEGs in a clinically relevant cohort, several databases with patient genomic data were used to analyze these DEGs. Patients in these databases can be classified based on their tumor type and subtype. Such databases include the Breast Cancer Gene-Expression Miner v4.7 and the Kaplan-Meier plotter (https://doi.org/10.1016/j.csbj.2021.07.014, accessed on 1 October 2021) to identify the survival of patients based on the expression of selected DEGs.

The cancer genomic database cBioPortal (https://www.cbioportal.org/, accessed on 1 October 2021) was used to identify the survival of patients based on mutations in these common DEGs, as well as to visualize the expression of these genes. cBioPortal hosts multiple cancer databases and/or datasets, and for the basis of this analysis, the TCGA PanCancer atlas was used.

The webserver "TIMER", an inclusive reserve that analyzes immune infiltrates across various cancer types, was used to evaluate the diagnostic and prognostic value of those specific genes, as well as identify the top immune infiltrates in the breast cancer datasets in relation to these genes.

3. Results

Our search yielded nine datasets that met our criteria, with a total of 1350 patient samples across the nine datasets. The re-analysis of these datasets revealed a total of 1217 DEGs in all datasets (Supplementary Files S1–S9), 34 of which are consistent across five of the nine datasets (50%) (Figure 2, Table 2, Supplementary File S10), with these genes being either up- or down-regulated. Of the significant and common 34 genes, 26.4% of the genes (n = 9) were up-regulated, and the remaining 73.5% (n = 25) were down-regulated. FOXA1 was the only consistently down-regulated gene across all nine datasets. The log fold change of the 34 DEGs is represented in Figure 3.

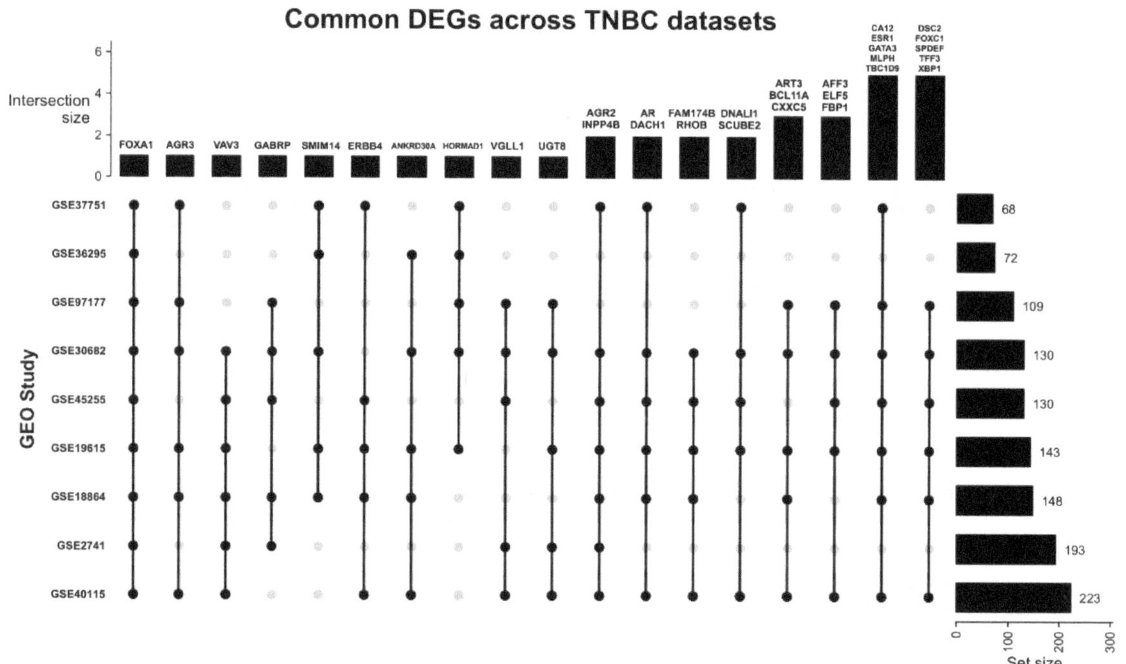

Figure 2. Upset plot illustrating the common DEGs across all nine datasets. Most datasets share at least one DEG with other datasets. FOXA1 is the only DEG that is present in all datasets, and the study GSE30682 has 33 of the 34 common DEGs.

Table 2. Differentially expressed genes present in 50% or more of the datasets analyzed. DEG: differentially expressed gene, DE: differentially expressed.

Genes	Gene Name	Number of Datasets the Gene Was DE in	Percentage of Datasets the Gene Was DE in	Regulation of the DEG in TNBC
FOXA1	Forkhead box A1	9	100%	Down-regulated
AGR2	Anterior gradient 2	7	78%	Down-regulated
CA12	Carbonic anhydrase 12	7	78%	Down-regulated
ESR1	Estrogen receptor 1	7	78%	Down-regulated
GATA3	GATA binding protein 3	7	78%	Down-regulated
INPP4B	Inositol polyphosphate-4-phosphatase type II B	7	78%	Down-regulated
MLPH	Melanophilin	7	78%	Down-regulated
TBC1D9	TBC1 domain family member 9	7	78%	Down-regulated
AGR3	Anterior gradient 3	6	67%	Down-regulated
AR	Androgen receptor	6	67%	Down-regulated
DACH1	Dachshund family transcription factor 1	6	67%	Down-regulated
DSC2	Desmocollin 2	6	67%	Up-regulated
FOXC1	Forkhead box C1	6	67%	Up-regulated
SPDEF	SAM pointed domain containing ETS transcription factor	6	67%	Down-regulated
TFF3	Trefoil factor 3	6	67%	Down-regulated
VAV3	Vav guanine nucleotide exchange factor 3	6	67%	Down-regulated
XBP1	X-box binding protein 1	6	67%	Down-regulated
DNALI1	Dynein axonemal light intermediate chain 1	6	67%	Down-regulated
VGLL1	Vestigial-like family member 1	5	56%	Up-regulated
GABRP	Gamma-aminobutyric acid type A receptor subunit pi	5	56%	Up-regulated
AFF3	ALF transcription elongation factor 3	5	56%	Down-regulated
ANKRD3OA	Ankyrin repeat domain 30A	5	56%	Down-regulated
ART3	ADP-ribosyltransferase 3 (inactive)	5	56%	Up-regulated
BCL11A	BCL11 transcription factor A	5	56%	Up-regulated
CXXC5	CXXC finger protein 5	5	56%	Down-regulated
ELF5	E74-like ETS transcription factor 5	5	56%	Up-regulated
ERBB4	Erb-b2 receptor tyrosine kinase 4	5	56%	Down-regulated
FAM174B	Family with sequence similarity 174 member B	5	56%	Down-regulated
FBP1	Fructose-bisphosphatase 1	5	56%	Down-regulated
RHOB	Ras homolog family member B	5	56%	Down-regulated
SCUBE2	Signal peptide, CUB domain and EGF-like domain containing 2	5	56%	Down-regulated
UGT8	UDP glycosyltransferase 8	5	56%	Up-regulated
HORMAD1	HORMA domain containing 1	5	56%	Up-regulated
SMIM14	Small integral membrane protein 14	5	56%	Down-regulated

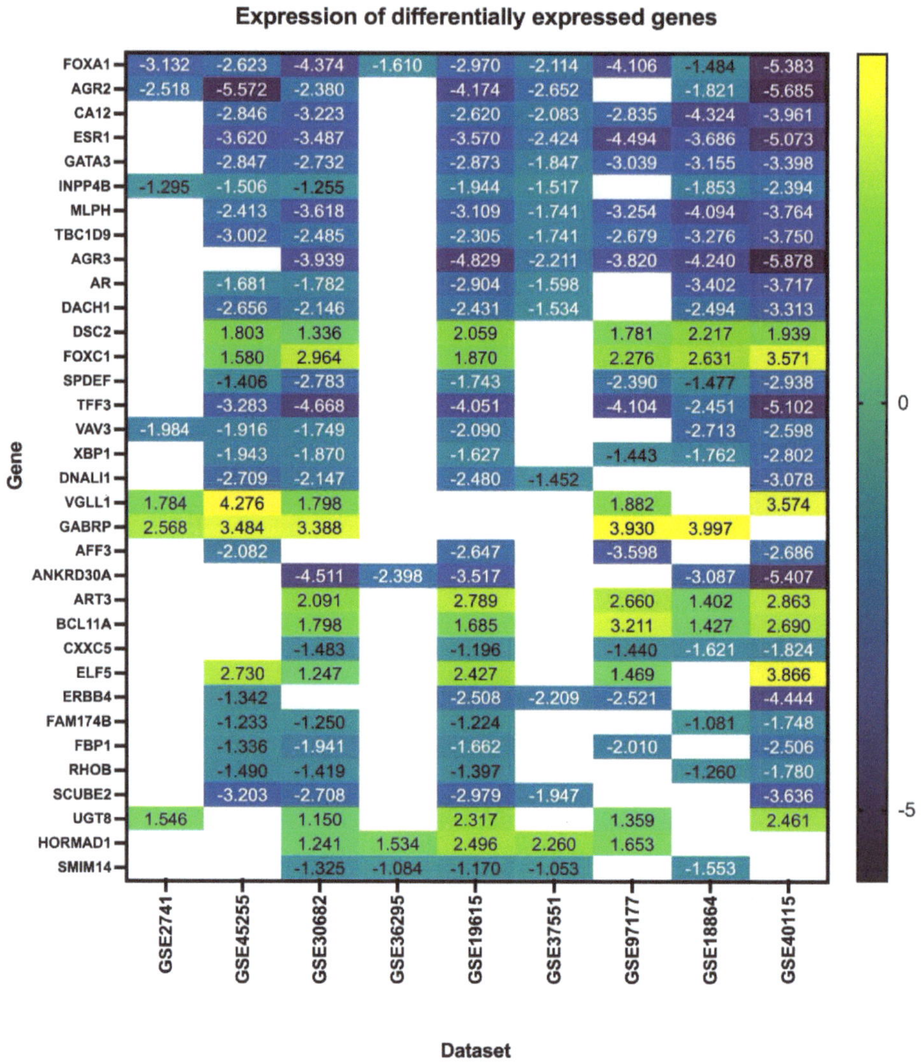

Expression of differentially expressed genes

Gene	GSE2741	GSE45255	GSE30682	GSE36295	GSE19615	GSE37551	GSE97177	GSE18864	GSE40115
FOXA1	-3.132	-2.623	-4.374	-1.610	-2.970	-2.114	-4.106	-1.484	-5.383
AGR2	-2.518	-5.572	-2.380		-4.174	-2.652		-1.821	-5.685
CA12		-2.846	-3.223		-2.620	-2.083	-2.835	-4.324	-3.961
ESR1		-3.620	-3.487		-3.570	-2.424	-4.494	-3.686	-5.073
GATA3		-2.847	-2.732		-2.873	-1.847	-3.039	-3.155	-3.398
INPP4B	-1.295	-1.506	-1.255		-1.944	-1.517		-1.853	-2.394
MLPH		-2.413	-3.618		-3.109	-1.741	-3.254	-4.094	-3.764
TBC1D9		-3.002	-2.485		-2.305	-1.741	-2.679	-3.276	-3.750
AGR3			-3.939		-4.829	-2.211	-3.820	-4.240	-5.878
AR		-1.681	-1.782		-2.904	-1.598		-3.402	-3.717
DACH1		-2.656	-2.146		-2.431	-1.534		-2.494	-3.313
DSC2		1.803	1.336		2.059		1.781	2.217	1.939
FOXC1		1.580	2.964		1.870		2.276	2.631	3.571
SPDEF		-1.406	-2.783		-1.743		-2.390	-1.477	-2.938
TFF3		-3.283	-4.668		-4.051		-4.104	-2.451	-5.102
VAV3	-1.984	-1.916	-1.749		-2.090			-2.713	-2.598
XBP1		-1.943	-1.870		-1.627		-1.443	-1.762	-2.802
DNALI1		-2.709	-2.147		-2.480	-1.452			-3.078
VGLL1	1.784	4.276	1.798				1.882		3.574
GABRP	2.568	3.484	3.388				3.930	3.997	
AFF3		-2.082			-2.647		-3.598		-2.686
ANKRD30A			-4.511	-2.398	-3.517			-3.087	-5.407
ART3			2.091		2.789		2.660	1.402	2.863
BCL11A			1.798		1.685		3.211	1.427	2.690
CXXC5			-1.483		-1.196		-1.440	-1.621	-1.824
ELF5		2.730	1.247		2.427		1.469		3.866
ERBB4		-1.342			-2.508	-2.209	-2.521		-4.444
FAM174B		-1.233	-1.250		-1.224			-1.081	-1.748
FBP1		-1.336	-1.941		-1.662		-2.010		-2.506
RHOB		-1.490	-1.419		-1.397			-1.260	-1.780
SCUBE2		-3.203	-2.708		-2.979	-1.947			-3.636
UGT8	1.546		1.150		2.317		1.359		2.461
HORMAD1			1.241	1.534	2.496	2.260	1.653		
SMIM14			-1.325	-1.084	-1.170	-1.053		-1.553	

Gene (y-axis) / **Dataset** (x-axis)

Figure 3. Log fold change of the differentially expressed genes in triple–negative breast cancer across nine datasets.

3.1. Survival Rates of TNBC Patients Are Affected by FOXA1 Expression

A Kaplan–Meier plot was used to test the FOXA1 regulation effect on the survival rate of all breast cancer patients (n = 2976). Patients with high and low FOXA1 expression were compared at a follow-up threshold of five years. Figure 4a reveals that high expression of FOXA1 is associated with a worse prognosis. This low prognosis was consistent when each subtype of breast cancer—TNBC (n = 126), ER/PR–positive (n = 2005), and HER2–positive (n = 30)—was analyzed individually, as shown in Figure 4b–d. Furthermore, another analysis of the TCGA breast cancer dataset from cBioPortal showed that patients with a mutated FOXA1 had a lower survival rate than those without a FOXA1 mutation (Figure 5). However, it is of importance that only one TNBC patient had a mutated FOXA1 in this dataset.

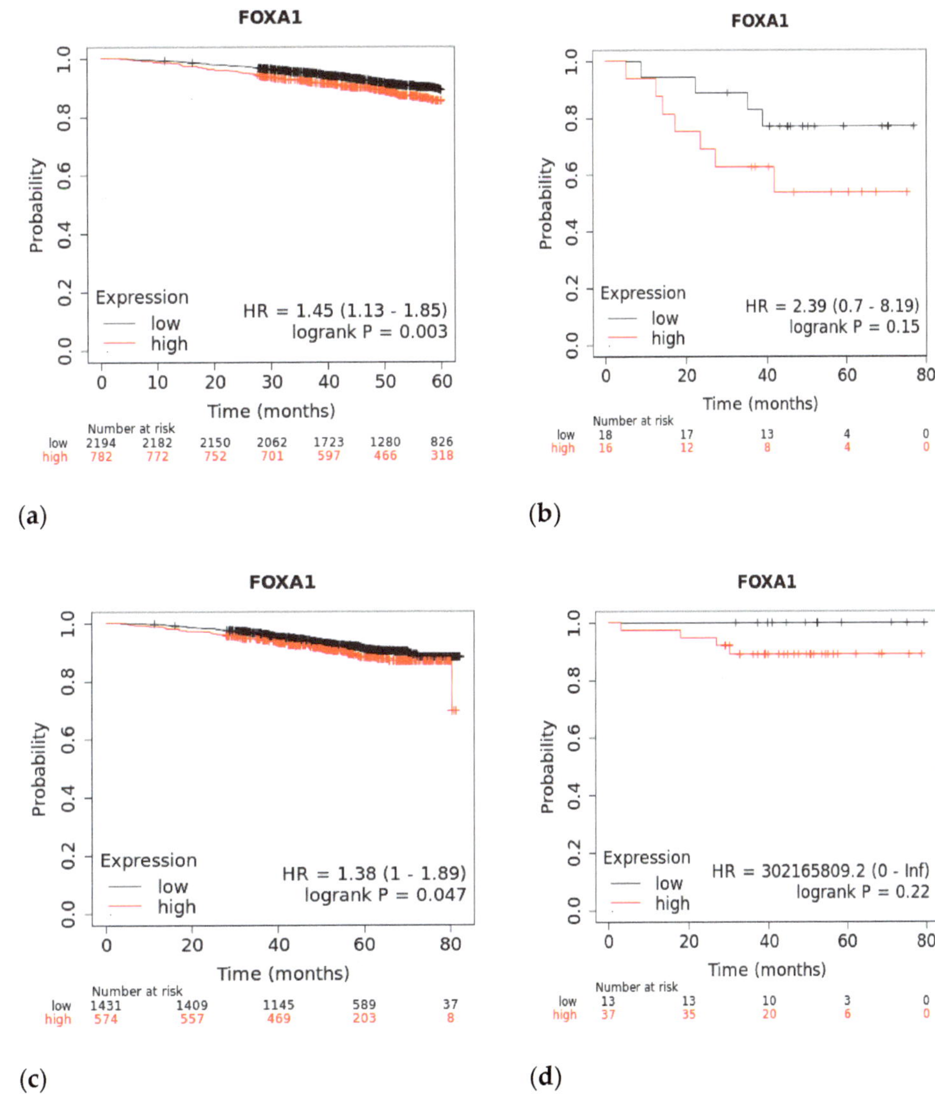

Figure 4. The effect of FOXA1 expression on survival rates. (**a**) Overall survival rate of high vs. low FOXA1 expression in BC patients for the first 60 months since diagnosis; those with high expression rates had lower survival rates. (**b**) Overall survival rate of high vs. low FOXA1 expression in TNBC patients. (**c**) Overall survival rate of high vs. low FOXA1 expression in ER+ and PR+ BC patients. (**d**) Overall survival rate of high vs. low FOXA1 expression in HER2+ BC patients. Created by Breast Cancer Gene-Expression Miner v4.7.

(a)   **(b)**

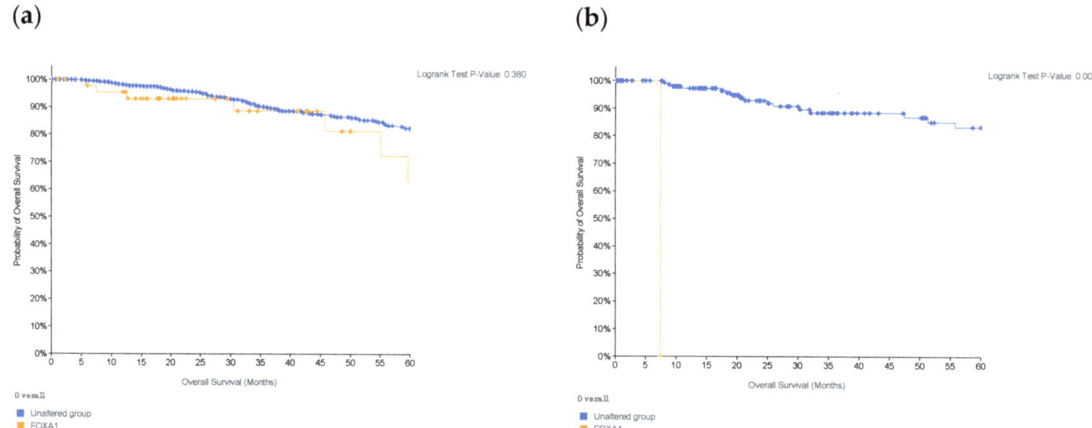

Figure 5. Survival of patients with FOXA1 mutations in (**a**) breast cancer, and in (**b**) patients who are classified as TNBC. Survival decreases in the presence of mutations in the gene; however, in TNBC patients, only one had the mutation and, therefore, this is not of statistical significance. Created by cBioPortal.

3.2. Functional Analysis of the Common DEGs Reveal the Involvement of Estrogen-Dependent Gene Expression Pathway and Related Genes

The functional pathway analysis of the 34 common DEGs performed via Metascape provided the pathways these genes were associated with (Figure 6). The most enriched pathway is that of estrogen-dependent gene expression followed by epithelial cell differentiation. Furthermore, the majority of these genes appear to be regulated by the transcription factor interferon regulatory factor 1 (IRF-1), as illustrated in Figure 7.

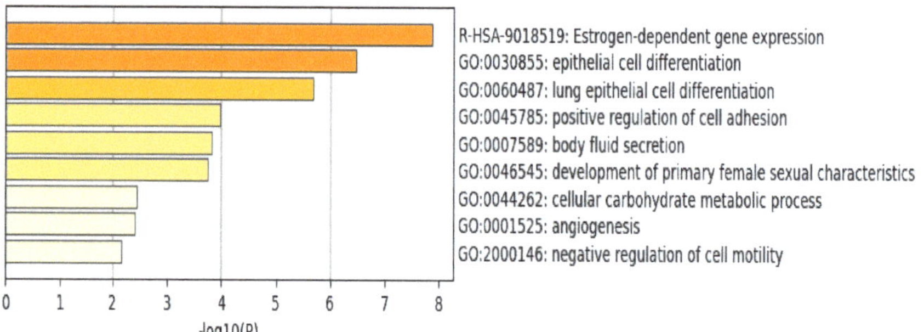

Figure 6. Enriched pathways of the 34 common DEGs using Metascape reveal that estrogen–dependent gene expression is the most significant and enriched pathway. Created by Metascape.

The Molecular Complex Detection (MCODE) algorithm was utilized to identify densely connected network components. Four genes (GATA3, FOXA1, TFF3, and ESR1) were found to be involved in protein–protein interactions (Figure 8a). Extended enrichment analysis showed that most of the genes were regulated primarily by GATA3, as shown in Figure 8b.

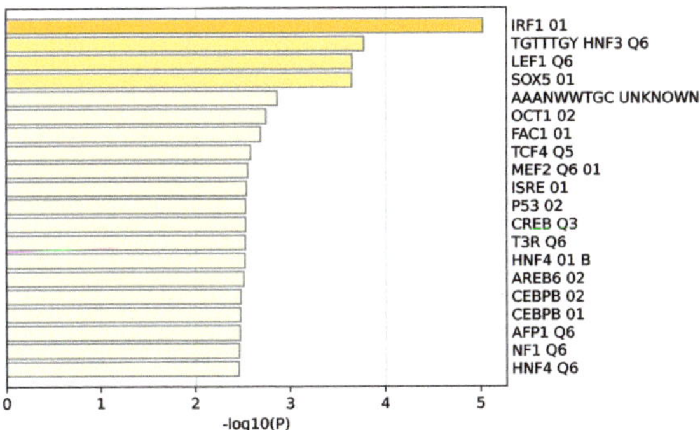

Figure 7. Enrichment analysis of the 34 common DEGS reveal 20 different transcriptomic target factors, with IRF1 being the most common target of these genes. Created by Metascape.

(a) **(b)**

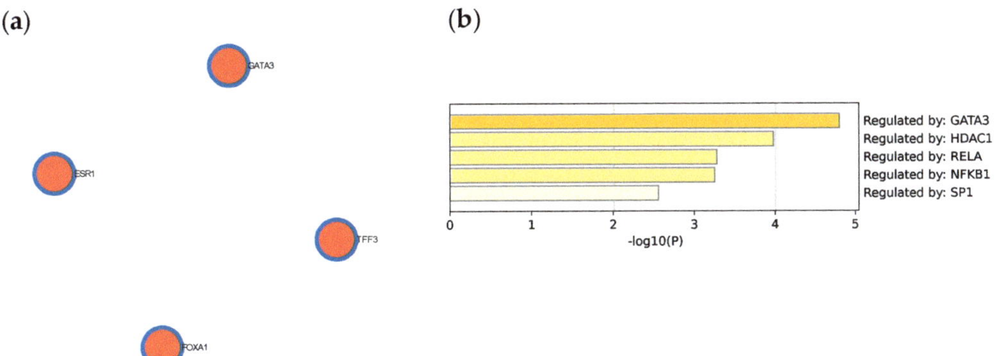

Figure 8. (**a**) Protein–protein interaction network and MCODE components identified in the TNBC significant genes list. (**b**) Summary of enrichment analysis according to gene regulatory functions of the 34 identified DEGs. Created by Metascape.

3.3. GATA3 Is Down-Regulated in TNBC Which Leads to Poor Survival

To evaluate the significance of GATA3 expression in terms of diagnostic and prognostic values across the four subtypes of breast cancer in a larger number of patient datasets, Breast Cancer Gene-Expression Miner v4.7 was used. Our analysis revealed a significant decrease in the expression of GATA3 ($p < 0.0001$) in TNBC patients in comparison to non-TNBC patients, as illustrated in Figure 9a. Furthermore, GATA3 expression and nodal involvement in breast cancer were not correlated with each other ($p = 0.4288$), as illustrated in Figure 9b. TNBC with low GATA3 mRNA expression also had a lower distant metastasis-free survival rate as well as a decreased overall survival rate, as shown in Figure 9c,d, respectively. Furthermore, TCGA patient data show that GATA3 is only mutated in 14% of the TNBC patients, with most of these mutations being amplification, while in non-TNBC patients, GATA3 mutations occur in 16% of patients and there are different types of mutations in these patients such as in-frame mutations, splice mutations, and truncating mutations. Consequently, GATA3 expression is higher in non-TNBC patients, followed by TNBC patients with GATA3 amplification and TNBC patients without any GATA3 mutations.

3.4. Three Immune Cell Types Are Found in the Tumor Site

Immune cell infiltrates analyzed with the web server "TIMER" reveal that there are three immune cell populations that are particularly involved with GATA3 expression in BRCA-Basal breast cancer are myeloid dendritic cells, neutrophils, and macrophages, as illustrated in Table 3. These immune infiltrates play a role in the innate immune response. The presence of antigen-presenting cells such as macrophages and dendritic cells in the immune microenvironment of the tumor plays a role in tumor progression.

Table 3. Immune cell involvement in basal-like breast cancer.

Cancer	Infiltrates	*p*-Value	Adjusted *p*-Value
BRCA-Basal (n = 191)	Myeloid dendritic cell activated	0.00151881	0.005368
BRCA-Basal (n = 191)	Neutrophil	0.019378582	0.049857
BRCA-Basal (n = 191)	Macrophage	0.006514787	0.019388

(**a**)

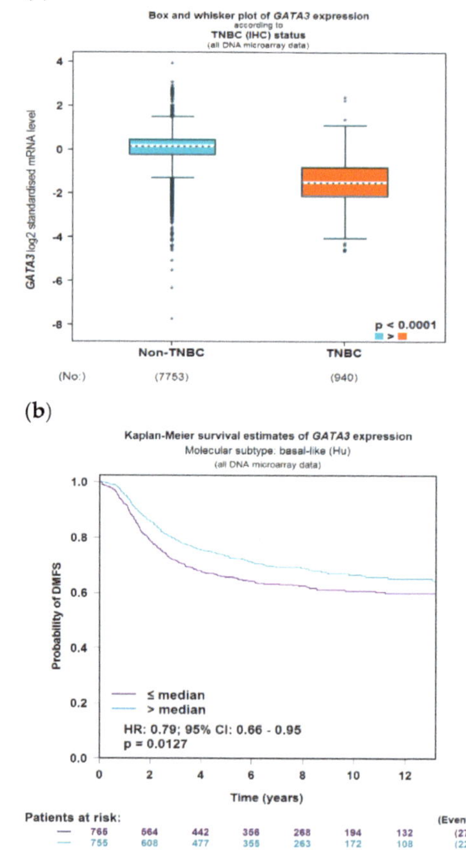

(**b**)

Figure 9. *Cont.*

(c)

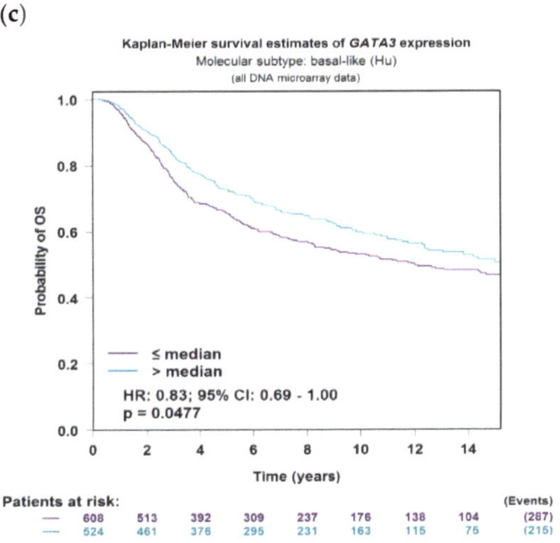

(d)

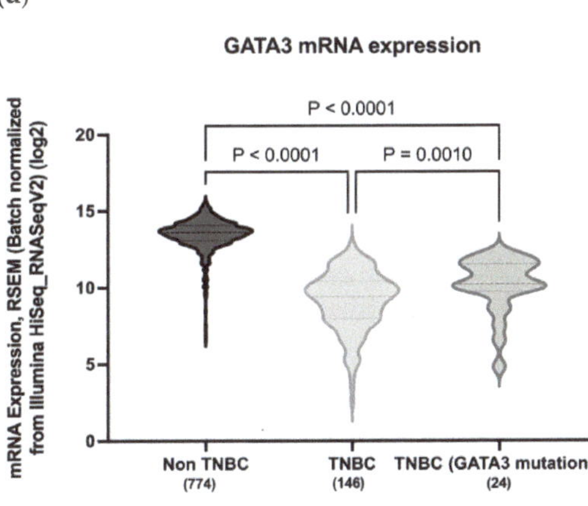

Figure 9. GATA3 expression between breast cancer subtypes. (**a**) GATA3 expression in TNBC vs. non-TNBC. (**b**) Distant metastasis-free survival (DMFS) in TNBC according to GATA3 expression. (**c**) Overall survival in TNBC according to GATA3 expression. (**d**) GATA3 expression in the Cancer Genome Atlas (TCGA) dataset patients obtained from cBioPortal, showing expression in non-TNBC patients, TNBC patients, and TNBC patients with a GATA3 mutation. Created by Breast Cancer Gene-Expression Miner v4.7.

4. Discussions

4.1. Most DEGs Are Down-Regulated in TNBC Which Can Be Attributed to Poor Prognosis

Within the nine TNBC datasets, there were 34 genes that were consistently differentially expressed, with the majority of these genes being down-regulated, and only nine genes being up-regulated. FOXA1 is the only gene that was down-regulated in all datasets analyzed and has been linked with poor prognosis.

Many of these genes have different roles and functions in breast cancer that affect tumor survival and response to therapy. The R-HSA-9018519 estrogen-dependent gene expression pathway involves the following DEGs: ERBB4, ESR1, GATA3, FOXA1, TFF3, CXXC5, ELF5, and VAV3, with four of these genes being involved in protein–protein interactions and regulated by GATA3. Furthermore, this pathway includes the reaction 'R-HSA-9018494: FOXA1 and GATA3 bind TFF genes' that utilizes both FOXA1 and GATA3 to aid oncogenesis and metastasis [14]. The most enriched transcription factor target is IRF1, a transcription regulator and tumor suppressor, which activates genes in both innate and acquired immune responses.

4.2. FOXA1 Can Increase Malignancy in Breast Cancer

FOXA1 is a transcriptional factor that plays an important role in hormone signaling in both breast cancer and normal breast tissues [15]. It has been shown that low expression of FOXA1 can increase malignancy and cancer stemness [15]. This gene has been used as one of many subtyping markers in the identification of triple-negative breast cancers [15]. Knock down studies have shown that reduction or deletion of FOXA1 decreases apoptosis and accelerates cell proliferation, which can explain the aggressive nature of TNBC and its worsening prognosis [15]. Furthermore, it has been shown that loss of FOXA1 expression is associated with worse survival and increased expression is an indicator of good prognosis [16,17]. Our re-analysis (Figure 4a) links increased expresion with lower survival rates in the first 60 months for breast cancer patients, which contradicts previous findings. However, when comparing the high and low expression of FOXA1 in TNBC patients, there is no statistical difference. On the other hand, the difference in FOXA1 expression is statistically significant in ER+/PR+ breast cancer. This indicates that FOXA1 might have a different mechanism in TNBCs compared to other sutypes.

4.3. GATA3 Is a Major Transcription Factor That Is Found in Many Breast Cancer Subtypes

GATA3 is a transcription factor that is involved in the embryonic development of different types of tissues as well as in inflammatory and humoral immune responses. It is a potent regulator of the tumor microenvironment and plays a role in the proper functioning of the endothelium layer in various types of blood vessels. GATA3 has been proven to be affected in multiple breast cancer subtypes, such as its high expression in the Luminal A subtype due to its strong association with estrogen receptor expression [18]. As seen in our analysis, GATA3 also regulates other DEGs that are affected in TNBC. Due to the strong link between GATA3 and ER expression, high GATA3 levels observed in immunohistochemistry can be used as a positive prognostic method and are linked with favorable pathological features such as positive ER status [19]. This is in line with our results (Figure 9), which link low GATA3 expression with a lower overall survival and distant metastasis-free rate. On the other hand, the lack of an ER receptor in TNBC also reduces GATA3, which is reflected in immunhistochemistry staining sensitivity and is linked with a worse prognosis, distant metastasis-free survival rate, and overall survival [20]. A study carried out in 205 TNBC samples that were divided into five molecular subtypes showed that GATA3 is categorized with a negative stain score (staining intensity x proportion) in 74.6% of all samples [21]. On the other hand, the rate of focal positivity was significantly higher in one of the molecular apocrine subtypes, at 73.9% [21].

However, GATA3 can still stain positive in TNBC, and can be useful when used as a diagnostic and prognostic measuring tool when characterizing metastatic tumors of unknown origin, which is also demonstrated in Figure 9 [22]. This has been supported with a systematic evaluation conducted by Ashley et al., which demonstrated a 44% stain positivity rate of GATA3 across 44 TNBC patients at a staining threshold of 5% [23], and another study revealing a 66% positivity rate when the staining threshold was 1% [24].

Furthermore, mRNA expression of TCGA patients reveals that non-TNBC patients had higher GATA3 expression compared to TNBC patients (Figure 9d). In TNBC patients, those with mutations in their GATA3 had a significantly higher expression ($p = 0.001$) than those without any mutation in their GATA3. These TNBC patients with the GATA3 mutation—all of which are amplification—had a higher survival rate than those without the mutation, which supports the idea that increased GATA3 expression increases survival.

4.4. Estrogen-Dependent Gene Expression Plays a Vital Role in Breast Cancer

The R-HSA-9018519 estrogen-dependent gene expression pathway has been shown to be involved in most of the 34 DEGs identified in this study (Figure 6). This is consistent with results from a study published by Treeck O et al., highlighting the effect of estrogen on TNBC, a breast cancer subtype that does not express the estrogen receptor, and yet plays a vital role in pathogenesis [25]. ERα—estrogen receptor alpha—is a major driver of about 70% of breast cancers, with TNBCs being responsive to ERα-independent pathways, which are involved in pathogenesis. A study using a TNBC experimental metastasis model comparing ovariectomy and estrogen supplementation showed that ovariectomy is 56% more efficient in decreasing the frequency of brain metastasis [25]. In addition to ovariectomy, the aromatase inhibitor letrozole reduced the frequency of large lesions by 14.4% in the estrogen control [25]. Another study demonstrated that elevated levels of circulating estrogens were enough to stimulate the development and progression of ERα-negative cancers [26].

The mentioned studies demonstrate that estrogen can act on cells that are distinct from cancer cells, promoting angiogenesis via a systemic pathway by promoting mobilization and recruitment of bone marrow stromal-derived cells in TNBC. This, along with our results (Figure 6), illustrates how these DEGs can serve as biomarkers for TNBC, and how estrogen plays a pivotal role in the pathophysiology of TNBC regardless of the estrogen receptor expression.

4.5. GATA3, ESR1, TFF3, FOXA1 Interaction

Of the genes involved in the estrogen-dependent pathway, four genes have been found to be involved in protein–protein interactions. These genes are GATA3, ESR1, TFF3, and FOXA1. ESR1, also known as NR3A1 (nuclear receptor subfamily 3, group A, member 1), is one of the two main types of estrogen receptors. TFF3 is a secretory protein that has multiple and diverse functions such as protection of the mucosa, thickening of the mucosa, and increasing epithelial healing rates [27]. TFF3 has not been well defined, yet closely resmbles the gene TFF1. Some of the properties of TFF1 include inhibition of cell growth, colony formation, and migration and invasion of breast cancer cells in vitro [28].

A study showed that FOXA1 enhances the response to estrogen due to its regulatory properties on the ER binding of the promoter region of its targets [29,30]. The expression of FOXA1 is regulated by GATA3, which, in turn, enhances the expression of the estrogen receptor in epithelial cells [31]. Therefore, if one gene is down-regulated, it will negatively influence genes downstream, which is seen in this study where GATA3 is down-regulated and, therefore, the genes affected by it are also down-regulated. This is indicated in cancer cells that have GATA3 depletion, where there is decreased ESR1-binding affinity, which, in turn, decreases the expression of FOXA1 [32].

While not being involved in the same pathway, TFF1 mRNA expression was correlated with that of FOXA1, GATA3, ESR1, XBP1, and MYB. Additionally, breast cancer patients with a positive ER expressed TFF1 higher than those who were negative for ER [33]. This shows a correlation between TFF1 and the status of estrogen receptor, as seen in the down-regulated TFF3 in TNBC patients. While possessing many genetic interactions, the novel variant TFF3 remains understudied, and further evaluation is required to validate its role and effect in TNBC.

4.6. IRF1 Is a Major Transcriptional Factor Target

IRF1 is a transcriptional factor regulator and tumor suppressor that involves immune responses to pathogens such as bacteria and viruses, as well as playing a role in cell proliferation and DNA damage response. This protein represses the transcription of other genes such as by regulating the transcription of INF and INF-induced genes (provided by RefSeq, August 2017). Many of the DEGs identified in this study appear to be associated with this transcription factor.

IRF1 exerts an anti-oncogenic and anti-proliferative effect by its ability to induce the expression of cell growth down-regulatory target genes [34]. Genes targeted by IRF1 include protein kinase R (PKR) and signal transducer and activator of transcription (STAT) and (STAT1) in the Janus kinase (JAK)-STAT pathway. The JAK-STAT pathway signaling is a pathway whose dysregulated activation is known and recorded in many types of tumors and is being studied as a molecular target for cancer therapeutics [35].

The enrichment analysis carried out in this study has shown that the DEGs identified in TNBC are linked to IRF1, a protein that has been considered a potential diagnostic and prognostic biomarker for recurrence-free survival in patients with colorectal cancer by some studies [36]. Therefore, not only can IRF1 be used as a biomarker, but these DEGs can be used as well.

4.7. Immune Cell Involvement in Triple-Negative Breast Cancer Can Lead to Better or Worse Prognosis

The tumor microenvironment is an important and variable aspect in the progression of breast cancer. Both the innate and adaptive immune systems with a variety of immune cells are involved in breast cancer [37].

In this study, 'TIMER' identified three types of immune cells that were involved in the pathogenesis of TNBC based on the gene expression of GATA3: myeloid dendritic cells, neutrophils, and macrophages. Myeloid dendritic cells were shown to be the most significant immune cell infiltrate. Dendritic cells are a major part of innate immunity and are linked to adaptive immunity through their antigen-presenting properties [38]. A study conducted by Gabrilovich et al. demonstrated the presence of a defect in the dendritic cells of cancer patients, citing that these cells were not effectively presenting antigens [39].

Furthermore, macrophages appear to be significantly involved in basal-like breast cancer based on the TIMER analysis. Macrophages, like dendtritic cells, are antigen-presenting cells and an important part of the innate system. There are two subtypes of macrophages, M1 and M2, which exhibit both inflammatory and anti-inflammatory properties [40]. Breast cancer polarizes macrophages to the M2 form, which is the subtype that promotes tumor growth and cell proliferation [40]. Tumors displaying this subtype of macrophages are often associated with unfavorable prognosis, and favoring features such as nodal involvement and metastasis [41]. Our results demonstrate three immune cell populations involved in breast cancer in relation to GATA3 expression, and the importance of this gene towards the immune microenvironment. Similarly, a study conducted by Dieci M et al. demonstrates the importance of immune infiltrations' involvement in breast cancer and the possibility of their use as potential biomarkers [38].

4.8. Clinical Implications

Significant key information about tumors can be obtained from the identified biomarkers, especially as a prognostic tool. Patient prognosis can be evaluated according to the biomarkers present in tumor DNA, which is an advisable screening option due to the lack of well-defined molecular targets that make cytotoxic chemotherapy the only treatment option for TNBC patients [42]. In order to avoid such harsh treatment plans, understanding genetic biomarkers could provide a platform for new diagnostic and therapeutic options specifically designed to target TNBC, with the expression of selected markers being used as identifiers for the ideal course of treatment and response to such treatments. For example, GATA3 can be used to evaluate response to hormonal treatments targeting the estrogen receptor pathways.

Another implication is that four of the genes identified in this study—AGR2, AGR3, TFF3, and SCUBE2—have protein products that are secreted in the blood by breast cancer. This can lead to the use of non-invasive methods such as blood tests for preliminary diagnosis before tissue biopsies, which can lead to more tests being conducted and earlier detection. Another benefit of using blood tests can include increased testing for more robust monitoring of the disease, such as taking a test before and after treatments.

4.9. Strengths and Limitations

This study was carried out using in silico methods and tools, which revolve around the use of publicly available transcriptomic data to mimic in vitro studies. A benefit to in silico analysis is the data accessibility and cost-effectiveness of this method, with a large body of patient data and databases available. This approach could reduce the time for the conceptualization of a hypothesis before going into in vitro testing, as well as identifying targets for in vitro testing and validation. Furthermore, this in silico study helped create a shortlist of potential genes that are involved in TNBC, which can be studied further and used to generate different hypotheses.

However, these in silico studies need in vitro validation to confirm any final conclusions. Furthermore, in silico data can result in contradicting results due to the varying patient types and accompanying clinical information in each dataset. This has been observed in some of our data, as cBioPortal has only one TNBC patient with a FOXA1 mutation, compared to Breast Cancer Gene-Expression Miner v4.7, which has several TNBC patients with high or low expression. Therefore, the use of one database is not enough and several tools need to be used, as we have in our re-analysis.

5. Conclusions

In conclusion, our study identified 34 DEGs in TNBC compared to the other subtypes of breast cancer. The generated shortlisted genes could be used in clinical settings as biomarkers to detect TNBC at an early stage and improve the overall prognosis of the patient as well as aiding in their treatment course. This in silico analysis study demonstrated the various physiological effects of the genes involved in TNBC such as estrogen-dependant pathways, which provide possible alternative targeted treatment options as compared to the standard non-specicific options currently available. Given that this study is an in silico analysis, we had access to a limited number of patients in certain subtypes of disease, and need further validation. We believe that our findings could provide advancements in the field of TNBC, and we encourage future in vitro and in vivo studies to further solidify the validity of these results.

Supplementary Materials: The following supporting information can be downloaded at: https://www.mdpi.com/article/10.3390/life13020422/s1. Supplementary File S1: GSE2741; Supplementary File S2: GSE18864; Supplementary File S3: GSE19615; Supplementary File S4: GSE30682; Supplementary File S5: GSE36295; Supplementary File S6: GSE37751; Supplementary File S7: GSE40115; Supplementary File S8: GSE45255; Supplementary File S9: GSE97177; Supplementary File S10: Common DEGs.

Author Contributions: R.K. and M.H. assisted with the study design, had full access to the study data, and assume responsibility for the integrity of the data and the accuracy of the analysis. M.A. and F.A. assisted with the interpretation of the data. R.K. and M.A. wrote the first draft, and M.H. and F.A. were responsible for the final editing of the manuscript. All authors have read and agreed to the published version of the manuscript.

Funding: This research received no external funding.

Institutional Review Board Statement: Not applicable.

Informed Consent Statement: Not applicable.

Data Availability Statement: The datasets supporting the conclusion of this article are available in the NCBI repository, (https://www.ncbi.nlm.nih.gov/geo/, accessed on 1 October 2021), with their accession codes in Table 1.

Acknowledgments: We would like to acknowledge the generous support of the Mohammed bin Rashid University of Medicine and Health Sciences (MBRU). We acknowledge Hanan Faraji for proofreading the manuscript.

Conflicts of Interest: The authors declare that the research was conducted in the absence of any commercial or financial relationships that could be construed as a potential conflict of interest.

Abbreviations

MBRU	Mohammed Bin Rashid University of Medicine and Health Sciences
ER	estrogen receptor
PR	progesterone receptor
HER2	human epidermal growth factor receptor 2
NCBI	National Center for Biotechnology Information
GEO	Gene Expression Omnibus
IRF-1	interferon regulatory factor 1
MCODE	Molecular Complex Detection
DMFS	distant metastasis-free survival
OS	overall survival
PKR	protein kinase R
STAT	signal transducer and activator of transcription
JAK	Janus kinase
NR3A1	nuclear subfamily 3, group A, member 1
HNF-3A	hepatocyte nuclear factor 3-alpha

References

1. Wild, W.E.; Stewart, B.W. (Eds.) *World Cancer Report: Cancer Research for Cancer Prevention*; International Agency for Research on Cancer: Lyon, France, 2020.
2. Eliyatkın, N.; Yalçın, E.; Zengel, B.; Aktaş, S.; Vardar, E. Molecular Classification of Breast Carcinoma: From Traditional, Old-Fashioned Way to A New Age, and A New Way. *J. Breast Health* **2015**, *11*, 59–66. [CrossRef]
3. Almansour, N.M. Triple-Negative Breast Cancer: A Brief Review About Epidemiology, Risk Factors, Signaling Pathways, Treatment and Role of Artificial Intelligence. *Front. Mol. Biosci.* **2022**, *9*, 836417. [CrossRef]
4. Ramadan, W.S.; Talaat, I.M.; Hachim, M.Y.; Lischka, A.; Gemoll, T.; El-Awady, R. The impact of CBP expression in estrogen receptor-positive breast cancer. *Clin. Epigenetics* **2021**, *13*, 72. [CrossRef]
5. Dass, S.A.; Tan, K.L.; Selva Rajan, R.; Mokhtar, N.F.; Mohd Adzmi, E.R.; Wan Abdul Rahman, W.F.; Tengku Din, T.; Balakrishnan, V. Triple Negative Breast Cancer: A Review of Present and Future Diagnostic Modalities. *Medicina* **2021**, *57*, 62. [CrossRef]
6. Hachim, M.Y.; Hachim, I.Y.; Talaat, I.M.; Yakout, N.M.; Hamoudi, R. M1 Polarization Markers Are Upregulated in Basal-Like Breast Cancer Molecular Subtype and Associated With Favorable Patient Outcome. *Front. Immunol.* **2020**, *11*, 560074. [CrossRef]
7. Dana-Farber Cancer Institute. "Liquid Biopsy" Blood Test Accurately Detects Key Genetic Mutations in Most Common Form of Lung Cancer, Study Finds. In *News Wise*; Dana-Farber Cancer Institute: Boston, MA, USA, 2016.
8. Kashyap, D.; Kaur, H. Cell-free miRNAs as non-invasive biomarkers in breast cancer: Significance in early diagnosis and metastasis prediction. *Life Sci.* **2020**, *246*, 117417. [CrossRef]
9. Abramson, V.G.; Mayer, I.A. Molecular Heterogeneity of Triple Negative Breast Cancer. *Curr. Breast Cancer Rep.* **2014**, *6*, 154–158. [CrossRef]
10. Zhou, J.; Tang, Z.; Gao, S.; Li, C.; Feng, Y.; Zhou, X. Tumor-Associated Macrophages: Recent Insights and Therapies. *Front. Oncol.* **2020**, *10*, 188. [CrossRef]
11. Petitprez, F.; Vano, Y.A.; Becht, E.; Giraldo, N.A.; de Reyniès, A.; Sautès-Fridman, C.; Fridman, W.H. Transcriptomic analysis of the tumor microenvironment to guide prognosis and immunotherapies. *Cancer Immunol. Immunother.* **2018**, *67*, 981–988. [CrossRef]
12. Zhou, Y.; Zhou, B.; Pache, L.; Chang, M.; Khodabakhshi, A.H.; Tanaseichuk, O.; Benner, C.; Chanda, S.K. Metascape provides a biologist-oriented resource for the analysis of systems-level datasets. *Nat. Commun.* **2019**, *10*, 1523. [CrossRef]
13. López-Ozuna, V.M.; Kogan, L.; Hachim, M.Y.; Matanes, E.; Hachim, I.Y.; Mitric, C.; Kiow, L.L.C.; Lau, S.; Salvador, S.; Yasmeen, A.; et al. Identification of Predictive Biomarkers for Lymph Node Involvement in Obese Women With Endometrial Cancer. *Front. Oncol.* **2021**, *11*, 695404. [CrossRef] [PubMed]
14. Busch, M.; Dünker, N. Trefoil factor family peptides–friends or foes? *Biomol. Concepts* **2015**, *6*, 343–359. [CrossRef]
15. Dai, X.; Cheng, H.; Chen, X.; Li, T.; Zhang, J.; Jin, G.; Cai, D.; Huang, Z. FOXA1 is Prognostic of Triple Negative Breast Cancers by Transcriptionally Suppressing SOD2 and IL6. *Int. J. Biol. Sci.* **2019**, *15*, 1030–1041. [CrossRef]
16. Rangel, N.; Fortunati, N.; Osella-Abate, S.; Annaratone, L.; Isella, C.; Catalano, M.G.; Rinella, L.; Metovic, J.; Boldorini, R.; Balmativola, D.; et al. FOXA1 and AR in invasive breast cancer: New findings on their co-expression and impact on prognosis in ER-positive patients. *BMC Cancer* **2018**, *18*, 703. [CrossRef]
17. Badve, S.; Turbin, D.; Thorat, M.A.; Morimiya, A.; Nielsen, T.O.; Perou, C.M.; Dunn, S.; Huntsman, D.G.; Nakshatri, H. FOXA1 expression in breast cancer–correlation with luminal subtype A and survival. *Clin. Cancer Res.* **2007**, *13*, 4415–4421. [CrossRef]
18. Byrne, D.J.; Deb, S.; Takano, E.A.; Fox, S.B. GATA3 expression in triple-negative breast cancers. *Histopathology* **2017**, *71*, 63–71. [CrossRef]
19. Voduc, D.; Cheang, M.; Nielsen, T. GATA-3 expression in breast cancer has a strong association with estrogen receptor but lacks independent prognostic value. *Cancer Epidemiol. Biomark. Prev.* **2008**, *17*, 365–373. [CrossRef]

20. Wasserman, J.K.; Williams, P.A.; Islam, S.; Robertson, S.J. GATA-3 expression is not associated with complete pathological response in triple negative breast cancer patients treated with neoadjuvant chemotherapy. *Pathol. Res. Pract.* **2016**, *212*, 539–544. [CrossRef]

21. Kim, S.; Moon, B.-I.; Lim, W.; Park, S.; Cho, M.S.; Sung, S.H. Expression patterns of GATA3 and the androgen receptor are strongly correlated in patients with triple-negative breast cancer. *Human Pathol.* **2016**, *55*, 190–195. [CrossRef]

22. Asif, A.; Mushtaq, S.; Hassan, U.; Akhtar, N.; Azam, M. Expression of Gata 3 in Epithelial Tumors. *PAFMJ* **2020**, *70* (Suppl. S1), S20–S25.

23. Cimino-Mathews, A.; Subhawong, A.P.; Illei, P.B.; Sharma, R.; Halushka, M.K.; Vang, R.; Fetting, J.H.; Park, B.H.; Argani, P. GATA3 expression in breast carcinoma: Utility in triple-negative, sarcomatoid, and metastatic carcinomas. *Hum. Pathol.* **2013**, *44*, 1341–1349. [CrossRef] [PubMed]

24. Krings, G.; Nystrom, M.; Mehdi, I.; Vohra, P.; Chen, Y.Y. Diagnostic utility and sensitivies of GATA3 antibodies in triple-negative breast cancer. *Hum. Pathol.* **2014**, *45*, 2225–2232. [CrossRef]

25. Treeck, O.; Schüler-Toprak, S.; Ortmann, O. Estrogen Actions in Triple-Negative Breast Cancer. *Cells* **2020**, *9*, 2358. [CrossRef] [PubMed]

26. Filardo, E.J.; Quinn, J.A.; Bland, K.I.; Frackelton, A.R., Jr. Estrogen-induced activation of Erk-1 and Erk-2 requires the G protein-coupled receptor homolog, GPR30, and occurs via trans-activation of the epidermal growth factor receptor through release of HB-EGF. *Mol. Endocrinol.* **2000**, *14*, 1649–1660. [CrossRef]

27. Hoffmann, W. Trefoil Factor Family (TFF) Peptides and Their Links to Inflammation: A Re-evaluation and New Medical Perspectives. *Int. J. Mol. Sci.* **2021**, *22*, 4909. [CrossRef]

28. Karakuş, F.; Eyol, E.; Yılmaz, K.; Ünüvar, S. Inhibition of cell proliferation, migration and colony formation of LS174T Cells by carbonic anhydrase inhibitor. *Afr. Health Sci.* **2018**, *18*, 1303–1310. [CrossRef] [PubMed]

29. Helzer, K.T.; Szatkowski Ozers, M.; Meyer, M.B.; Benkusky, N.A.; Solodin, N.; Reese, R.M.; Warren, C.L.; Pike, J.W.; Alarid, E.T. The Phosphorylated Estrogen Receptor α (ER) Cistrome Identifies a Subset of Active Enhancers Enriched for Direct ER-DNA Binding and the Transcription Factor GRHL2. *Mol. Cell. Biol.* **2019**, *39*, e00417-18. [CrossRef]

30. Laganière, J.; Deblois, G.; Lefebvre, C.; Bataille, A.R.; Robert, F.; Giguère, V. From the Cover: Location analysis of estrogen receptor alpha target promoters reveals that FOXA1 defines a domain of the estrogen response. *Proc. Natl. Acad. Sci. USA* **2005**, *102*, 11651–11656. [CrossRef]

31. Manavathi, B.; Samanthapudi, V.S.; Gajulapalli, V.N. Estrogen receptor coregulators and pioneer factors: The orchestrators of mammary gland cell fate and development. *Front. Cell. Dev. Biol.* **2014**, *2*, 34. [CrossRef]

32. Theodorou, V.; Stark, R.; Menon, S.; Carroll, J.S. GATA3 acts upstream of FOXA1 in mediating ESR1 binding by shaping enhancer accessibility. *Genome Res.* **2013**, *23*, 12–22. [CrossRef]

33. Yi, J.; Ren, L.; Li, D.; Wu, J.; Li, W.; Du, G.; Wang, J. Trefoil factor 1 (TFF1) is a potential prognostic biomarker with functional significance in breast cancers. *Biomed. Pharmacother.* **2020**, *124*, 109827. [CrossRef] [PubMed]

34. Ohsugi, T.; Yamaguchi, K.; Zhu, C.; Ikenoue, T.; Takane, K.; Shinozaki, M.; Tsurita, G.; Yano, H.; Furukawa, Y. Anti-apoptotic effect by the suppression of IRF1 as a downstream of Wnt/β-catenin signaling in colorectal cancer cells. *Oncogene* **2019**, *38*, 6051–6064. [CrossRef] [PubMed]

35. Rah, B.; Rather, R.A.; Bhat, G.R.; Baba, A.B.; Mushtaq, I.; Farooq, M.; Yousuf, T.; Dar, S.B.; Parveen, S.; Hassan, R.; et al. JAK/STAT Signaling: Molecular Targets, Therapeutic Opportunities, and Limitations of Targeted Inhibitions in Solid Malignancies. *Front. Pharmacol.* **2022**, *13*, 821344. [CrossRef] [PubMed]

36. Xu, X.; Wu, Y.; Yi, K.; Hu, Y.; Ding, W.; Xing, C. IRF1 regulates the progression of colorectal cancer via interferon-induced proteins. *Int. J. Mol. Med.* **2021**, *47*, 104. [CrossRef] [PubMed]

37. Burugu, S.; Asleh-Aburaya, K.; Nielsen, T.O. Immune infiltrates in the breast cancer microenvironment: Detection, characterization and clinical implication. *Breast Cancer* **2017**, *24*, 3–15. [CrossRef] [PubMed]

38. Dieci, M.V.; Miglietta, F.; Guarneri, V. Immune Infiltrates in Breast Cancer: Recent Updates and Clinical Implications. *Cells* **2021**, *10*, 223. [CrossRef]

39. Gabrilovich, D.I.; Ishida, T.; Nadaf, S.; Ohm, J.E.; Carbone, D.P. Antibodies to vascular endothelial growth factor enhance the efficacy of cancer immunotherapy by improving endogenous dendritic cell function. *Clin. Cancer Res.* **1999**, *5*, 2963–2970.

40. Saqib, U.; Sarkar, S.; Suk, K.; Mohammad, O.; Baig, M.S.; Savai, R. Phytochemicals as modulators of M1-M2 macrophages in inflammation. *Oncotarget* **2018**, *9*, 17937–17950. [CrossRef]

41. Tiainen, S.; Tumelius, R.; Rilla, K.; Hämäläinen, K.; Tammi, M.; Tammi, R.; Kosma, V.M.; Oikari, S.; Auvinen, P. High numbers of macrophages, especially M2-like (CD163-positive), correlate with hyaluronan accumulation and poor outcome in breast cancer. *Histopathology* **2015**, *66*, 873–883. [CrossRef]

42. Palma, G.; Frasci, G.; Chirico, A.; Esposito, E.; Siani, C.; Saturnino, C.; Arra, C.; Ciliberto, G.; Giordano, A.; D'Aiuto, M. Triple negative breast cancer: Looking for the missing link between biology and treatments. *Oncotarget* **2015**, *6*, 26560–26574. [CrossRef]

MDPI AG
Grosspeteranlage 5
4052 Basel
Switzerland
Tel.: +41 61 683 77 34

Life Editorial Office
E-mail: life@mdpi.com
www.mdpi.com/journal/life